Optical Access Networks and Advanced Photonics:
Technologies and Deployment Strategies

Ioannis P. Chochliouros
Hellenic Telecommunications Organization S.A., Greece

George A. Heliotis
Hellenic Telecommunications Organization S.A., Greece

T0345428

INFORMATION SCIENCE REFERENCE

Hershey · New York

Director of Editorial Content:	Kristin Klinger
Senior Managing Editor:	Jamie Snavely
Assistant Managing Editor:	Michael Brehm
Publishing Assistant:	Sean Woznicki
Typesetter:	Sean Woznicki
Cover Design:	Lisa Tosheff
Printed at:	Yurchak Printing Inc.

Published in the United States of America by
Information Science Reference (an imprint of IGI Global)
701 E. Chocolate Avenue
Hershey PA 17033
Tel: 717-533-8845
Fax: 717-533-8661
E-mail: cust@igi-global.com
Web site: http://www.igi-global.com/reference

Library of Congress Cataloging-in-Publication Data

Optical access networks and advanced photonics : technologies and deployment strategies / Ioannis P. Chochliouros and George A. Heliotis, editors.
    p. cm.
  Includes bibliographical references and index.
  Summary: "This book presents a comprehensive overview of emerging optical access network solutions to efficiently meet the anticipated growth in bandwidth demand"--Provided by publisher.

  ISBN 978-1-60566-707-2 (hardcover) -- ISBN 978-1-60566-708-9 (ebook) 1. Optical communications. 2. Broadband communication systems. I. Chochliouros, Ioannis P., 1966- II. Heliotis, George A., 1975-
  TK5103.59.O653 2010
  621.382'7--dc22
                                    2009004395

British Cataloguing in Publication Data
A Cataloguing in Publication record for this book is available from the British Library.

All work contributed to this book is new, previously-unpublished material. The views expressed in this book are those of the authors, but not necessarily of the publisher.

List of Reviewers

C.C.K. Chan, *The Chinese University of Hong Kong, Hong Kong*
I.P. Chochliouros, *Dr. Telecommunications Engineer, Greece*
C.-W. Chow, *National Chiao Tung University, Taiwan*
G. Fountakos, *Telecommunications Engineer, Greece*
G. Heliotis, *Dr. Telecommunications Engineer, Greece*
A. Huang, *University of California, USA*
T. Kamijoh, *Oki Electric Industry Co., Ltd. Japan*
V. Kyriakidou, *University of Athens, Greece*
G.K. Lalopoulos, *Hellenic Telecommunications Organization S.A.-OTE, Greece*
N. Merayo, *University of Valladolid, Spain*
M. Molnar, *Institut de Recherche en Informatique et Systèmes Aléatoires (IRISA-INSA), France*
H.T. Mouftah, *University of Ottawa, Canada*
G.A. Pagiatakis, *School for Pedagogical and Technological Education-ASPETE, Greece*
A. Rissons, *Université de Toulouse, France*
J. Schussmann, *Carintia University of Applied Sciences, Austria*
A.S. Spiliopoulou, *Lawyer, LL.M., Member of Athens Bar Association, Greece*
C. Vassilopoulos, *Hellenic Telecommunications Organization S.A.-OTE, Greece*
T. Yokotani, *Mitsubishi Electric Corp., Japan*
W. Yue, *Fujitsu Network Communications Inc., USA*

Table of Contents

Section 1
Current Market View: Trends, Challenges and Opportunities for Development

Chapter 1
Ioannis P. Chochliouros, Dr. Telecommunications Engineer, Greece
George A. Heliotis, Dr. Telecommunications Engineer, Greece
Anastasia S. Spiliopoulou, Lawyer, LL.M., Member of Athens Bar Association, Greece

Chapter 2
William Yue, Fujitsu Network Communications Inc., USA
Brian Hunck, Fujitsu Network Communications Inc., USA

Section 2
Modern Optical Technologies and Future Architectures for Broadband Access

Chapter 3
Gerasimos C. Pagiatakis, School for Pedagogical and Technological Education
(ASPETE), Greece

Chapter 4
Calvin C. K. Chan, The Chinese University of Hong Kong, Hong Kong

Section 5
The Way Forward

Detailed Table of Contents

Section 1
Current Market View: Trends, Challenges and Opportunities for Development

> *Ioannis P. Chochliouros, Dr. Telecommunications Engineer, Greece*
> *George A. Heliotis, Dr. Telecommunications Engineer, Greece*
> *Anastasia S. Spiliopoulou, Lawyer, LL.M., Member of Athens Bar Association, Greece*

With the current continuously growing bandwidth demand, it is apparent that conventional broadband access solutions will quickly become bottlenecks in terms of bandwidth provision. In this chapter, the authors analyze the present global challenge for extended bandwidth provision in the scope of the fast developing electronic communications sector, creating a fully converged environment. In particular, first the authors examine several potential options imposed by distinct technologies, as they are currently applied in the marketplace. Then they present a comprehensive review of the emerging optical access solutions, focusing mainly on passive optical network (PON) technologies that promise to efficiently meet the anticipated growth in bandwidth demand and at the same time be economically viable and future-proof from an operator's perspective, and evaluate their capabilities to the conventional copper-based broadband solutions. They also survey the current deployment efforts and relevant policies in the European Community area, as well as discuss why Europe is lagging with regard to deployment pace when compared to Asia and the USA. Specific and detailed analysis is given for recent developments performed in the European Union, where the authors identify current trends, potential hurdles and/or difficulties, as well as perspectives for further growth and development.

Chapter 2

The access network is the last loop, or last mile, in the provider network between the central office (CO) or point of presence (PoP) and the customer premises. Competitive pressure to provide high-bandwidth services (such as video) to consumers, and Ethernet transport to enterprises, is forcing service providers to rebuild their access networks. More optical fibers are being added in the last mile to meet these new bandwidth demands since legacy access networks have not been sufficient to support bandwidth-intensive applications. This chapter reviews the multiple definitions of "optical access" and the migration from direct copper loops to a variety of optical architectures, including Synchronous Optical Networking (SO-NET), Synchronous Digital Hierarchy (SDH), Fiber to the x (FTTx), Ethernet and wavelength delivery. Key business drivers such as carrier competition, bandwidth needs, and the reliability and service level agreement issues of optical technology are covered. The chapter concludes by considering the near future of optical access product trends and key optical deployment options in applications such as cellular backhaul. The data presented in this chapter is mainly based on our recent deployment experience in the North American optical access market segment.

Section 2
Modern Optical Technologies and Future Architectures for Broadband Access

In this chapter, active optical access networks (AONs) are examined. AONs are a special type of optical access networks in which the sharing of optical fibers among users is implemented by means of active equipment (as opposed to passive optical networks –PONs– where sharing is achieved by using multiple passive splitters). In active optical access networks, user-side units, known as Optical Network Units (ONUs), are usually grouped in access Synchronous Digital Hierarchy (SDH) rings and fiber-interconnected to a local exchange unit, known as Optical Line Termination (OLT). In AONs (as well as in PONs) the optical fiber (originally used in the trunk network) is introduced in the access domain, namely between the customer and the local exchange. Practically, this means that the huge bandwidth provided by the optical fiber becomes directly available to the normal user. Despite the obvious financial and techno-economical issues related to the massive deployment of optical access networks, the possibilities and challenges created are enormous. This chapter examines the various units and modules composing an active optical access network and presents the basic procedures for implementing such a network.

Calvin C. K. Chan, The Chinese University of Hong Kong, Hong Kong

Wavelength division multiplexing passive optical network has emerged as a promising solution to support a robust and large-scale next generation optical access network. It offers high-capacity data delivery and flexible bandwidth provisioning to all subscribers, so as to meet the ever-increasing bandwidth requirements as well as the quality of service requirement of the next generation broadband access networks. The maturity and reduced cost of the WDM components available in the market are also among the major driving forces to enhance the feasibility and practicality of commercial deployment. In this chapter, the author will provide a comprehensive discussion on the basic principles and network architectures for WDM-PONs, as well as their various enabling technologies. Different feasible approaches to support the two-way transmission will be discussed. It is believed that WDM-PON is an attractive solution to realize fiber-to-the-home (FTTH) applications.

Chi-Wai Chow, National Chiao Tung University, Taiwan

Passive optical network (PON) is considered as an attractive fiber-to-the-home (FTTH) technology. Wavelength division multiplexed (WDM) PON improves the utilization of fiber bandwidth through the use of wavelength domain. A cost-effective solution in WDM PON would use the same components in each optical networking unit (ONU), which should thus be independent of the wavelength assigned by the network. Optical carriers are distributed from the head-end office to different ONUs to produce the upstream signals. Various solutions of colorless ONUs will be discussed. Although the carrier distributed WDM PONs have many attractive features, a key issue that needs to be addressed is how best to control the impairments that arise from optical beat noise induced by Rayleigh backscattering (RB). Different RB components will be analyzed and RB mitigation schemes will be proposed. Finally, some novel PONs including signal remodulation PONs, long reach PONs and wireless/wired PONs will be highlighted.

Section 3
Technical Challenges and Determinants for Further Growth

Noemí Merayo, University of Valladolid, Spain
Patricia Fernández, University of Valladolid, Spain
Ramón J. Durán, University of Valladolid, Spain
Rubén M. Lorenzo, University of Valladolid, Spain
Ignacio de Miguel, University of Valladolid, Spain
Evaristo J. Abril, University of Valladolid, Spain

Passive Optical Networks (PONs) are very suitable architectures to face today's access challenges. This technology shows a very cost saving architecture, it provides a huge amount of bandwidth and efficiently supports Quality of Service (QoS). In PON networks, as all subscribers share the same uplink channel, a medium access control protocol is required to provide a contention method to access the channel. As the performance of Time Division Multiplexing Access (TDMA) protocol is not good enough because traffic nature is heterogeneous, Dynamic Bandwidth Allocation (DBA) algorithms are proposed to overcome the problem. These algorithms are very efficient as they adapt the bandwidth assignment depending on the updated requirements and traffic conditions. Moreover, they should offer QoS by means of both class of service and subscriber differentiation. Long-Reach PONs, which combine the access and the metro network into only one by using 100 km of fibre, is an emergent technology able to reach a large number of far subscribers and to decrease the associated costs.

Chapter 7
Jun Zheng, Southeast University, China
Hussein T. Mouftah, University of Ottawa, Canada

Bandwidth allocation is one of the critical issues in the design of Ethernet passive optical networks (EPONs). In an EPON system, multiple optical network units (ONUs) share a common upstream transmission channel for data transmission. To efficiently utilize the limited bandwidth of the upstream channel, a system must dynamically allocate the upstream bandwidth among multiple ONUs based on the instantaneous bandwidth demands and quality of service requirements of end users. This chapter gives an introduction of the fundamental concepts on bandwidth allocation in an EPON system, discusses the major challenges in designing a polling protocol for bandwidth allocation, and presents an overview of the state-of-the-art dynamic bandwidth allocation (DBA) algorithms proposed for EPONs.

Chapter 8
Miklós Molnár, IRISA-INSA, France
Fen Zhou, IRISA-INSA, France
Bernard Cousin, IRISA-Université de Rennes I, France

Widely available broadband services in the Internet require high capacity access networks. Only optical networking is able to efficiently provide the huge bandwidth required by multimedia applications. Distributed applications such as Video-Conferencing, HDTV, VOD and Distance Learning are increasingly common and produce a large amount of data traffic, typically between several terminals. Multicast is a bandwidth-efficient technique for one-to-many or many-to-many communications, and will be indispensable for serving multimedia applications in future optical access networks. These applications require robust and reliable connections as well as the satisfaction of QoS criteria. In this chapter, several access network architectures and related multicast routing methods are analyzed. Overall network performance and dependability are the focus of our analysis.

The proposal chapter aims at highlighting the tremendous emergence of the Vertical-Cavity Surface-Emitting Laser (VCSEL) in the FTTX systems. The VCSEL is probably one of the most important and promising components of the "last-leg" Optical Access Networks. To satisfy the bandwidth rise as well as the inexpensive design constraints, the VCSEL has found its place between the Light-Emitting-Diode (LED) and the Edge-Emitting-Laser (EEL) such as the DFB (Distributed-Feedback) laser. Hence, the authors dedicate a chapter to the promising VCSEL technology that aims to give an overview of the advances, the physical behavior, and the various structures regarding VCSELs. They discuss the VCSEL features and performance to weigh up the specific advantages and the weaknesses of the existing technology. Finally, diverse potentials of Optical Access Network architectures are discussed.

Section 4
Business Models and Techno-Economic Evaluations

In Europe initiatives towards the development of optical networks infrastructures' have been undertaken in order to address the need of faster and more telecommunications services. Consequently, the implementation of appropriate business model seems to be meaningful tool for infrastructures as it could ensure commercial viability and limit investment's risks. Although, the variety of forms characterizes a business model, there are some aspects that networks owners or decision makers should consider in each case. The purpose of this book chapter is to highlight the main issues that should be considered from the main actors involved in the business models like telecom operators, fiber optical constructors as well as municipalities that own the main part of the infrastructure.

In the near future, broadband access networks will be required with data rates of over 1Gbit/s per customer. Currently, time-division multiple access passive optical networks (TDMA-PONs) are deployed. However, TDMA-PONs cannot keep up with the requirements for the broadcasting of a great number of HDTV channels and the unicasting of several triple-play services (voice, data and video). In contrast, wavelength-division-multiplexed PONs (WDM PONs) will be able to provide these required high data rates per user causing higher costs than with TDM-PONs. The introduced paradigm shift, at least one wavelength per service and user, leads to the introduction of new aspects in the design of future WDM PON access networks. In techno-economic evaluations, new network architectures with the highest potential concerning economic considerations have been identified. Access to these newly identified network architectures will prompt market introduction as well as market penetration helping Fiber-to-the-Home (FTTH) to become reality.

Section 5
The Way Forward

Chapter 12

Christos Vassilopoulos, Hellenic Telecommunications Organization S.A., Greece

This chapter offers a qualitative approach towards the development of the new generation access network, based on FTTx implementations. After a brief description of the current state of traditional access networks and an estimation of the expected data rate per household in terms of services, the chapter examines all the available Network Technologies (FTTx), Access Technologies (xDSL, Ethernet and PON) for both P2P and P2MP development schemes and their relevant implementations. The prospects of NGA are also strategically examined in view of the complicated multi-player environment, involving Telco (ILEC and CLEC), regulators and pressure interest groups, all striving to serve their individual, often conflicting interests. The chapter concludes with an outline of the different deployment strategies for both ILEC and CLEC Telco.

Chapter 13

Tetsuya Yokotani, Mitsubishi Electric Corporation, Japan

As optical broadband access networks have been popularized, triple play services using IP technologies, such as Internet access, IP telephony, and IP video distributing services, have been also popularized. However, consumers expect new services for a more comfortable life. Especially, when QoS guarantee and high reliable services are provided in NGN (Next Generation Network) era, various home network services over NGN are deployed. For this purpose, the home gateway has been installed in consumer houses for the connection between access and home networks, and providing various services to consumers. Even though, the broadband router currently plays a role similar to the home gateway, this home gateway should comprehend functionalities of the broadband router, and should have additional features. The functional requirements of such home gateway have been discussed in standard bodies. That is, the next generation home gateway in NGN era generally should have four features as follows; High performance for IP processing, Compliance with the interface of carrier grade infrastructure including NGN, Flexible platform for various services, and Easy management and maintenance. This chapter describes the standardization of the home gateway and, proposes its evolution scenario the present to the future. Then, it also proposes these four requirements, and technologies to comply with features described above.

Preface

We are currently witnessing an unprecedented growth in bandwidth demand, which is mainly driven by the development and proliferation of advanced broadband multimedia applications, including video-on-demand, high-definition digital television (HDTV), multi-party videoconferencing, online gaming and many others. These multimedia digital services are highly bandwidth intensive, and therefore require underlying access network infrastructures that are capable of supporting the very high-speed data transmission rates that are needed. It is apparent that today's conventional broadband access communications networks (such as xDSL networks) will quickly become "bottlenecks" in terms of bandwidth provision. Hence, telecom providers are currently focusing on developing new network infrastructures that will constitute future-proof solutions in terms of the anticipated growth in bandwidth demand, but at the same time be economically viable.

Indeed, while most users currently enjoy relatively high speed communication services mostly through xDSL networks, purely copper-based access networks can be viewed as relatively short-term solutions since this aging infrastructure is rapidly approaching its fundamental speed limits. In contrast, fiber optics-based technologies offer tremendously higher bandwidth, a fact that has long been recognized by all telecom providers at international level, which have upgraded their core networks to optical technologies. Having won the battle in core networks, optical networks are now entering the access networks arena, reducing or even totally eliminating the need for data transfer over conventional telephone wires.

It should be emphasized that optical access networks are not a new concept, but instead have been considered as a solution for the subscriber access network for quite some time. Early proposals and developments can even be traced back to the early 80s. However, these were abandoned due to the technology not being mature enough, and most importantly, due to prohibitively high costs coupled with the fact that there was no actual bandwidth demand to necessitate the deployment of such networks at the time. Nevertheless, since then, photonic and fiber-optic components technology has progressed remarkably and volume production has dramatically reduced costs. The cost of deploying such networks is also falling steadily, and coupled with the ever-increasing bandwidth demand, mass worldwide optical access networks deployment is now reality. Fiber access networks are capable of delivering extremely high bandwidth at large distances and can cater for all current and predicted future voice, data and video services requirements. In this sense, it can be said that "once fiber is installed, no significant further investments or re-engineering is likely to be required for decades".

Also referred to as "Fiber-to-the-X" (FTTX, where X can be Home, Node, Building etc. depending on the degree of optical fiber penetration), such architectures offer a viable solution to the access network bottleneck problem, and promise extremely high bandwidth to the end-user. In addition, they future-proof the telecomunications operator's investment, as they offer relatively easy upgrade and upscale, should such a need arises in the future. Deployment projects are now underway in Asia, the USA and Europe, creating an international diversity of distinct business cases, which all provide major contribution to the global effort for promoting broadband and creating opportunities for an "Information Society for All".

This book presents a comprehensive overview of the emerging optical access network solutions that promise to cater for current and predicted future voice, data and video services needs, and ultimately change the way we experience broadband. With a thematically wide selection of chapters that are written by renowned researchers of the field, the book provides a well-rounded and thorough overview of the current and upcoming state-of-the-art optical access networks, discussing technological, regulatory, business and deployment aspects, based on current international trends, policies and applied measures.

The core objective of the book is two-fold: It intends to discuss the challenge imposed by fibre-optic access networks worldwide as a strategic choice (or even as a "necessity" due to rapid technology and market development) of market operators, in order to provide immediate "responses" to the fast evolutionary progress in bandwidth demand. To this aim, apart from the exact analysis of all underlying strategic initiatives, we discuss and evaluate specific technology trends in order to realize the expected beneficial use of optical access infrastructures, as a "modern" and viable network solution. On the other hand, we deal with current market cases and assess several potential opportunities observed at the global level, by discussing and analyzing specific cases (as currently appear in Asia, America and Europe). The current diversity among international markets can also reflect other trends explaining why in some cases such networks have been very rapidly deployed (and have already affected growth by contributing to the national economies) while in others there are still "limitations" and uncertainties, preventing from further growth. This option is quite important, especially in the scope of liberalized and competitive markets, as any further delay may result to negative results for further economy progress.

The book presents a comprehensive overview of the emerging optical access network solutions, that promise to efficiently meet the anticipated growth in bandwidth demand by delivering extremely high data rates to end-users, and offering truly unprecedented levels of performance that can cater for all current and predicted future voice, data and video services.

The book provides comprehensive coverage of the area of optical access networks, covering not only technology, but also real-world deployment, business and economical issues. Hence, a very well-rounded source of information is concentrated on a single book

Written primarily from a general perspective and at the same time not aimed at the experienced engineer, a very wide audience can benefit from this book. Most technical background is included and explained within it, focusing only on the most important issues and avoiding unnecessary specialised technicalities. Overall, the book provides up-to-date authoritative information in a single-source, and is useful for anyone who wishes to introduce himself or deepen and update his knowledge in this fascinating field. Specifically, the book is intended to serve as a valuable source of information to readers coming from a wide range of different professional disciplines including postgraduate science and engineering students, academics and researchers, telecommunication engineers and technicians, communication network planners and designers, ICT business development strategists, telecommunication market analysts and many others.

The book is organized into thirteen distinct chapters, which together capture a very wide range of timely issues regarding optical access networks.

Chapter 1, by Ioannis P. Chochliouros, George A. Heliotis and Anastasia S. Spiliopoulou is entitled *An Introduction to Optical Access Networks: Technological Overview and Regulatory Issues for Large-Scale Deployment* and gives comprehensive review of the emerging optical access solutions, focusing mainly on passive optical network (PON) technologies and contrasting their capabilities to conventional copper-based broadband solutions. The chapter discusses the present global challenge for extended bandwidth provision in the scope of the fast developing electronic communications sector, creating a fully converged environment and it examines several potential options imposed by distinct technologies, as currently adopted in the marketplace and experienced by the various categories of the "market actors". The chapter also analyzes, in detail, the regulatory issues that surround the deployment of such next generation networks, and surveys current deployment efforts and applied policies, as well as potential perspectives for further growth and development, on a worldwide level (with distinct paradigms chosen from the international experience).

Chapter 2, by William Yue and Brian Hunck of the Fujitsu Network Communications Inc. (USA) is entitled *Optical Access Comes of Age in a Packet-Delivery World* and reviews the current and upcoming "flavors" of optical access networks and the migration from direct copper loops to such optical architectures. Competitive pressure to provide high-bandwidth services (such as video) to consumers, and Ethernet transport to enterprises, is forcing service providers to rebuild their access networks. To this aim, more optical fibers are being added in the last mile to meet these new bandwidth demands since legacy access networks have not been sufficient to support bandwidth-intensive applications. Key business drivers such as carrier competition, bandwidth needs, and reliability and service level agreement issues of optical technology are discussed. The chapter also considers the near future of optical access product trends and some key optical deployment options.

Chapter 3, by Gerasimos C. Pagiatakis of the School for Pedagogical and Technological Education, Greece, is entitled *Active Optical Access Networks*, and focuses on optical access networks in which the sharing of fibers among end-users is implemented by means of active equipment. In active optical access networks (AONs), user-side units, known as optical network units, are usually grouped in access synchronous digital hierarchy (SDH) rings and fiber-interconnected to a local exchange unit, known as optical line termination. In AONs (as well as in passive optical networks) the optical fiber (originally used in the trunk network) is introduced in the access domain, namely between the customer and the local exchange. Practically, this means that the huge bandwidth provided by the optical fiber becomes directly available to the normal user. Despite the obvious financial and techno-economical issues related to the massive deployment of optical access networks, the possibilities and challenges created are enormous. The chapter examines in detail the various units and modules that comprise an active optical access network as well as their respective role, and it presents the fundamental procedures for implementing such a network.

Chapter 4, by Calvin C.K. Chan of the Chinese University of Hong Kong, Hong Kong, is entitled *Wavelength Division Multiplexed Passive Optical Networks: Principles, Architectures and Technologies* and presents a very comprehensive treatment of the principles and network architectures of Wavelength Division Multiplexed Passive Optical Networks (WDM-PONs), as well as a thorough discussion of their enabling technologies. Wavelength division multiplexing passive optical network has emerged as a promising solution to support a robust and large-scale next generation optical access network. It offers high-capacity data delivery and flexible bandwidth provisioning to all subscribers, so as to meet the ever-increasing bandwidth requirements as well as the quality of service requirement of the next generation broadband access networks. The maturity and reduced cost of the WDM components available in the market are also among the major driving forces to enhance the feasibility and practicality of commercial deployment. Different feasible approaches to support the two-way transmission are discussed, as WDM-PON seems as an "attractive solution" to realize fiber-to-the-home (FTTH) applications.

Chapter 5, by Chi-Wai Chow of the National Chiao Tung University, Taiwan, is entitled *Broadband Optical Access using Centralized Carrier Distribution* and discusses the main characteristics, advantages and attractive features of WDM-PONs with centralized carrier distribution. Wavelength division multiplexed (WDM) PON improves the utilization of fiber bandwidth through the use of wavelength domain. A cost-effective solution in WDM PON would use the same components in each optical networking unit (ONU), which should thus be independent of the wavelength assigned by the network. Optical carriers are distributed from the head-end office to different ONUs to produce the upstream signals. Various solutions of colorless ONUs are discussed. The chapter also analyzes the technical challenges that these implementations present, focusing on Rayleigh backscattering (RB) induced impairments and proposing attractive solutions for their mitigation, with some novel PONs highlighted.

Chapter 6, by Noemí Merayo, Patricia Fernández, Ramón J. Durán, Rubén M. Lorenzo, Ignacio de Miguel and Evaristo J. Abril by the University of Valladolid, Spain, is entitled *Bandwidth Allocation Methods in Passive Optical Access Networks (PONs)* and thoroughly discusses the various PON standards that exist as well as the technology for long-reach implementations, focusing on the main challenges that these present relative to bandwidth allocation methods, quality of service (QoS) and service level agreement (SLA) issues. PONs appear as quite suitable architectures to face today's access challenges; this technology "shows" a very cost saving architecture, it provides a huge amount of bandwidth and efficiently supports QoS. In PON networks, as all subscribers share the same uplink channel, a medium access control protocol is required to provide a contention method to access the channel. As the performance of Time Division Multiplexing Access (TDMA) protocol is not good enough because traffic nature is heterogeneous, Dynamic Bandwidth Allocation (DBA) algorithms are proposed to overcome the problem. These algorithms are very efficient as they adapt the bandwidth assignment depending on the updated requirements and traffic conditions.

Chapter 7, by Jun Zheng and Hussein T. Mouftah of the University of Ottawa, Canada, is entitled *Dynamic Bandwidth Allocation for Ethernet Passive Optical Networks*, and gives an extensive introduction to the critically important bandwidth allocation mechanisms in Ethernet-PON (EPON) systems. In an EPON system, multiple optical network units (ONUs) share a common upstream transmission channel for data transmission. To efficiently utilize the limited bandwidth of the upstream channel, a system must dynamically allocate the upstream bandwidth among multiple ONUs based on the instantaneous bandwidth demands and quality of service requirements of end users. The chapter discusses the major challenges in designing protocols for bandwidth allocation, and presents an overview of the state-of-the-art dynamic bandwidth allocation (DBA) algorithms that are proposed for EPON implementations.

Chapter 8, by Miklós Molnar, Fen Zhou and Bernard Cousin of the Institut de Recherche en Informatique et Systèmes Aléatoires (INISA) and of the Université de Rennes-I, France, is entitled *Multicast Routing in Optical Access Networks* and discusses multicasting as an instrumental feature for the efficient delivery of multimedia services in future optical access networks. Widely available broadband services in the Internet require high capacity access networks and optical networking is able to efficiently provide the huge bandwidth required by multimedia applications, producing large amounts of data traffic, typically between several distinct types of terminals. Multicast is a bandwidth-efficient technique for one-to-many or many-to-many communications, and will be instrumental in serving multimedia applications in future optical access networks, requiring robust and reliable connections together with compliance to QoS criteria. The chapter gives an extensive analysis of multicast routing methods in a variety of optical access network architectures, focusing on performance and dependability issues.

Chapter 9, by Angélique Rissons and Jean-Claude Mollier of the Université de Toulouse, France, is entitled *The Vertical-Cavity Surface-Emitting Laser: A Key Component in Future Optical Access Networks* and relates to photonic instrumentation that is utilized in optical access networks, focusing on the emerging and promising technology of Vertical Cavity Surface-Emitting Lasers (VCSELs). For more than ten years, the access network market is "attracted" by the VCSEL technology. This chapter aims at highlighting the tremendous emergence of the VCSEL technology in the FTTX systems, especially by providing an overview of the advances, the physical behavior, and the various structures regarding VCSELs. The technology, features and physical behaviour of these laser sources are thoroughly analyzed, highlighting their advantages and potential for use in optical access networks systems. Diverse system configurations to generate a signal for the optical access networks and potentials of network architectures are discussed. The VCSEL is probably one of the most important and promising components of the "last-leg" optical access networks.

Chapter 10, by Vagia Kyriakidou, Aristidis Chipouras, Dimitris Katsianis and Thomas Sphicopoulos of the University of Athens, Greece, is entitled *Business Models for Municipal Metro Networks: Theoretical and Financial Analysis*, and gives a solid understanding of techno-economic evaluation of optical access networks. Broadband penetration is continuously increasing and operators have to face end users' demand for more bandwidth. More specifically, in Europe, initiatives towards the development of optical networks infrastructures' have been undertaken to address the need of faster and more telecommunications services. New infrastructures and equipment are required to deliver real broadband services to end users and, consequently, the implementation of appropriate business model seems to be meaningful tool for infrastructures as it could ensure commercial viability and limit investment's risks. After reviewing important current cases of worldwide deployment, the chapter discusses and analyzes a variety of business models for the emerging optical access network markets, highlighting the main issues that should be addressed by the main actors involved, including telecommunication operators, infrastructure constructors, and so forth.

Chapter 11, by Jürgen Schussmann and Thomas Schirl of the Carinthia University of Applied Sciences, Austria, is entitled *Modeling and Techno-Economic Evaluations of WDM-PONs*, and considers the bandwidth and capacity requirements that are posed by future services, emphasizing and analyzing in detail the solutions that WDM-PON networks can offer. In the near future, broadband access networks will be required with data rates of over 1Gbit/s per customer. Currently, time-division multiple access passive optical networks (TDMA-PONs) are deployed; however, these networks cannot keep up with the requirements for the broadcasting of a great number of HDTV channels and the unicasting of several triple-play services (i.e., voice, data and video). In contrast, WDM PONs will be able to provide these required high data rates per user causing higher costs than with TDM-PONs. The chapter presents detailed techno-economic evaluations of single- and multi-stage WDM-PONs, also discussing design issues and identifying and analyzing the most promising network architectures in terms of cost effectiveness. Access to these newly identified network architectures will prompt market introduction as well as market penetration helping Fiber-to-the-Home (FTTH) to become reality.

Chapter 12, by Christos Vassilopoulos of the Hellenic Telecommunications Organization S.A., Greece, is entitled *The New Generation Access Network*, and gives an extensive review of current and upcoming access networks implementations, focusing on a strategic analysis of their prospects in view of the complicated multi-player telecommunications environment, involving telecommunications Operators (both Incumbent Local Exchange Carriers-ILECs and Competitive Local Exchange Carriers-CLECs), regulators and other pressure interest groups. The chapter offers a qualitative approach towards the development of the new generation access network, based on FTTx implementations. After a brief description of the current state of traditional access networks and an estimation of the expected data rate per household in terms of services, the chapter examines all the available network technologies (FTTx), access technologies (xDSL, Ethernet and PON) for both point-to-point (P2P) and point-to-multipoint (P2MP) development schemes and their relevant implementations. The chapter also outlines different deployment strategies that could be followed for optimal development.

Chapter 13, by Tetsuya Yokotani of the Mitsubishi Electric Corporation, Japan, is entitled "*Next Generation Home Network and Home Gateway associated with Optical Access*", and discusses the importance of home gateways for the efficient provisioning of services to the end-users of next generation access networks. As optical broadband access networks have been popularized, triple play services using IP technologies, such as Internet access, IP telephony, and IP video distributing services, have been also popularized. However, consumers expect new services for a more comfortable life. Especially, when QoS guarantee and high reliable services are provided in NGN (Next Generation Network) era, various home network services over NGN are deployed. For this purpose, the home gateway has been installed in consumer houses for the connection between access and home networks. The chapter analyzes in detail the functional characteristic, as well as the service features that home gateways should possess, that is: high performance for IP processing; compliance with the interface of carrier grade infrastructure including NGN; flexible platform for various services, and; easy management and maintenance. The chapter also discusses their standardization efforts and progress, and proposes evolution scenarios from the present to the future.

The combination of the selected chapters included in the book (all composed by recognised experts in the relevant thematic "fields") provides a detailed analysis of the relevant scope, bringing together ideas and experience from the academic and the business sector, thus "joining together" current market trends and research priorities, from a worldwide scope. As the proposed book covers a great variety of issues (i.e. technical, strategic, social, regulatory and business) all relevant to a major challenge for promoting broadband-based market evolution and deployment, it will provide essential facilities to any potential reader (mainly by offering exact and updated information to a great audience of potential recipients, of multiple origin, i.e. from researchers to market players, from students to authorities, etc.). Thus, its "value" for new scientists, for investors and for market players will be extremely high.

As can be seen from the above short chapter summaries, the book provides a solid foundation on optical access networks, and also captures the great diversity of current issues affecting their real-world deployment and future perspectives. We trust that the book will prove valuable for those who will read it, and hope it will allow a thorough understanding of the current and upcoming FTTx implementations and will promote strategic thinking for future issues.

Ioannis P. Chochliouros
George A. Heliotis
The Editors
Athens, January 2009

Acknowledgment

We want to recognize the expertise, enthusiasm, and cooperative spirit of the authors of the present book. Without their commitment to the actual multidisciplinary effort, we would not have succeeded. *"Optical Access Networks and Advanced Photonics: Technologies and Deployment Strategies"*, like all fieldbooks, would not exist without a great deal of time and attention from many contributors and/ or participants, originating from multiple sectors of the international electronic communications market (including recognized representatives from the academia, the research sector and the business communities). The efforts we wish to acknowledge took place over the last 15 months first at the structuring of the project, then the challenges faced, and finally the book itself took shape.

We are particularly indebted to all the authors involved in this book who provided us the great opportunity to interact and work with the best and the brightest from around the world, thus collecting approaches from the international experience and depicting the actual state-of-the-art for a quite interesting and many promising issue/challenge, like the objective of the present book. We would like to thank all of them: Anastasia S. Spiliopoulou (Lawyer, LL.M., Member of the Athens Bar Association, Greece), William Yue and Brian Hunck (Fujitsu Network Communications Inc., USA), Gerasimos C. Pagiatakis (School for Pedagogical and Technological Education-ASPETE, Greece), Calvin C.K. Chan (The Chinese University of Hong Kong, Hong Kong), Chi-Wai Chow (National Chiao Tung University, Taiwan), Noemí Merayo, Patricia Fernández, Ramón J. Durán, Rubén M. Lorenzo, Ignacio de Miguel and Evaristo J. Abril (University of Valladolid, Spain), Jun Zheng and Hussein T. Mouftah (University of Ottawa, Canada), Miklós Molnár and Fen Zhou (Institut de Recherche en Informatique et Systèmes Aléatoires, IRISA-INSA, France) and Bernard Cousin (IRISA- Université de Rennes I, France), Angélique Rissons and Jean-Claude Mollier (Université de Toulouse, France), Vagia Kyriaki-dou, Aristidis Chipouras, Dimitris Katsianis and Thomas Sphicopoulos (University of Athens, Greece), Jürgen Schussmann and Thomas Schirl (Carinthia University of Applied Sciences, Austria), Christos Vassilopoulos (Hellenic Telecommunications Organization S.A.-OTE, Greece) and Tetsuya Yokotani (Mitsubishi Electric Corporation, Japan).

We want to thank all of the authors for their insights and excellent contributions to this book. Working with them was a really extraordinary and quite fascinating experience.

A further special note of thanks goes to Heather A. Probst (for her initial support and guidance) and, most importantly, to Tyler Heath (Administrative Editorial Assistant) at IGI Global whose contributions throughout the whole process from the formation of the initial book's scope to the final publication have been invaluable. Furthermore, IGI Global has been extremely helpful and supportive every step of the way.

We would like to acknowledge the help of all involved in the structuring and review of the book, without whose support the entire project could not have been satisfactorily completed. Most of the authors of chapters included in this book also served as referees/reviewers for articles written by other authors. Extended and sincere editors' thanks go to all those experts who provided constructive, detailed and comprehensive reviews, in order to facilitate our effort to depict current results, experience gained at international level and to delimit further challenges and areas of expected development, thus creating a very much promising future.

Ioannis P. Chochliouros
George A. Heliotis
The Editors
Athens, January 2009

Section 1
Current Market View:
Trends, Challenges and Opportunities for Development

Chapter 1
An Introduction to Optical Access Networks
Technological Overview and Regulatory Issues for Large-Scale Deployment[1]

Ioannis P. Chochliouros
Dr. Telecommunications Engineer, Greece

George A. Heliotis
Dr. Telecommunications Engineer, Greece

Anastasia S. Spiliopoulou
Lawyer, LL.M., Member of Athens Bar Association, Greece

ABSTRACT

With the current continuously growing bandwidth demand, it is apparent that conventional broadband access solutions will quickly become bottlenecks in terms of bandwidth provision. In this chapter, the authors analyze the present global challenge for extended bandwidth provision in the scope of the fast developing electronic communications sector, creating a fully converged environment. In particular, first the authors examine several potential options imposed by distinct technologies, as they are currently applied in the marketplace. Then they present a comprehensive review of the emerging optical access solutions, focusing mainly on passive optical network (PON) technologies that promise to efficiently meet the anticipated growth in bandwidth demand and at the same time be economically viable and future-proof from an operator's perspective, and evaluate their capabilities to the conventional copper-based broadband solutions. They also survey the current deployment efforts and relevant policies in the European Community area, as well as discuss why Europe is lagging with regard to deployment pace when compared to Asia and the USA. Specific and detailed analysis is given for recent developments performed in the European Union, where the authors identify current trends, potential hurdles and/or difficulties, as well as perspectives for further growth and development.

DOI: 10.4018/978-1-60566-707-2.ch001

INTRODUCTION

We are currently witnessing an unprecedented growth in bandwidth demand, mainly driven by the development of advanced broadband multimedia applications, including video-on-demand (VoD), interactive high-definition digital television (HDTV) and related digital content, multi-party video-conferencing etc, as most of them are offered by the Internet in a converged global environment. These IP-based services are bandwidth intensive and require an underlying network infrastructure that is capable of supporting high-speed data transmission rates (Hellberg, Greene, & Boyes, 2007). Hence, telecom providers are currently focusing on developing new network infrastructures that will constitute future-proof solutions in terms of the anticipated growth in bandwidth demand, but at the same time be economically viable (Chochliouros & Spiliopoulou, 2005).

Most users currently enjoy relatively high speed communication services mostly through DSL (Digital Subscriber Line) access technologies (Starr, Sorbara, Cioffi, & Silverman, 1999). What really fuelled DSL's deployment is the fact that it allowed network operators to use their already laid copper infrastructure to provide broadband connectivity services to their customers, without actually needing to make large investments in access infrastructure. However, DSL schemes may be considered as not so future-proof solutions, since the aging copper-based infrastructure is rapidly approaching its fundamental speed limits. For example, the most recent DSL variant, VDSL2 (Very high speed DSL, version 2), represents the current state-of-the-art and can theoretically offer up to 100 Mbps symmetric data transfer rates (though much less in real conditions), but only for very small distances (~300 meters). New services will possibly push data rates beyond the capabilities of such networks: for instance a multi-channel HDTV service will pose strong challenges in order to operate efficiently. In addition, the fact that high speed and large distances cannot be achieved

simultaneously, results in solutions that are not economically favorable (for example requiring the installation of a large number of new neighborhood nodes). Furthermore, the emerging requirement for capacity symmetry for certain applications or businesses constitutes a significant challenge.

It is evident that these copper-based access networks create a bottleneck in terms of bandwidth and service provision between the operator and the (corporate or residential) end-user. Contrary to that, optical access architectures enable communication via optical fibers that extend all the way from the telecom operator premises to the customer's home or office (or at least to close proximity), thus eliminating the need for data transfer over telephone wires. Such architectures offer a viable solution to the access network bottleneck problem (Green, 2006), and promise extremely high, symmetrical bandwidth to the end-user (Prat, Balaquer, Gene, Diaz, & Fiquerola, 2002). In addition, they future-proof the operator's CAPEX (capital expenditures) investment, as they offer easy and low-cost speed upscale, should such a need arises in the future. While the cost of deploying optical access networks has been prohibitively high in the past (Frigo, Iannone, & Reichmann, 2004), this has been falling steadily, and such networks are now likely to be the dominant broadband access technology within the next decade.

EXISTING BROADBAND SOLUTIONS AND THE NEED FOR MORE BANDWIDTH

Today, the most widely deployed broadband access solutions are DSL and Cable TV (CATV) networks. DSL is truly the current "king" of broadband with more than 200 million lines provisioned worldwide as of June 2007 (DSL Forum, 2007), accounting for more than 65% of the total broadband installations. DSL deployment has been a major political issue and priority in many countries that view it as a critical ingredient for their efficient transition into the modern knowledge-based economies and the future competitiveness of their industries. Indeed, DSL is now available in just about all developed countries, as well as in many countries that are now transitioning into market economies. In addition, DSL's constantly dropping prices are greatly helping its further adoption (OECD, 2007). What really fuelled DSL's deployment is the fact that it makes it possible for an operator to re-use its existing telephone wires so that they can deliver high bandwidth data services to end-users. In particular, since voice telephony is restricted to ~ 4 KHz, DSL operates at frequencies above that (for example up to 1 MHz for ADSL or 30 MHz for VDSL2), with different regions of this range allocated for upstream (US) or downstream (DS) traffic. This has enabled operators to more fully utilize the capabilities of their already laid (for telephone purposes) copper infrastructure, and to provide relatively high-speed data services to their customers, without actually needing to make large investments in access infrastructure. Currently, there are many DSL connection variants, with ADSL offering a maximum upstream rate of 1 Mbps and a maximum downstream rate of 8 Mbps, and operating over a maximum distance of ~ 5.5 Km (ITU-T 1999a; 1999b). Most operators have already boosted the available data transfer rates by upgrading to ADSL2+, a major improvement over the standard ADSL that offers up to 3.5Mbps and 24 Mbps for the upstream and downstream directions respectively (ITU-T, 2003b). Finally, the most recently standardized DSL variant, VDSL2 (ITU-T, 2006), represents the current state-of-the-art and can theoretically offer data rates of more than 200 Mbps (though much less in real conditions, especially in the upstream direction), but only for very small distances (Androulidakis, Kagklis, Doukoglou, & Skenter, 2004). For instance, VDSL2's highest speed can only be achieved in loops shorter than ~ 300 meters.

CATV networks are currently the second most popular broadband access technology, accounting for about 22% of the total broadband connections (Multimedia Research Group Inc., 2007). These are operated mostly by the cable TV industry, and they typically feature a hybrid fiber-coaxial (HFC) architecture. In particular, a fiber runs from the cable operator's central office to a neighborhood's optical node, from where the final drop to the subscriber is through a coaxial cable. The coaxial part of the network usually uses amplifiers, and splits the signal among many subscribers. The main limitation of CATV networks lies on the fact that they were not designed for data communications but merely for TV broadcasting. As such, they allocate most capacity to downstream traffic (for streaming channels) and only a small amount of bandwidth for upstream communications (as upstream traffic is minimal for TV distribution purposes), which also has to be shared among a large number of subscribers. Currently, a lot of effort is being devoted to the transformation of CATV networks for efficient data communications, so that they can compete with the newest DSL variants. Today most cable internet service is delivered via DOCSIS (Data Over Cable Service Interface Specifications) 1.1 (Data Over Cable Service Interface Specifications Group, 1999; ITU-T, 2004), while many operators are upgrading to DOCSIS 2.0 (ITU-T, 2002), which can offer up to 38 and 27 Mbps downstream and upstream respectively, but shared between many users (usually ~100 users). The recently agreed DOCSIS 3.0 can significantly boost the data transmission rates of CATV networks but deployment has not started yet, though its introduction is much anticipated in the coming years.

Although both DSL and cable networks have been evolving rapidly over the years, and some of the progress has truly been outstanding, they cannot always -and under all potential circumstances- be seen as definitive and future-proof access network solutions (de la Rosa, 2005). Currently, data traffic is increasing at phenomenal rates and we are now entering a new era of "triple-play" service bundles, where telecom operators or service providers offer data, video and voice services in a single package (Hellberg, Greene, & Boyes, 2007). This bundling of services over a converged IP network will present a multitude of bandwidth challenges to the existing copper infrastructure. New services such as video-on-demand (VoD), interactive IPTV, HDTV, 3DTV, multi-party videoconferencing, telemedicine, online gaming and other bandwidth intensive applications will definitely drive data rates beyond the capabilities of our aging access networks (Hirosaki, Emura, Hayano, & Tsutsumi, 2003). For instance, a multi-channel HDTV service may require up to ~100 Mbps to operate efficiently. With the current bandwidth growth rate of 42% per year, by 2020 we may need an astonishing 2000 Mbps per home (George, 2006). Even if this growth rate slows down, it is evident that neither DSL nor cable networks have the *"adequate"* and/or the *"full"* ability to keep up with increasing bandwidth demand and meet the challenges for future broadband applications. In addition, high speed and large distances cannot be achieved simultaneously with these technologies, resulting in solutions that are not always economically favourable (for example requiring the installation of a large number of new neighborhood nodes). Furthermore, the emerging requirement for capacity symmetry for certain applications or businesses constitutes a significant challenge for both technologies. Hence, multiple telecom operators in the international electronic communications arena have come to the point to consider that a "novel" solution is needed, able to adequately address users' future needs as well as to constitute a future-proof investment of their resources (European Commission, 2006). Fibers are suited to this job. Having already won the battle for the backbone networks, it seems natural for fibers to be considered as a reliable solution for constituting some of the future access networks (Personick, 2002).

In this work we present an overview of the technology and innovative characteristics of optical access networks, focusing mainly on passive optical network (PON) implementations. We discuss the already established PON standards (namely BPON, GPON and EPON) as well the emerging wavelength division multiplexing-based PONs (WDM-PONs). We then survey the current PON deployment efforts that are taking place worldwide, focusing mainly on the European Community area, and we highlight the regulatory and policy obstacles that need to be overcome in order for Europe to tackle its slow FTTH (fiber-to-the-home) growth rate and stay on the broadband race, when compared to Asia and the USA (Abrams, Becker, Fujimoto, O'Byrne, & Piehler, 2005).

OPTICAL ACCESS NETWORKS

Optical access networks are not a new concept, but instead have been considered as a solution for the subscriber access network for quite some time. Early proposals and developments can even be traced back to the late 70s - early 80s (Shimada, Hashimoto, & Okada, 1987; Stern, Ballance, Faulkner, Hornung, Payne, & Oakley, 1987). However, these were abandoned due to the technology not being mature enough, and most importantly, due to prohibitively high costs coupled with the fact that there was no actual bandwidth demand to necessitate the deployment of such networks at the time. However, regarding deployment costs, photonic and fiber-optic components have progressed remarkably in more recent years and volume production has dramatically reduced their cost. As a result, the cost of deploying optical access networks has fallen steadily to the point that worldwide deployment is now reality (Rashid, 2004). Fiber access networks are capable of delivering extremely high bandwidth at large distances beyond 20km and can cater for all current and predicted future voice, data and video services requirements (Bayvel, 2000). In this sense, it can be said that once fiber is installed, no significant further investments or re-engineering is likely to be required for decades. In addition, such networks offer quick and simple repair, low-cost maintenance and easy upgrade.

Optical access networks come in many "flavours" such as fiber-to-the-curb (FTTC), fiber-to-the-building (FTTB), fiber-to-the-home (FTTH) etc., depending on how close is the optical fiber terminated to the end-user. A FTTN (fiber-to-the-node) scheme, where a fiber is terminated close to the subscriber in a local node and the final drop is made through conventional copper wires (perhaps using VDSL2 for high speed transmission) may sound more cost effective and adequate for the current and near-future predicted needs (Gillespie, Orth, Profumo, & Webster, 1997). However, it might be that such solutions will not always be able to accommodate the future bandwidth requirements of subscribers, imposed by demanding services such as HDTV, 3DTV, and that further penetration of fibers will be needed, to a certain extent. Hence, under certain circumstances, a fully optical access network (i.e. FTTH) may appear as a quite "challenging" implementation in the local loop, and a promising future-wise solution (Keiser, 2006). Under this scope, we further discuss the FTTH perspective (Payne & Davey, 2002) in the following paragraphs.

We can distinguish three main FTTH deployment architectures, which are illustrated in Figure 1. The simplest and most straightforward way to deploy fiber in the local access loop is to use a point-to-point topology (Figure 1a), in which single, dedicated fibers link the network operator's local exchange with each end-user. It is evident that, though simple, this topology requires a very large number of fibers, as well as significant fiber termination space at the local exchange, making this solution quite expensive. In this sense, to reduce fiber deployment, it is possible to use a remote switch, which will act as a concentrator, close to the subscribers (Figure 1b). Assuming negligible distance from the switch to the subscribers, the deployment costs are lower, since now only a single fiber link is needed (the

link between the network operator's premises and the switch). This configuration does, however, have the drawback of the added extra expense of an active switch that needs to be placed in every neighborhood, as well as the extra operational expenditure of providing the electrical power needed for the switch to function. Therefore, a logical progression of this architecture would be one that does not require any active elements (like switches and concentrators) in the local loop. Such a point-to-multipoint architecture is shown in Figure 1c, and it is termed as Passive Optical Network (PON). In PONs, the neighborhood switches are replaced by inexpensive passive (i.e. requiring no electric power) splitters, whose only function is to split an incoming signal into many identical outputs (Effenberger, Ichibangase, & Yamashita, 2001]. PONs are viewed as possibly a very attractive solution for bringing fiber to the home, since they are comprised of only passive elements (fibers, splitters, splicers, etc.) and are therefore less costly. In addition to being capable of very high bandwidths, a PON can operate at distances of the order of 20km, significantly higher than the distances supported by high-speed DSL variants (Girard, 2005). Furthermore, PONs reduce installation and maintenance cost, whilst also allowing for easy upgrades to higher speeds, since upgrades need only be done centrally at the network operator's central office where the relevant active equipment is housed (Lin, 2006).

We can distinguish two main network elements in a PON implementation: The Optical Line Termination (OLT) and the Optical Network Unit (ONU). The OLT resides at the network operator's premises, while each user has its own ONU. Besides the basic star topology depicted in Figure 1c, addition of further splitters in a PON network allows the easy formation of many different point-to-multipoint topologies to suit particular network needs, nicely illustrating the high flexibility of PON architecture.

The main drawback of PONs is the need for complex mechanisms to allow shared media access to the subscribers so that data traffic collisions are avoided. This arises from the fact that although a PON is a point-to-multipoint topology from the OLT to the ONU (i.e. downstream direction), it is multipoint-to-point in the reverse (upstream) direction (Gumaste & Anthony, 2004). This simply means that data from two ONUs transmitted simultaneously will enter the main fiber link at the same time and collide. The OLT will, therefore, be unable to distinguish them. Hence, it is evident that there needs to be a shared media access mechanism implemented in the upstream direction, so that data from each ONU can reach the OLT without colliding and getting distorted. Note that an elegant, though expensive, way to achieve this would be to use a Wavelength Division Multiplexing (WDM) scheme, in which each ONU is allocated a particular wavelength, so that all ONUs can use the main fiber link simultaneously (Park, Lee, Jeong, Park, Ahn, & Song, 2004). The current preferred solution is based on a Time Division Multiplexing (TDM) scheme, in which each ONU is allowed to transmit data only at a particular time window dictated by the OLT. This means that only a single upstream wavelength is needed, which considerably lowers the associated costs as a universal type of ONU can be employed in every site.

Finally, should the need for more bandwidth is required in the future, the PON architecture allows for easy upscale. For example, this can be accomplished either through the use of statistical multiplexing or through a combination of WDM and TDM schemes in the access fiber (where different wavelengths carry different TDM PON streams) or even by using a WDM access scheme, where each subscriber is allocated a different wavelength as noted above (ITU-T, 2001). The latter is actually a very exciting proposition that promises to bring truly enormous bandwidth to the end-users by fully utilizing the fiber's spectral windows. Such a proposition currently attracts a significant amount of research interest, concentrated mostly on lowering the high costs of such

Figure 1: FTTH deployment architectures: (a) point-to-point; (b) curb-switched active Ethernet, and; (c) passive optical Network (PON).

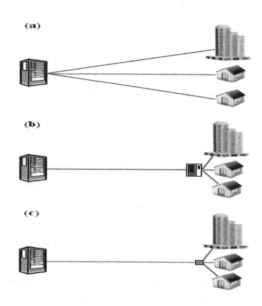

a scheme by replacing the need for an expensive light source (usually a tuneable distributed feedback laser) at the side of the end-user (Wagner & Kobrinski, 1989).

TYPES OF PASSIVE OPTICAL NETWORKS

There are currently three main PON standards: Broadband PON (BPON), Gigabit PON (GPON) and Ethernet PON (EPON). The first two are ratified by the ITU (International Telecommunications Union), while the third from the IEEE (Institute of Electrical and Electronic Engineers). The initial specification for a PON-based optical network was actually called APON and described PON operation using Asynchronous Transfer Mode (ATM) encapsulation. This was first developed in the 1990s by the Full Service Access Network group (FSAN), an alliance between several major worldwide network operators. APON has since further developed, and its name was later changed to the current BPON (broadband PON) (ITU-T, 2005) to reflect the support for additional services other than ATM data traffic such as Ethernet access, video content delivery etc.

The final BPON version provides speeds of up to 1244 and 622 Mbps downstream and upstream respectively, though the most common variant supports 622 Mbps of downstream traffic and 155 Mbps of upstream bandwidth. As already stated, upstream and downstream BPON traffic uses ATM encapsulation. Since all cells transmitted from the OLT reach all ONUs, to provide security BPON uses an algorithm-based scheme to scramble downstream traffic. The upstream transmission is governed by a TDM scheme.

The GPON standard, first released in 2003 (ITU-T, 2003a), allows for a significant bandwidth boost and can provide symmetric downstream/upstream rates of up to 2488 Mbps, although the asymmetric variant of 2488/1244 Mbps downstream/upstream is the most common. The split ratio can be up to 1:128. The significant advantage of GPON is that it abandons the legacy ATM encapsulation, and instead utilizes the new GPON or Generic Encapsulation Mode (GEM) that allows framing of a mix of TDM cells and packets like Ethernet. ATM traffic is also still possible, since a GPON frame can carry ATM cells and GEM data simultaneously. Overall, GEM is a very well designed encapsulation method that can deliver delay-sensitive data such as video traffic with high efficiency.

Finally, the EPON standard (IEEE, 2004) was finalized in 2004 as part of the IEEE "Ethernet in the First Mile Project" and is the main competitor to GPON. It uses standard Ethernet (802.3) framing and can offer symmetric downstream/upstream rates of up to 1Gbps and split ratios of up to 1:64, while work on a 10 Gbps version has started recently. Since Ethernet is not meant for point-to-multipoint architectures, to account for the broadcasting downstream nature of PONs, EPON defines a new scheduling protocol, the Multi-Point Control Protocol (MPCP), that al-

locates transmission time to ONUs so that data collisions are avoided (Kramer, 2005).

It should be noted that all three PON standards are optically similar. They all use a simple WDM scheme to offer full duplex operation (i.e. simultaneous downstream and upstream traffic) over a single fiber, in which the 1310nm-centred band is used for upstream data, the 1490nm-centred band is used for downstream traffic, while the 1550nm-centred band is reserved for future TV broadcasting.

The three PON standards are summarized in Table 1. A quick comparison with the features of VDSL2 illustrates how significantly higher are both the bandwidth offered by a PON implementation as well as the distance this bandwidth can be attained for.

WDM-PONS

The PON architecture allows for easy service upscale should the need for more even more bandwidth arises. This can be elegantly accomplished through a combination of WDM and TDM schemes in the access fiber (where different wavelengths carry different TDM PON streams) or even better by using a WDM access scheme, where each subscriber is allocated a different wavelength (ITU-T, 2001). Indeed, WDM-PONs are a very exciting proposition that represents the next generation in PON development, and promise to bring truly enormous bandwidth to the end-users by fully utilizing the fiber's spectral windows (Zhang, Lin, Huo, Wang, & Chan, 2006).

As already noted, standard PONs (i.e. BPON, GPON and EPON) utilize single wavelengths for data communication. In particular, 1310 nm is used for upstream communication, whereas 1490 nm is used in the downstream direction. Thus the bandwidth in these two wavelengths is inherently shared between all end-users. It is evident that the full bandwidth capacity of fibers is not utilized in such schemes. In contrast, in a WDM-PON each

Table 1. Passive Optical Networks Standards

	BPON	GPON	EPON
Standard	ITU-T G.983	ITU-T G.984	IEEE 802.3ah
Upstream rate	622 Mbps	2488 Mbps	1250 Mbps
Downstream rate	1244 Mbps	2488 Mbps	1250 Mbps
Distance (Km)	20	20	20
Upstream λ (nm)	1310	1310	1310
Downstream λ (nm)	1490/1550	1490/1550	1490/1550
Upstream efficiency	87%	92%	65%
Downstream efficiency	87%	92%	72%
Transmission	ATM	Ethernet and/or ATM via GEM	Ethernet
Voice support	TDM over ATM (VoATM), TDM over IP (VoIP)	native TDM, TDM over ATM (VoATM), TDM over IP (VoIP)	TDM over IP (VoIP)
Video support	RF overlay (over 1550nm), IPTV	RF overlay (over 1550nm), IPTV	IPTV
Security (encryption)	AES	AES	none specified
Network protection	Standard	Standard	none specified

* Maximum values of transmission rates and operating distance are quoted

ONU uses a dedicated separate wavelength to communicate with the appropriate OLT. In this way, a point-to-point connection is established between each ONU and the OLT, resulting in a dedicated communication channel for each end user, and hence a dedicated fixed bandwidth. The different wavelengths are routed from the OLT to the appropriate ONUs and backward using a wavelength multiplexer/demultiplexer (e.g. a grating) at the splitting point. A simple illustration of a WDM-PON is shown in Figure 2. It should be noted that there are no time synchronization requirements between the different wavelength channels, and that even the same wavelength could be used for both upstream and downstream traffic between each ONU-OLT connection, provided that noise from back-reflections (e.g. at fiber imperfections or connection points) are minimal.

Figure 2. A simple WDM-PON implementation where multiple wavelengths are assigned for communication between the various Optical Network Units (ONUs) and the Optical Line Termination (OLT).

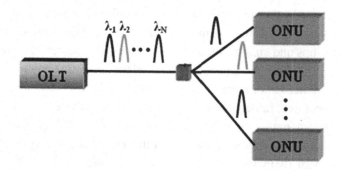

Although WDM-PONs truly have the potential of supporting massive bandwidth rates as well as scalable operation and ease of customization, their major disadvantage is associated with the high costs of the required equipment. Hence, there is currently a significant amount of research interest on finding ways to lower the high costs of such schemes, mostly by addressing the need for expensive broadband light sources (Wagner & Kobrinski, 1989).

In particular, in the most straightforward WDM-PON implementation a dedicated wavelength is assigned to each ONU in a fixed manner. As a result, for upstream communication each ONU needs to possess its own wavelength-specific light source. This necessitates the use of expensive widely tuneable laser diodes, which are usually high-cost Distributed Feedback (DFB) lasers that offer excellent performance since they support tuneable purely single-mode operation. Vertical-Cavity Surface-Emitting Lasers (VCSELs), may provide a more attractive alternative as they offer lower-cost manufacturing. Importantly, and in addition to the high cost of the associated light sources, the requirement for a wavelength-specific source at each ONU in WDM-PONs also dramatically increases the cost and the complexity of administration and maintenance of such networks (e.g. manual reconfiguration of the ONU is required each time a change or upgrade is performed).

To lower the cost of WDM-PONs it is, therefore, highly desirable to use a so-called "colorless" (i.e. not wavelength-specific) ONU. This means that identical ONUs can be distributed to all end-users, thus decreasing the ONU cost through mass production and also greatly reducing the operation, maintenance and upgrade costs. Several approaches have been proposed to achieve colorless ONUs. For example a spectral-slicing approach (Frigo, Iannone, & Reichmann, 1998) proposes the use of a broadband light source (e.g. a superluminescent LED) whose output is appropriately spectrally filtered at the remote node. This method, however, has a negative effect on the available optical power budget, which in turn decreases the operating network distance. Alternatively, a centralized light source at the OLT can be coupled with injection-locked Fabry-Perot lasers at the ONUs (Park, Lee, Jeong, Park, Ahn, & Song, 2004). Another very interesting approach is to re-use the downstream light emitted by the OLT as the upstream light that is provided by each ONU (Kani, Teshima, Akimoto, Takachio, Suzuki, Iwatsuki, & Ishii, 2003). In this approach a part of the original downstream light is re-modulated by the ONU to provide the upstream signal and sent back to the ONU. The major advantage of this approach is that it eliminates the need for the existence of a light source at the ONUs, hence greatly reducing the equipment and maintenance costs. Re-modulation at the ONU side can be achieved through the use of low cost reflective optical modulators, but again there is a penalty in the available optical budget achieved. Higher powers can be achieved if re-modulation is performed through reflective semiconductor optical amplifiers (R-SOAs) (Healey, Townsend, Ford, Johnston, Townley et *al.*, 2001).

DEPLOYMENT OF OPTICAL ACCESS NETWORKS IN ASIA AND THE US: A SUCCESS STORY

Remarkable progress has been achieved recently with regard to PON deployment, and mass worldwide installations are now reality. Asia is clearly the leader in the area of very high-speeds deployments, with large scale deployments having been rolled out by Japanese and Korean incumbents, and more recently by alternative operators and power utilities in Japan. NTT (Nippon Telegraph & Telephone, Japan's dominant network operator) pushed fibre optic cable deep into its network early on, with its two regional subsidiaries providing FTTH access based primarily on PON technologies.

Currently, Japan is the world leader with regard to FTTH adoption, even though cities like Tokyo are amongst the most densely populated and, hence, DSL could prove a strong opponent (Shinohara, 2005). By the end of 2006, NTT had over 7 million FTTH subscribers. Currently the FTTH annual growth rate (about 60%) is larger than that of DSL, and NTT expected the number of FTTH subscribers to exceed the number of DSL subscribers for the first time by the end of 2007 (FTTH Council, 2007a). NTT presently has fibre coverage in 80% of its distribution area and says that it expects to have about 30 million FTTH subscribers by 2010 (Wieland, 2006). NTT, whose deployments include both EPONs and BPONs, also faces competition from very active alternative providers offering higher-speed FTTH connections but at a smaller scale. As of the end of 2007, FTTH subscribers exceeded 10 million households in Japan, with nearly 300,000 new subscribers per month. Recent studies have demonstrated that nearly 60,000 subscribers per month switch from DSL to FTTH in the national Japanese market (FTTH Council Europe, 2007). Currently, it is quite remarkable that 50% of Japanese end users have access either to VDSL or FTTH, with the Government's "e-Japan" national plan aiming to establish Japan as a "fibre optic nation" by 2010; the local Government's objective is to bring fibre to 30 million homes at a rate of 3 million per year.

Apart from Japan, FTTH adoption has been very quick in other Asian countries as well, with South Korea and Hong-Kong leading the way. In particular, in South Korea, the incumbent telecom operator KT and other operators rely chiefly on bringing fibre optic cable to the basement of buildings, then extended by LAN (Ethernet) or VDSL. Japanese and Korean figures as "pioneers" can be explained by the support of public authorities, along with certain particular characteristics such as high population density in the large metropolitan areas of Tokyo and Seoul, and the prominence of high-rise buildings and aerial connections, all

of which make for cost-effective FTTx projects. In total, Asia today accounts for about 70% of worldwide FTTH installations (FTTH Council, 2007b).

In the US, the national regulatory authority made clear in 2001 that it would forebear from regulating fibre deployments for at least five years. This was followed by considerable fibre deployment to the home, especially by Verizon (www.verizon.com), and to the cabinet by other Bell companies, in the context of competition between those companies and cable. In particular, FTTH deployments were initially confined to trials in a few municipalities. The USA accounts for about 10% of the current worldwide FTTH installations, with the majority of them being GPON. The US FTTH growth rate has currently reached an astonishing 100% per year, with about 1.3 million subscribers and 8 million connected homes on March 2007. The number of subscribers has overcome the limit of 2 million household subscribers as of the end of 2007 (FTTH Council Europe, 2007). The deployment is now advancing rapidly, being greatly helped by the Federal Communications Commission (FCC) decision to hand the incumbent local-exchange carriers (ILECs) a virtual monopoly on broadband services, provided that those services reach customers over new fiber-optic infrastructure, and the progressive elimination of unbundling local loop policy, that has in the past presented obstacles for large operators to invest in new optical-access technologies (Federal Communications Commission-FCC, 2003). Currently, the US FTTH deployment is being led mostly by Verizon and AT&T (ex-SBC) who deploy GPONs, while other operators are following as well.

At the end of 2006, Verizon's base could reach 900,000 subscribers, for 6 million homes passed. AT&T (Project Lightspeed), in the meantime, relies on FTTN solutions (for 17 million homes) based on ADSL2+ or VDSL, except in greenfield areas (1 million FTTH households).

Out-distanced in the broadband access market by cable modem operators, and handicapped compared to European telecommunication operators by long copper loops, Verizon and AT&T are banking on the deployment of their new infrastructures to increase their market share and allow them to roll out services like HDTV. Their projects are benefiting from recent decisions from the courts and the FCC, absolving them of having to share the optical infrastructure with third parties (i.e. avoiding them the requirement to unbundle their new developed fiber-based infrastructures). However, in most parts of the country, they are nonetheless facing one sizeable obstacle, namely the obligation to negotiate franchise rights with each municipality (Montagne, 2006).

EUROPEAN DEVELOPMENTS

Compared with the FTTH growth in Asia and the USA, evidently Europe seems to be lagging in related developments. Growth is relatively slow, but encouraging progress has been achieved in some countries recently (Young, 2007). However, fortunes differ country by country. The annual FTTH growth rate in Europe is currently at about 13% (significantly lower than that of the USA and Asia), and there were about 780,000 subscribers in Europe by the end of 2006, though the number of homes passed is about 3 million (FTTH Council Europe, 2007). However, 97% of these subscribers were located in just five countries, namely Sweden, Italy, Norway, Netherlands and Denmark that show the most mobility towards FTTH (Tauber, 2007). The Nordic countries in particular all have comparatively large fibre rollouts achieved over a sustained period of network building. The rest of the continent is far behind, with a lot of countries not even having formulated plans for future deployment, due to certain reasons.

However, drivers for corresponding deployment exist -varying from country to country- including, *for example*, cable TV penetration,

encroachment of VoIP traffic on telecommunications providers' traditional voice business, the existence of national plans for broadband development, client's willingness to pay for broadband services and the level of intermodal competition in the market(s), the maturity of deregulation/ unbundling, the ease of construction-related issues (i.e. cost and facility of civil works, in-house wiring, and so forth) etc. Some other factors affecting further development relate to the state and the "age" of existing network infrastructure, the length of the local loop, the distribution of number of users and the number of street cabinets for local exchange.

The challenge becomes of remarkable importance as certain studies (Heavy Reading, 2006) propose that about 16.3 million homes are expected to be connected to FTTH in Europe by the end of 2011, just over 8% of all European homes (by considering the 25 EU Member States at that time).

Recent approaches (IDATE, 2006) have demonstrated that civil engineering works can account for more than 70% of relevant network deployment costs. Furthermore, construction costs appear to be higher in Europe, than in other territories (because in much of the US and in Japan it is possible to use aerial fiber). In order to reduce such costs, some private operators have decided to use existing passive infrastructures (for example, Milan and Paris sewer systems) to deploy fibre as close as possible to the buildings (Amendola & Pupillo, 2008). An alternative option might be the adoption of nationwide urban planning laws which require greater sharing of existing or new buried ducts (especially in Europe where there are no longer aerial deployments in urban areas). However, the latter option considers a "maturity" at the level of the market and cannot be easily applied at all cases, as there is a risk to "affect" investments in the market. Another difficulty when deploying fibre to the home is how to gain access to the building. Current approaches (i.e. negotiating installations one-by-one) are not always cost-effective for operators involved. This suggests the possibility to find a "harmonized" solution either at the national level or in each municipality (by reviewing urban planning laws and creating a system of "*Building Certification*").

Europe's slow FTTH adoption rate can be attributed mainly to the reluctance of the incumbent operators to invest in such developments. Most European incumbents & alternative operators are currently heavily investing in large DSL roll-outs and are focusing their (often limited) financial resources on this technology, which is well suited to less densely populated areas compared to Asian countries. In fact, very few market operators have the appropriate resource to invest massively in FTTH deployments. Indeed, as Table 2 shows, the FTTH European market is dominated by municipality and utility installations (Ecobilan S.A., 2008), which have a very large share reaching about 70%, while incumbent operators have only a minor share of only about 6.6% (IDATE, 2006), (Houbby, 2006).

Concerning a more detailed European depiction for the year 2007, several initiatives have been deployed in the scope of FTTH-related activities: For example, in France, four operators started to deploy fibre in Paris; in Netherlands, the Amsterdam CityNet company has started mass-deployment; in Denmark, several utility-companies continued to deploy their FTTH-based networks; in Switzerland, EWZ, the power-utility-company in Zurich, started to deploy fibre after more than 2/3 of the citizens have voted for FTTH in a referendum; in Germany, municipalities, utility-companies and several operators started to deploy FTTH; in Spain, Italy, Slovenia, the

national incumbent operators have announced initiation of specific FTTH plans.

The reluctance of European incumbent operators to deploy optical access networks can be traced mostly to legislation issues arising from European policy (Chochliouros & Spiliopoulou, 2003), but also to a lack of vision and motivation (Commission of the European Communities, 2006). The latter originate from the fact that European incumbents have not yet been adequately convinced about the necessity of fiber-optic access networks, at least for today's uses. Indeed, triple-play services, IPTV, VoD and other bandwidth intensive digital applications are relatively new in the European telecommunications arena. Most incumbents have either almost just started provisioning them or plan to do so in the near future. Hence, they are only now beginning to comprehend the explosion of bandwidth demand that will occur as such services proliferate and evolve, and the impact this will have on their aging copper networks. To remain competitive and profitable, there is an urgent need to upgrade their infrastructure (Distaso, Lupi, & Manenti, 2006).

However, the main deterrent for incumbent operators to deploy fiber in the access part has been the current relevant European policy and legislation environment (J.P. Morgan Securities, 2006). In order to fulfill the vision of being the most dynamic knowledge-based economy by 2010, the European Commission has initiated a number of relevant action plans (European Commission, 2005a) that are based on the argument that enhanced development and innovation in the communication sector can only be achieved through increased market competition (European

Table 2. European Investors in FTTx Installations (data collected for June 2006)

Municipalities/Utilities	69.4%
Housing Companies	12.4%
Alternative Operators/ISPs	11.6%
Incumbent Operators	6.6%

Commission, 2005b). Hence, as dictated by these action plans, incumbent operators in all European countries were "forced" to open their network infrastructure to all new entrants and provide them access at highly attractive prices that are dictated and monitored by newly established governmental authorities at national level. This, clearly, has had a major impact in the European telecom landscape. Indeed, multiple players have now entered the market and competition has increased significantly, although often at the expense of losses from the incumbents' side due to the current policies being much more favorable to new entrants. In the broader European context, facilities-based or infrastructure-based competition implicates competition between providers of the same or "comparable" services, but delivered by different kinds of networks. So far, such form of competition has mainly relied on the unbundling of the legacy copper network. Investment in other infrastructures like fiber-based ones are hampered by lack of investor confidence, due to regulatory uncertainty and its impact on pricing and the sustainability of new business models.

The part of the network that has been affected the most is, as expected, the access part (Cave, 2007). This is due to the copper access networks being the most costly and difficult to deploy. Hence, under the recent policies, new entrants have gained a golden opportunity to use the already laid infrastructure at highly competitive prices. This enables them to directly reach the final end-user, where large profit margins exist through the provisioning of a variety of data, voice and video services. In essence, new entrants have had the opportunity to obtain revenues without the burden of sunk costs and the risk of expensive infrastructure investments. The relevant access-network legislative action has been called "Unbundling of the Local Loop - ULL" (European Commission, 2000; European Parliament & Council of the European Union, 2000) and has been employed by new entrants throughout Europe at varying degrees (Chochliouros & Spiliopoulou, 2002). The most

common approach has been the bitstream access scheme, where a new entrant provisions DSL services through access infrastructure that is totally owned by the incumbent operator. However, few new operators have installed their own DSLAMs (DSL Access Multiplexers) or switching equipment, and consequently switched to shared- or full-access schemes, illustrating, unfortunately, the hesitation for significant investments by a large number of new entrants. It must be stressed, though, that the ULL policy has ultimately been a success: the price for a broadband DSL connection has fallen steadily over the recent years, and as a result broadband penetration in European households has significantly increased.

If, however, any deployed network infrastructure (independently of its specific nature) is completely new, the European regulation of December 2000 (European Commission, 2000) that imposes *ex-ante* obligations does not apply.

It remains that services supported by the new specific infrastructure, as well as prices charged, should "differentiate" themselves significantly from services supported by ADSL access in the case where the operator is considered as "powerful" (i.e. having significant market power-"SMP") in the relevant broadband market (European Parliament and Council of the European Union, 2002). If this occurs, the operator's investment could be managed as an emerging market and not be subject to an ex-ante regulation imposing access at a controlled price and requiring cost-oriented tariffs. As it currently happens, *however*, there is no clear evidence to indicate that potential changes being made by an expected "review"/update to the current European regulatory framework will give European telecommunications operators greater flexibility, although a clearer regulatory stance on the issues will help steer the various players' future decisions regarding various forms of FTTx deployments. The new European framework for regulation of electronic communication services came into force in 2003 (Chochliouros & Spiliopoulou, 2003) and is already subject to a debate over how it should be reformed from 2010 (Cave, 2007).

A Recommendation on relevant markets (European Commission, 2003) identifies those markets which, in the European Commission's view, may warrant ex-ante regulation. Unlike the previous regime, markets must be defined in accordance with the principles of competition law. The Recommendation practically "identifies" three cumulative criteria for describing those markets which are suitable for ex-ante regulation: (i) high and non-transitory barriers to entry over the period of application of remedies; (ii) the expected persistence of such barriers to entry beyond that period, making the prospect of effective competition unlikely, and; (iii) the inability of competition law adequately to address the particular issue.

The first version of the Recommendation (originally issued on 2003) identified 18 markets. The second version published in 2006, reduces it by a third, notably eliminating most retail markets (Commission of the European Communities, 2006).

A stable and predictable regulatory and public policy environment is "key" to enabling any FTTx investment and is an important asset for all market players, competitors as well as incumbents. Regulatory certainty may improve the conditions for investment and innovation and is, therefore, conducive to increase the competitive dynamics of the market.

In fact, under the current circumstances, the incumbent operators face regulatory uncertainty as the specifics of the actually applied ex-ante European regulation may significantly damage their business case(s), while the alternative operators that have invested in the unbundling of the local loop are disadvantaged by the passive infrastructure assets already owned by incumbents, and their ULL investments might become obsolete as incumbent pursue FTTC or FTTH strategies. In fact, next generation access (NGA) deployment, able to fulfil all future market requirements, is still

at an "early" stage in many EU Member States. Therefore, it is important to gain momentum the right moment for prospectively analysing the developments ongoing and any expected regulatory activity needs to reflect these developments. Regulatory clarity and predictability can so be considered as a prerequisite for an efficient investment in NGA networks, by restricting any risky nature (Amendola, & Pupillo, 2008).

The current EU ULL regulations have, thus, being positive overall for DSL access schemes, but are, *however*, proving a major obstacle for incumbent operators to seriously consider deploying fiber access networks (Hogan & Hartson LLP, & Analysys Consulting Ltd., 2006). These operators have already suffered a significant market share loss due to ULL (Squire, Sanders, & Dempsey L.L.P., 2002). There is, indeed, little incentive now for them to embark on FTTH projects: if they are forced to share their new costly fiber access infrastructure with their competitors, the return on their investment will be much more challenging. Although the ULL policy does not apply to new fiber infrastructure (as already mentioned above), rather primarily in the copper access loop, the European Commission has yet to release clear regulation directions with regard to access fibers and, hence, the situation remains vague. Under these conditions, incumbent operators are largely hesitant in embarking into FTTH deployment (as also evidenced by their very low involvement rate illustrated in Table 2), which in turn greatly affects broadband development in Europe.

Taking advantage of the incumbent operators' reluctance to deploy fiber, European municipalities and utility companies have rushed to fulfill the need for high-speed data access by deploying their own FTTH networks (notably in the Nordic countries) (Fuller, 2006). Most of these networks operate on an open-access scheme that comprises three basic horizontal components, illustrated in Figure 3.

In particular, in such a scheme, the network is built and owned by a municipality or utility

Figure 3. A carrier-neutral open-access FTTH scheme illustrating the three major operational roles

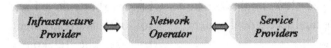

company that acts as an infrastructure provider (Chochliouros, Spiliopoulou, & Lalopoulos, 2005). An operator runs the network without being employed into any end-user services. The operator just offers access to various service providers that finally provide data and voice services to the subscribers. This can stimulate interest in the corresponding technology. Some representative and particularly successful citywide FTTH open-access scheme implementations are the Stokab in Sweden's Stockholm (www.stokab. se), the Reykjavik FTTH project in Iceland, and the Amsterdam's CityNet FTTH installation in the Netherlands (www.citynet.nl). Although such open-access FTTH schemes have proven particularly successful in municipality owned networks, there is no evidence that they will be a viable economic model for operators and it is still an open question whether they can significantly affect the European market as a whole.

It needs to be stressed that such open-access municipality and power utility built networks, involving public finance and governmental support, are not only based around financial considerations, but also have a strong public community character (Foxley, 2002). In particular, a strong objective of such implementations has been the local stimulation of economic development, as high-speed interconnected cities are likely to attract much more businesses and industries (Arnaud, 2001).

Although utility and municipality installations have provided a good starting paradigm for optical access networks in Europe, it is evident that such efforts are not able to lead to mass fiber deployment, as they implicate a "specific" infrastructure deployment, not always able to efficiently cover any form of needs and/or requirements implicated

by the fast growth of the global information society. There is no doubt that only a kind of "robust" involvement of incumbent operators is able to achieve such an important perspective and, *consequently,* Europe urgently needs to motivate them into embarking into such projects, by offering them appropriate and competitive incentives or terms/conditions for market deployment, especially by providing a clear regulatory framework. Otherwise, broadband development in Europe risks falling behind that of the USA or Asia, in turn affecting the future competitiveness of the region. A new (and properly updated) regulatory environment is needed now (Cave, Stumpf, & Valetti, 2006), one that will encourage investment and promote broader FTTx deployment by incumbents, but also able to stimulate alternative operators to participate in such developments as well (Jones, 2007). USA and Asia may provide good regulatory examples and/or (best) practices, where actions such as investment promotion and deregulation of new infrastructures, and restriction of ULL policies, have motivated large FTTx developments without affecting the healthy competition in the market.

In any case, several international approaches have so far demonstrated that deployment of fiber cannot be justified by short-term financial ratios, but rather by longer-term concerns about the shape of user demand and competitive pressures. The European Commission has already realized the need for an efficient new/updated regulatory framework, and the first rulings are expected very soon, as the current regulatory framework is actually under extended review. There is currently confidence that these rules will be drafted so as to efficiently enable Europe to enter the worldwide optical access race and tackle its slow FTTx growth rate. The year 2009 might then be a period where further optical access deployment across Europe may occur.

Another positive note is that even in today's "unclear" environment, some major European incumbents (WIK-Consult GmbH, 2007b) such as France Telecom[2], Deutsche Telekom[3] (WIK-Consult GmbH, 2007a), and others[4], have already embarked into pilot fiber-based projects (Montagne, 2006) and started deploying several types of optical access networks of various architectures (Ruderman, 2007), possibly feeling the pressure from the increased competition by the alternative operator-owned networks, as well as realizing the inevitable bandwidth demand rise that is expected. These operators are now deploying or making plans to deploy combined FTTN and VDSL solutions. It is important, here, to make an explicit distinguish between FTTN solutions and FTTH options. The former case deploys fibre optic cable all the way to an intermediate node located between the operator's central office (CO) and the user's terminal (The "node" is typically the street cabinet). Last mile connection termination is generally supplied by the phone line's copper infrastructure and appropriate xDSL equipment, particularly with VDSL2 options.

By "shortening" the copper infrastructure (i.e., up to the street cabinet instead of the MDF (Main Distribution Frame)) it becomes possible to achieve transmission speeds of up to 50 Mbit/s upstream (5 Mbit/s downstream) with VDSL access (for distances of 1,000 m or less from the street cabinet to the customer).

This type of configuration can guarantee and deliver very high speeds, but still depends on how close the node is to the user's building (up to 100 Mbps at less than 300 meters from the distribution node for VDSL2). When the intermediate node is a street cabinet, problems of coexistence can occur (signal disruption) with ADSL equipment installed by other operators, and lead to complex scenarios if the VDSL is unbundled. Also, FTTN+VDSL

architectures remain asymmetrical. VDSL2 has begun to be deployed, with one of the largest scale roll-outs, as already mentioned, being undertaken by Deutsche Telekom in Germany. There are three main considerations for incumbents to roll out FTTN-based solutions (Amendola & Pupillo, 2008): (i) First, due to the expected revenue upside, as incumbent operators believe that the superior capability of VDSL2 will allow them to charge premium prices; (ii) Secondly, due to cost savings from reduction in operating expenses and from the planned closure of many central offices when the VDSL2 network is completed, and; (iii) Thirdly, due to the responding strategic evaluation, as deployment of VDSL2 is a rational way to face competition from cable companies and defend market share against triple play cable offering.

As to FTTH, it is a solution that uses fibre optic cable from end-to-end between the central office and the user's premises. FTTH networks can be deployed using two distinct configurations: point-to-multipoint (fibre optic cable not dedicated to users but shared upstream of an intermediate node) or point-to-point (where each user has dedicated fibre optic cable from end-to-end). Concerning point-to-multipoint configurations, a distinction needs to be made between passive and active networks such as double star Ethernet.

Furthermore, FTTH deployment is strongly hindered by the cost of construction, which currently appears to be higher in Europe than in other territories. The main reason for this is that in much of the US and in Japan, it is possible to use aerial fiber. In many cities in Europe, cost of digging in fiber, including the final drop, is very high, and regulation in some cases may be onerous, increasing this cost further.

There is still, obviously, the issue of which optical platform to deploy, and a common trend has not emerged yet. Most open-access municipality networks have opted for simple point-to-point active Ethernet as it provides a number of manageability advantages that are attractive for such shorter-scale carrier-neutral networks (Van den Hoven, 2007). Incumbents, on the other hand, may possible follow the example of their US and Japanese counterparts, and choose a point-to-multipoint PON solution, where the battle will be between GPON and EPON (Rubenstein, 2005). There is also the question of whether to temporarily follow a FTTN hybrid solution (i.e. one that features deployment of fiber deep into the access network and the last drop to the subscriber is through the existing copper plant, possibly through VDSL) as a shorter-term less expensive solution.

In any case, further market growth and the expected exponential demand for bandwidth (mainly for video-related services) will affect deployment of all related infrastructures. The European regulation will play a very important role in determining the speed of the build-out, and decisions at both local and European level could have a major impact.

CONCLUSION

In this work, we have presented an overview of the emerging optical access network implementations, which promise to offer an innovative solution to the access network bottleneck problem. Such implementations can deliver exceptionally high bandwidth to the end-user, tremendously higher than any existing copper-wire based solution. They can, therefore, create a plethora of new business opportunities and pave the way for the introduction of advanced services that will pertain to a truly broadband society. Mass deployment projects are underway in Asia and the USA (with both continents featuring astonishingly high FTTH adoption rates), with passive optical networks being the preferred format. Europe is lagging in relevant developments, mainly due to the currently unclear European regulatory framework that is unable to predict the regulatory treatment of FTTH and consequently discourages incumbent operators in investing in these exciting

technologies. Currently most deployed networks in Europe are owned by municipalities or utility companies and operate on open-access schemes. They are typically based on point-to-point active Ethernet. There are also a number of trials being conducted by some incumbents, which are based on PON formats similar to USA and Asia. However, a new regulatory environment that relates to new access fiber installations is urgently needed now, in order to foster investment and promote large-scale deployment if Europe is to tackle its slow FTTH growth rate and stay on the broadband race, benefiting industry, commerce, public services and the future competitiveness of its economy in general.

This regulatory environment should consider the nature of the innovative service markets that are likely to emerge due to these ultra high-speed access networks, and, very importantly, the high levels of upfront deployment CAPEX required from telecommunication operators. In this sense, it may be beneficial for EU policy to promote and encourage public investment in passive infrastructures (trenches, ducts, pipes etc) so as to lower the initial construction costs and provide incentives for a network operator wanting to enter the FTTx arena, who, in turn, should focus in providing the opto-electronic and other network related infrastructure.

Overall, the regulatory framework for FTTx should provide certainty to investing companies, stimulate innovation and investment, and increase competition. With regard to the latter, a decision should be made whether to promote infrastructure competition or services competition (or both) among telecommunication players, as this should determine also the degree to which publicly-funded facilities infrastructure building will be utilized. Another issue to consider will be the entrance of new players once initial FTTx installations are deployed. While continuous infrastructure investment and service competition should be promoted, this should not be done in a way that discourages initial large-scale installation efforts, risking their long-term economic viability.

Obviously, the complexity of the issues surrounding the FTTx market, makes the task of compiling an efficient EU regulation framework far from an easy one. However, it is of outmost importance that the European Commission indicates fast which regulation direction will be taken, the type of competition that will be encouraged and the improvements to be undertaken, as this will shape Europe's future networks and significantly affect its economy and business.

REFERENCES

Abrams, M., Becker, P. C., Fujimoto, Y., O'Byrne, V., & Piehler, D. (2005). FTTP deployments in the United States and Japan – Equipment Choices and Service Provider Imperatives. *Journal of Lightwave Technology, 23*(1), 236–246. doi:10.1109/JLT.2004.840340

Amendola, G. B., & Pupillo, L. M. (2008). The Economics of Next Generation Access Networks and Regulatory Governance: Towards Geographic Patterns of Regulation. *Communications & Strategies, 69*, 85-105. Retrieved on November 17, 2008, from: http://mpra.ub.uni-muenchen.de/8823/

Androulidakis, S., Kagklis, D., Doukoglou, T., & Skenter, S. (2004). ADSL2: A sequel better than the original? *IEE Communications Engineering Magazine, June/July Issue* (pp. 22-27).

Arnaud, B. S. (2001). *Telecom issues and their impact on FTTx architectural designs (FTTH Council)*. CANARIE, Inc. Retrieved on August 22, 2004, from: http://www.canarie.ca/canet4/library/customer.html

Bayvel, P. (2000). Future high-capacity optical telecommunications networks. *Philosophical Transactions of the Royal Society of London. Series A: Mathematical and Physical Sciences, 358*, 303–329. doi:10.1098/rsta.2000.0533

Cave, M. (2007). *The regulation of access in telecommunications: A European perspective. (Revised, April 2007)*. Warwick, UK: Warwick Business School, University of Warwick. Retrieved on October 28, 2008, from: http://www.econ.upf.edu/docs/seminars/cave.pdf.

Cave, M., Stumpf, U., & Valetti, T. (2006, June). *A Review of certain markets included in the Commission's Recommendation on Relevant Markets subject to ex ante Regulation. Report for the European Commission.* Brussels, Belgium: European Commission.

Chochliouros, I. P., & Spiliopoulou, A. S. (2002). Local loop unbundling policy measures as an initiative factor for the competitive development of the European electronic communications markets. [TCN]. *The Journal of the Communications Network, 1*(2), 85–91.

Chochliouros, I. P., & Spiliopoulou, A. S. (2003). Innovative Horizons for Europe: The New European Telecom Framework for the Development of Modern Electronic Networks and Services. [TCN]. *The Journal of The Communications Network, 2*(4), 53–62.

Chochliouros, I.P., & Spiliopoulou, A.S. (2005). Broadband Access in the European Union: An Enabler for Technical Progress, Business Renewal and Social Development. *The International Journal of Infonomics (IJI), 1*, 05-21.

Chochliouros, I. P., Spiliopoulou, A. S., & Lalopoulos, G. K. (2005). Dark Optical Fiber as a Modern Solution for Broadband Networked Cities. In M. Pagani (Ed.), *The Encyclopedia of Multimedia Technology and Networking,* (pp. 158-164). Hershey, PA: IRM Press, Idea Group Inc.

Commission of the European Communities. (2006). *Communication on Market Reviews under the EU Regulatory Framework - Consolidating the internal market for electronic communications [COM(2006) 28 final, 06.02.2006].* Brussels, Belgium: Commission of the European Communities.

Data Over Cable Service Interface Specifications (DOCSIS) Group. (1999). *DOCSIS 1.1 Interface Specification.* (www.cablemodem.com).

de la Rosa, P. (2005, October). *The Access Network in the XXI Century.* Broadband World Forum, Madrid, Spain.

Distaso, W., Lupi, P., & Manenti, F. M. (2006). Platform Competition and Broadband Uptake. Theory and empirical evidence from the European Union. *Information Economics and Policy, 18,* 87–106. doi:10.1016/j.infoecopol.2005.07.002

DSL Forum (2007, June 5). *More than 200 million customers chose DSL.* News Release. Retrieved on July 23, 2007 from: http://www.dslforum.org/dslnews/pdfs/prbbwfasia060507.pdf

Ecobilan S.A. (2008, February). *Developing a generic approach for FTTH solutions using LCA methodology. Methodological Guide - Final Version.* Neuilly Sur Seine, France: FTTH Council Europe.

Effenberger, F. J., Ichibangase, H., & Yamashita, H. (2001). Advances in broadband passive optical networking technologies. *IEEE Communications Magazine, 39*(12), 118–124. doi:10.1109/35.968822

European Commission. (2000). *Communication on unbundled access to the local loop: Enabling the competitive provision of a full range of electronic communication services, including broadband multimedia and high-speed Internet [COM(2000) 394, 26.07.2000].* Brussels, Belgium: European Commission.

European Commission (2003). *Commission Recommendation 2003/311/EC of 11 February 2003 on relevant product and service markets within the electronic communications sector susceptible to ex ante regulation in accordance with Directive 2002/21/EC on a common regulatory framework for electronic communication networks and services. [Official Journal (OJ) C114 (pp.45-49)].* Brussels, Belgium: European Commission.

European Commission. (2005a). *Communication on i2010 - A European Information Society for growth and employment [COM(2005) 229 final, 01.06.2005].* Brussels, Belgium: European Commission.

European Commission. (2005b). *Commission Staff Working Paper on Communication from the Commission "i2010 - A European Information Society for growth and employment - Extended Impact Assessment" [SEC(2005) 717/2, 01.06.2005].* Brussels, Belgium: European Commission.

European Commission. (2006). *Communication to the Council, the European Parliament, the European Economic and Social Committee and the Committee of the Regions, on Bridging the Broadband Gap [COM(2006) 129 final, 20.03.2006].* Brussels, Belgium: European Commission.

European Parliament & Council of the European Union. (2000). *Regulation (EC) 2887/2000 on unbundled access to the local loop" [Official Journal (OJ) L336, 30.12.2002, pp.04-08].* Brussels, Belgium: European Parliament & Council of the European Union.

European Parliament and Council of the European Union (2002). *Directive 2002/21/EC of March 7 2002 on a common regulatory framework for electronic communications networks and services (Framework Directive), (Official Journal (OJ) L108, 24.04.2002, pp.33-50).* Brussels, Belgium: European Parliament and Council of the European Union.

Federal Communications Commission (FCC). (2003). *Triennial Review Order, document FCC-03-36A1: Review of the Section 251 Unbundling Obligations of Incumbent Local Exchange Carriers.* Washington, DC: FCC.

Foxley, D. (2002). *Dark fiber.* TechTarget. com, Inc. Retrieved on August 11, 2004, from: http://searchnetworking.techtarget.com// sDefinition/0,sid7_gci21189,00.html

Frigo, N. J., Iannone, P. P., & Reichmann, K. C. (1998, September 20-24). Spectral slicing in WDM passive optical networks for local access. In *Proceedings of European Conference on Optical Communications (ECOC)* (pp. 119-120). Madrid: Spain.

Frigo, N. J., Iannone, P. P., & Reichmann, K. C. (2004). A view of fiber to the home economics. *IEEE Optical Communications, 42*(8), S16–S23. doi:10.1109/MCOM.2004.1321382

FTTH Council. (2007a). *FTTH/FTTP Update April 2007 - Study by RVA LLC.* Retrieved on June 06, 2008 from: http://www.ftthcouncil.org/ documents/800832.pdf.

FTTH Council. (2007b). *Asia leads the world in Fiber-To-The-Home penetration.* Retrieved on June 06, 2008 from: http://www.ftthcouncil. org/?t=231.

FTTH Council Europe. (2007). *Europe at the Speed of Light - Presentation by J. M. Van Bogaert.* Retrieved on August 15, 2008 from: http://www. ftthcouncil.com.

Fuller, M. (2006). European municipalities lead FTTH charge. *Lightwave FTTX Direct Newsletter.* Retrieved on July 24, 2007, from: http:// www.localret.es/localretnews/bandaampla/num9/ docs/5num9.pdf.

George, J. (2006). Start thinking about 3 to 30 Gbps by 2030! *Broadband Properties, September 2006,* 42-47. Retrieved on July 14, 2007 from: *www.broadbandproperties.com/2006issues/sep-06issues/george_sep.pdf*

Gillespie, A., Orth, B., Profumo, A., & Webster, S. (1997). Evolving Access Networks: A European Perspective. *IEEE Communications Magazine, 35*(3), 47–54. doi:10.1109/35.581306

Girard, A. (2005). *FTTx PON Technology and Testing.* Tukwila, WA: EXFO Electro-Optical Engineering Inc.

Green, P. E. (2006). *Fiber To The Home: The New Empowerment.* New Jersey: John Wiley & Sons Inc.

Gumaste, A., & Anthony, T. (2004). *First Mile Access Networks and Enabling Technologies.* Indianapolis, IN: Cisco Press.

Healey, P., Townsend, P., Ford, C., Johnston, L., Townley, P., Lealman, I. , et *al.* (2001). Spectral slicing WDM-PON using wavelength-seeded reflective SOAs. *IEE Electron. Lett., 37*(19), 1181–1182. doi:10.1049/el:20010786

Heavy Reading (2006). *FTTH in Europe: Forecast & Prognosis, 2006-2011.* New York: Heavy Reading. Retrieved on November 18, 2008, from: http:// www.ftthcouncil.eu/documents/studies/Heavy_ Reading_FTTH_Europe_2006_final.pdf

Hellberg, C., Greene, D., & Boyes, T. (2007). *Broadband Network Architectures: Designing and Deploying Triple Play Services.* Prentice-Hall.

Hirosaki, B., Emura, K., Hayano, S., & Tsutsumi, H. (2003). Next-Generation Optical Networks as a Value Creation Platform. *IEEE Communications Magazine, 41*(9), 65–71. doi:10.1109/ MCOM.2003.1232238

Hogan & Hartson LLP, & Analysys Consulting Ltd. (2006). *Preparing the next steps in regulation of electronic communications. A contribution to the review of the electronic communications regulatory framework. Final Report for the European Commission, July 2006 - Service Contract No.05/48622*. Brussels, Belgium: European Commission.

Houbby, R. (2006, September 24-25). FTTH: Where, When, Why and How Much? In *Proceedings of Optical Network Europe (ONE) 2006 - Vision Strategies and Execution*. Cannes, France.

Institut de l'Audiovisuel et des Télécommuncations en Europe (IDATE) (2006, August). *FTTH Deployment - When and why?* Montpellier, France: IDATE.

Institute of Electrical and Electronic Engineers (IEEE). (2004). *IEEE 802.3ah Standard Ethernet in the First Mile*. New York: IEEE.

International Telecommunication Union - Telecommunications Standardization Sector (ITU-T). (1999a). *ITU-T Recommendation G.992.1: Asymmetric digital subscriber line (ADSL) transceivers*. Geneva, Switzerland: ITU-T.

International Telecommunication Union - Telecommunications Standardization Sector (ITU-T). (1999b). *ITU-T Recommendation G.992.2: Splitterless asymmetric digital subscriber line (ADSL) transceivers*. Geneva, Switzerland: ITU-T.

International Telecommunication Union - Telecommunications Standardization Sector (ITU-T). (2001). ITU-T Recommendation G.983.3: *A Broadband Optical Access System with Increased Service Capability by Wavelength Allocation*. Geneva, Switzerland: ITU-T.

International Telecommunication Union - Telecommunications Standardization Sector (ITU-T). (2002). *ITU-T Recommendation J.122: Second-generation transmission systems for interactive cable television services - IP cable modems*. Geneva, Switzerland: ITU-T.

International Telecommunication Union - Telecommunications Standardization Sector (ITU-T). (2003a). *ITU-T Recommendation G.984.1: Gigabit-capable Passive Optical Networks (GPON): General characteristics*. Geneva, Switzerland: ITU-T.

International Telecommunication Union - Telecommunications Standardization Sector (ITU-T). (2003b). *ITU-T Recommendation G.992.5: Asymmetric Digital Subscriber Line (ADSL) transceivers - Extended bandwidth ADSL2 (ADSL2plus)*. Geneva, Switzerland: ITU-T.

International Telecommunication Union - Telecommunications Standardization Sector (ITU-T). (2004). *ITU-T Recommendation J.112 Annex B: Data-over-cable service interface specifications: Radio-frequency interface specification*. Geneva, Switzerland: ITU.

International Telecommunication Union - Telecommunications Standardization Sector (ITU-T). (2005). *ITU-T Recommendation G.983.1: Broadband optical access systems based on Passive Optical Networks (PON)*. Geneva, Switzerland: ITU-T.

International Telecommunication Union - Telecommunications Standardization Sector (ITU-T). (2006). *ITU-T Recommendation G.993.2: Very high speed digital subscriber line transceivers 2 (VDSL2)*. Geneva, Switzerland: ITU.

Jones, R. (2007, February). *Opportunities in Fibre to the Home (FTTH) and How to Make a First Assessment*. London, UK: Ventura Team LLP. Retrieved on December 06, 2008 from: http://www.localret.es/localretnews/bandaampla/num18/docs/4num18.pdf.

J. P. Morgan Securities (2006). *The fibre battle – Changing dynamics in European wireline*. European Equity Research.

Kani, J., Teshima, M., Akimoto, K., Takachio, N., Suzuki, H., Iwatsuki, K., & Ishii, M. (2003). A WDM based Optical Access Network for Wide-Area Gigabit Access Services. *IEEE Communications Magazine, 41*(2), 43–48. doi:10.1109/MCOM.2003.1179497

Keiser, G. (2006). *FTTX Concepts and Applications*. New Jersey: Wiley Interscience.

Kramer, G. (2005). *Ethernet Passive Optical Networks*. New York: McGraw-Hill Communications Engineering.

Lin, C. (2006). *Broadband Optical Access Networks and Fiber-to-the-Home. Systems Technologies and Deployment Strategies*. John Wiley & Sons, Ltd.

Montagne, R. (2006, August). *The market's evolution to very high-speed*. Montpellier, France: Institut de l'Audiovisuel et des Télécommuncations en Europe (IDATE). Retrieved on October 06, 2008, from: http://www.idate.org.

Multimedia Research Group (MRG). Inc. (2007). *CATV Infrastructure: Assessing Strategies & Forecast - March 2007*. San Jose, CA: MRG, Inc.

Organization for Economic Coordination and Development (OECD) (2007). *OECD Communications Outlook 2007*. Paris, France: OECD. Retrieved on July 30, 2007, from: *http:www//213.253.134.43/oecd/pdfs/browseit/9307021E.PDF*

Park, S.-J., Lee, C.-H., Jeong, K.-T., Park, H.-J., Ahn, J.-G., & Song, K.-H. (2004). Fiber-To-The-Home Services based Wavelength-Division-Multiplexing Passive Optical Network. *Journal of Lightwave Technology, 22*(11), 2582–2590. doi:10.1109/JLT.2004.834504

Payne, D. B., & Davey, R. P. (2002). The future of fibre access systems? *BT Technology Journal, 20*(4), 104–114. doi:10.1023/A:1021323331781

Personick, S. D. (2002). Evolving toward the Next-Generation Internet: Challenges in the Path Forward. *IEEE Communications Magazine, 40*(7), 72–76. doi:10.1109/MCOM.2002.1018010

Prat, J., Balaquer, P. E., Gene, J. M., Diaz, O., & Fiquerola, S. (2002). *Fiber-to-the-home Technologies*. Boston: John Wiley & Sons Inc.

Rashid, S. (2004). *PON Delivers Optical Access to the Masses*. Alcatel Communications. Retrieved on June 06, 2007, from: www.alcatel.com/bnd/fttu/18282_FTTU_article_final.pdf

Rubenstein, R. (2005). GPON versus EPON: the battle lines are drawn. *Lightwave Europe, November 2005 issue*. Retrieved on July 24, 2007, from: http://fibers.org/articles/news/7/12/5/1

Ruderman, K. (2007, May). European trends favour FTTX. *Lightwave Europe*. Retrieved on July 24, 2007, from: http://lw.pennnet.com/display_article/293410/63/ARTCL/none/none/European-trends-favour-FTTX/

Shimada, S., Hashimoto, K., & Okada, K. (1987). Fiber optic subscriber loop systems for integrated services: the strategy for introducing fibers into the subscriber network. *Journal of Lightwave Technology, 5*(12), 1667–1675. doi:10.1109/JLT.1987.1075480

Shinohara, H. (2005). Broadband Access in Japan: Rapidly Growing FTTH Market. *IEEE Communications Magazine, 43*(9), 72–78. doi:10.1109/MCOM.2005.1509970

Squire, Sanders, & Dempsey L.L.P. (2002). *Legal study on part II of local loop unbundling sectoral inquiry (Contract No. Comp.IV/37.640)*. Brussels, Belgium: European Commission.

Starr, T., Sorbara, M., Cioffi, J., & Silverman, P. (1999). *Understanding Digital Subscriber Line Technology*. Upper Saddle River, NJ: Prentice Hall.

Stern, J. R., Ballance, J. W., Faulkner, D. W., Hornung, S., Payne, D. B., & Oakley, K. (1987). Passive optical local networks for telephony applications and beyond. *Electronics Letters, 23*(24), 1255–1256. doi:10.1049/el:19870872

Tauber, H. (2007). *FTTH: Europe must open its eyes*. Retrieved on July 10, 2007, from: http://optics.org/cws/article/industry/26992.

Van den Hoven, G. (2007). Why Europe is choosing point-to-point. *Lightwave Europe*. Retrieved on July 24, 2007, from: http://lw.pennnet.com/display_article/293398/63/ARTCL/none/none/Why-Europe-is-choosing-point-to-point/

Wagner, S. S., & Kobrinski, H. (1989). WDM Applications in Broadband Telecommunication Networks. *IEEE Communications Magazine, 27*(3), 22–30. doi:10.1109/35.20264

Wieland, K. (2006). *Voice and Low Prices Drive FTTH in Japan*. Retrieved on July 24, 2007, from: http://www.telecommagazine.com/newsglobe/article.asp?HH_ID=AR_2570

WIK-Consult GmbH. (2007a, June). *Possibilities and Prerequisites of a FTTx Strategy in Germany*. Bad Honnef, Germany: WIK-Consult GmbH. Retrieved on October 06, 2008, from: http://www.comreg.ie/_fileupload/publications/ComReg0795a.pdf.

WIK-Consult GmbH. (2007b, November). *Next Generation Bit Stream Access - Final. Study for the Commission for Communications Regulation (ComReg)*. Bad Honnef, Germany: WIK-Consult GmbH. Retrieved on October 06, 2008, from: http://www.comreg.ie/_fileupload/publications/ComReg0795a.pdf.

Young, G. (2007). Europe at the speed of light: Report of the FTTH Council of Europe Annual Conference. *Broadband Magazine, 291*, 18-19.

Zhang, B., Lin, C., Huo, L., Wang, Z., & Chan, C.-K. (2006, March). *A simple high-speed WDM PON utilizing a centralized supercontinuum broadband light source for colorless ONUs*. Paper presented at OFC'06, Anaheim, CA.

ENDNOTES

[1] This chapter reflects the personal views and opinions of the authors, and does not represent the official views of their respective employing institution.

[2] France Telecom began testing a pure FTTH solution (based on GPON) in Paris, the surrounding area and the main cities as well (Lille, Lyon, Toulouse, Poitiers, Marseille) to cover several thousand subscribers in June 2006. The solution was based on GPON technology. The original budget estimation was 270M€ till 2008, with a perspective of reaching 3 to 4.5 billions € till 2012. France Telecom is currently rolling out fibre in its ducts, which were inherited by the former monopoly. A certain number of these ducts are not occupied and can be used to roll out FTTH networks. At the time of the initiation of the project (2006), the company expected to pass 1 million homes by the end of 2008.

[3] Deutsche Telekom (DT) was the first to announce a large scale plan (investment

of 3 billion Euros (with € 500 million of these for the first stage of the project) and objective to deploy very high speed access infrastructure based on VDSL2 in 10 cities by the end of 2006, and 50 cities by the end of 2007. The company intended to invest in a VDSL network (combining fibre optic to the curb/node and VDSL on the copper sub-loop to the subscriber), having exclusive access to the infrastructures deployed. This project has, however, created conflicts with the European Commission in Brussels which, running contrary to an agreement previously negotiated with German authorities, required that DT make its infrastructures available to third parties. The European Commission refused the original company's request, citing application of the regulatory framework that requires access to be provided to any interested third party, even for partial use of the copper loop. The European Commission ordered the German government to "reverse" its initial decision to give DT exclusive use of its new FTTN-VDSL2 network. Negotiations are nonetheless still possible over the tariffs that DT can set for access to the network, which will not necessarily be cost-oriented, and could factor in the financial risks being shouldered by the operator. In the same scope, although France Telecom has announced a new infrastructure deployment based on VDSL2 technology, only a limited number of sites have been realized, on an experimental basis, until the European unbundling debate becomes clear.

4

Along with Deutsche Telekom, Swisscom in Switzerland was due to launch commercial VDSL2 services in summer 2006, and planned on having half of all Swiss households covered by its VDSL network by 2007. KPN in Netherlands and Belgacom in Belgium have also chosen FTTN + VDSL solutions to deliver very high-speed access to the Internet. Among the major European nations, the UK, *however*, demonstrates the least inclination to "move" toward fiber, and although British Telecom (BT) has been experimenting with FTTH for almost two decades, it does not currently consider a specific case or scenario to deploy it. In fact, BT considers the "joined FTTC + VDSL" case as a "potential near-term solution", apart from some specific direct FTTH deployments (for some green-field builds). As municipal government is weak and depends on central funding, there are no significant local initiatives for fiber deployment.

Chapter 2
Optical Access Comes of Age in a Packet–Delivery World

William Yue
Fujitsu Network Communications Inc., USA

Brian Hunck
Fujitsu Network Communications Inc., USA

ABSTRACT

The access network is the last loop, or last mile, in the provider network between the central office (CO) or point of presence (PoP) and the customer premises. Competitive pressure to provide high-bandwidth services (such as video) to consumers, and Ethernet transport to enterprises, is forcing service providers to rebuild their access networks. More optical fibers are being added in the last mile to meet these new bandwidth demands since legacy access networks have not been sufficient to support bandwidth-intensive applications. This chapter reviews the multiple definitions of "optical access" and the migration from direct copper loops to a variety of optical architectures, including Synchronous Optical Networking (SONET), Synchronous Digital Hierarchy (SDH), Fiber to the x (FTTx), Ethernet and wavelength delivery. Key business drivers such as carrier competition, bandwidth needs, and the reliability and service level agreement issues of optical technology are covered. The chapter concludes by considering the near future of optical access product trends and key optical deployment options in applications such as cellular backhaul. The data presented in this chapter is mainly based on our recent deployment experience in the North American optical access market segment.

INTRODUCTION

Optical access is the use of optical fiber to span the "local loop," or "last mile" of network transmission between a central office or point of presence and the customer premises. The idea of using optical fiber to connect last-mile equipment to the customer edge has been pursued for decades, but critical economic factors have only recently aligned to facilitate large-scale deployment.

Bandwidth demand is driven by new consumer and business applications such as High-Definition TV (HDTV), Web advertising, e-commerce, tele-medicine, high-quality videoconferencing and

DOI: 10.4018/978-1-60566-707-2.ch002

Figure 1. Multimedia Service Offering via Broadband Access Network

interactive gaming. This demand creates new revenue opportunities for service providers, who are seeking ways to compensate for the steadily eroding revenue stream from traditional voice applications.

The traditional copper-based access network has gone through a number of technology upgrades in recent decades. Advances were made to drive capacity per loop from a single voice channel up to a full Digital Signal 1 (DS1), also known as T-carrier (T1), of 28 channels (ANSI, 1991). Copper span distances also have consistently been improved via repeated T1 and various High bit-rate Digital Subscriber Line (HDSLx) technologies. Finally, the embedded copper plant has been stretched to provide significant data services by the use of a family of Asymmetric Digital Subscriber Line (ADSL) and Very High Bit-rate DSL (VDSL) solutions. These advancements have extended the useful life of the copper-based access network, but they are not able to scale to the much greater bandwidth levels now being targeted for new data services (Yue & Mocerino, 2007, p.1).

Optical access technologies have also been undergoing generational improvements in the last ten years, as technology has migrated from equipment platforms, which were deployed previ-

ously in the core network. Optical access products have become more affordable, more feature-rich, and more broadly available from a wide range of vendors. Advancements in the manufacturing of fiber optic cable have also seen steady progress, which has led to higher quality and lower prices. This has virtually eliminated any price premium when a service provider considers copper or fiber installation in new access construction. Today fiber deployment is the strategic choice when service providers choose to invest in new infrastructure for both residential and business access.

The optical access network has finally delivered on the promise of becoming an effective and economical choice for network operators who are driving innovative new services to the customer edge. Figure 1 shows the multiservice offering delivered via the broadband access network.

BACKGROUND

In the last decade, unprecedented levels of Internet usage and the need for new broadband applications have created exploding demand for emerging digital video and high-speed data services. In the U.S., the Federal Communications Commission

(FCC) mandated a plan for television stations to convert from traditional, analog National Television Standards Committee (NTSC) to Digital TV (DTV). This was completed mid-2009.

The world is entering the new broadband era; people across the globe are beginning to experience true broadband services. The ultimate goal is to deliver high-quality multimedia services, including voice, data and video, to users at home, in transit, or at their workplace. This universally accessible content will deliver a completely new user experience, which has never been possible before.

The increasing demand from users for this new experience affords service providers opportunities for expansion and increased revenue from digital video and data services. However, there are a number of hurdles that must be overcome before the access network infrastructure can provide a consistent, reliable, high-speed network between the customer and the application service centers. Service providers have already invested heavily in upgrading the transport bandwidth of their core networks. Reconfigurable Optical Add Drop Multiplexer (ROADM) and Dense Wavelength Division Multiplexer (DWDM) products have greatly expanded the capacity of core networks and simultaneously improved the flexibility of configuring transport paths. The process of upgrading the access network has lagged behind; relying instead on incremental improvements provided by changes in the use of existing copper-based access technologies.

Enterprise LANs, Wi-Fi hotspots, or campus Private Branch Exchange (PBX) networks can consume megabits to gigabits of transport throughput. Existing copper-based technologies such as DSL or cable modem, over a standard 12,000-foot access loop, can only provide several megabits. This mismatch of required service to offered rate becomes the bandwidth bottleneck, limiting the service provider's ability to deliver new broadband services, like HDTV, which require very high bandwidth to deliver good signal quality.

Fiber has been available for years, but the cable was expensive and there was limited customer demand for the few high-speed services offered by carriers. Fiber deployment was generally limited to dense metropolitan areas where a number of customers could be grouped together to amortize the high cost of fiber installation. This situation has changed because numerous factors have lowered the cost of fiber installation. For example, the price of fiber optic cable has been significantly reduced to eliminate any premium over copper cable; a wider variety of fiber cable types exist to fit application needs specifically; and fiber splicing and termination procedures have improved, eliminating specialized training and reducing the time required for site installation. These advances have made fiber the media type of choice for new builds in a variety of industries like Telco plant, intra-building office, and residential neighborhoods. Fiber is future-proof and can be scaled and upgraded to meet future demand in the same strand of glass, which protects the infrastructure investment.

Fiber penetration in the U.S. access network extends to only 10-12% overall. Remaining locations still have no fiber access facilities and represent a large opportunity for future optical fiber deployment to deliver broadband Ethernet and IP-based services.

Description of Optical Access

Optical Access is the use of optical fiber to span the "local loop," or "last mile," of a network. The access network is implemented with a wide variety of technologies and products for business and residential services like Passive Optical Network (PON), Wave Division Multiplexing (WDM), SONET/SDH (Telcordia, 2001) and Ethernet. Figure 2 shows samples of optical access products. This section discusses various optical access architectures and their impact.

Fiber-To-The-Home (FTTH), also known as Fiber-To-The-Premises (FTTP), is being de-

Figure 2: Optical Access Products

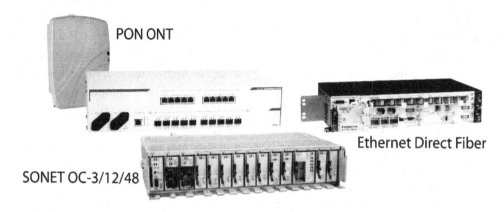

ployed by many service providers to deliver a new generation of combined voice, video and data applications, or groups of services known as "triple play" over a single service connection. Typically a FTTH network is implemented using one of a variety of PON technologies (FTTH Council, 2006).

In a PON network an optical line termination (OLT) product is placed in the central office and connects to the service provider's voice, video, and data transport facilities. The OLT extends a number of fiber optic connections to the access network where each fiber connects to a passive splitter. The signal is split onto 32 or more fiber optic feeder lines, each of which connects to an Optical Network Terminal (ONT) device typically located on the side of a home or business.

PON networks have many advantages. Their all-optical design allows for high bandwidth delivery. Passive devices at intermediate network points reduce operational issues associated with the deployment of power systems and correction of power interruptions. Centralized switching and management control at the OLT reduces operational complexity. Lastly, the ability of the OLT to scale to support hundreds of remote ONT devices reduces overall space, power and per-line costs.

A variety of PON technologies have evolved which are driving bandwidth ever higher and driving down the deployment costs per bit transported. Initial deployments of PON focused on ATM-based broadband PON (APON), which led to Broadband PON (BPON) (ITU-T, 2002; 2005a; 2005b) and Ethernet PON (EPON) (Institute of Electrical and Electronic Engineers, 2004, p.25). BPON simplified operations and increased downlink stream bandwidth to 622 Mbps but itself was exceeded by EPON, which is designed to provide an Ethernet-based system with 1.25 Gbps connections. However, when split into 32 or more ONTs, even these high-downlink bandwidths may not have the capacity to provide enough throughput to support a full assortment of triple play services. Specifically, the demands of HDTV services may quickly exhaust the available bandwidth. To address this concern, the Full-Service Access Network (FSAN/ITU-T) committee has defined the Gigabit PON (GPON) standard (ITU-T, 2003, p.2). The GPON standard increases downlink bandwidth up to 2.5 Gbps, increases the possible number of ONTs per splitter, and improves Quality of Service (QoS) compared to earlier generations. In North America, service providers are moving toward GPON as the preferred PON technology. However in Asia, EPON established substantial early deployment and is continuing to expand its footprint.

Looking forward, service providers are discussing options to provide 100 Mbps residential access, which would enable delivery of premium bandwidth services and richer applications to residential customers. Both the IEEE and FSAN have begun studies of 10G PON systems as possible future enhancements to achieve these service provider goals.

However, even with the advantages provided by PON, the technology has shortcomings. Most notable is that it requires a full optical fiber connection to every residential or business node. These connections are not commonly available today and require new construction and all the associated costs of acquiring right-of-way, trenching and installation, which can be very high.

A number of service providers are addressing this construction issue by taking a pragmatic approach and developing hybrid solutions that leverage the existing, installed residential copper access lines and pairing them with an optical access concentrator such as a DSLAM. These carriers are implementing a Fiber-To-The-Curb (FTTC) or similar Fiber-To-The-Node (FTTN) architecture. The FTTC network uses optical access solutions from a central office to a DSLAM node in the residential area that terminates the optical path. The DSLAM is responsible for differentiating the voice signal, which is sent to a voice switch, from the data signal, which is high-speed Internet access, Internet Protocol TV (IPTV) or Video On Demand (VOD), and is directed to a Layer 2 switch[1] or Layer 3 router.

On the subscriber side, the DSLAM uses xDSL to drive service connections over the existing right-of-way copper residential lines to a DSL modem inside the residence or business. With advances in ADSL2, VDSL, and VDSL2 technologies the FTTC architecture is able to provide very high bandwidth connections by extending the optical access to nodes within neighborhoods and thereby shortening the length of the final copper connection.

A business case study was done to assess the rate of return on capital investment in GPON versus copper infrastructure (Han, Yue, & Smith,

2006, p.3). The study reviews three competitive broadband access architectures (see Figure 3) and then analyzes the economic perspectives that include comparisons of revenue streams, cash flows and cumulative discounts and respective payback years. The study confirmed that the FTTC DSL required lower up-front investment expenses and generated a faster payback period. However, the FTTH PON solution exhibited a faster growth in cash flow and overcame the higher initial expenses in only a slightly longer payback period.

Another key technology in the optical access area is DWDM which can provide direct wavelength service to business users who require very high bandwidth services. DWDM technology allows service providers to add hundreds of signals to the same fiber by sending the signals at different wavelengths. The technology has traditionally been used for inter-office, national long haul, and transoceanic optical networks. However, vendors are adapting these large DWDM products into smaller product versions, at lower price points, for use in access networks. A variant of DWDM is Coarse Wave Division multiplexing (CWDM). This technology is similar to DWDM but multiplexes fewer optical wavelengths onto the optical fiber at wider frequencies, and thus allows for lower product cost while still delivering adequate high bandwidth for access applications. CWDM is specified by ITU-T G694.2 and can support as many as 20 wavelengths on the same fiber. The use of the same fiber to carry multiple signals with different wavelengths can offer a scalable, low-cost, high-bandwidth solution to avoid deploying extra fiber during future upgrades (Hinderthur & Friedric, 2003, p.1).

The prevailing solution for optical access, however, remains SONET and SDH. SONET and SDH are mature (Telcordia, 2001), established circuit-based technologies and have been the industry optical transport standard for a number of years. These technologies dominate access and transport infrastructure for business and wireless backhaul applications. They have a large installed

Figure 3. FTTx Network Architectures

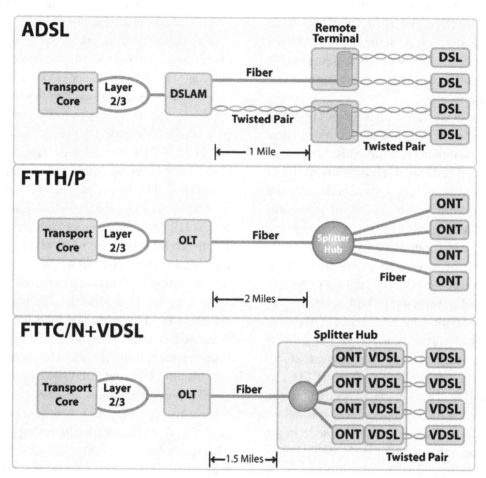

base because of their reliability, low cost, and large pool of trained technicians in the market place. Additionally, the SONET/SDH standard has been resilient against replacement, given that the standard is highly scalable with in-service upgrades possible and because it has already been integrated with service providers' Operation Support Systems (OSS). SONET and SDH are similar technologies with slight bandwidth and overhead structure differences for use in different parts of the world. These optical solutions are optimized for prevailing Time-Division Multiplexing (TDM)/ Plesiochronous Digital Hierarchy (PDH) voice circuits but are being adapted to support new high bandwidth "Ethernet over SONET/SDH" (EoS) services with new pluggable optical modules (Mocerino, 2006)

A continuation of this hybrid TDM/Ethernet hardware configuration approach will be desirable as service providers transition their networks to a fully Ethernet- or packet-based architecture (Yue & Hunck, 2005, p.6). The recent Multiservice Provisioning Platform (MSPP) class of product is designed to be an optimized, cost-saving solution. MSPPs consolidate video, data and voice services on the existing deployed SONET/SDH-based network infrastructure, to provide all-TDM and Ethernet services, such as DS1/E1, DS3/E3, Optical Carrier-N (OC-N) and Synchronous Transfer Mode (STM-N) (Yue & Gutierrez, 2003a, p.7).

MSPPs are the current standard in transport equipment, combining the functions of Add/Drop Multiplexers (ADMs), switching/grooming, and

Figure 4. Various Optical Access Architectures

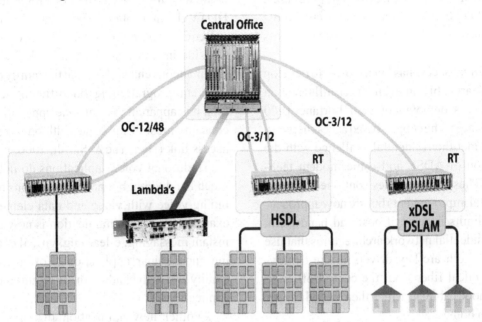

data interfaces into integrated platforms. These are much more compact and energy-efficient than previous generations that use separate pieces of equipment. Equipment vendors have also been extending their product families and developing "mini" and "micro" MSPPs in very small form factors to target access applications, providing highly reliable direct SONET/SDH delivery to business and cellular tower locations (Yue, 2006a, p.4).

The diverse nature of access networks means that a single technology or architecture will not be ideal for all applications. However, it seems clear that the copper-based access network is slowly being replaced with optical access in its many forms. SONET/SDH is still a key metro and access technology, which is excellent for low-speed aggregation with ring-based protection and bandwidth sharing. DWDM/CWDM has begun to move into metro and access applications as prices for solutions continue to drop. These technologies will likely gain additional application space as end-user bandwidth demands continue to rise, since these xWDM solutions are ideal for high-bandwidth

point-to-point applications such as storage area networking, Fortune 1000[2] private networking, and full Gigabit Ethernet service. Figure 4 illustrates various optical access architectures involving direct SONET or wavelength services.

REASONS FOR DEPLOYING OPTICAL ACCESS ARCHITECTURES

Bandwidth demand and the proliferation of new, high-bandwidth applications are the main reasons for deploying an optical access architecture.

Bandwidth Demand

Bandwidth demand has significantly increased in three connectivity areas: Fortune 1000 business access, residential access, and cellular tower backhaul.

Fortune 1000 business access has been the traditional driving force behind optical access deployments. Large companies demanded optical access to their office locations and were willing to pay a premium for dedicated, highly reliable, high-bandwidth connections, which were impossible to provide with the copper-based access network.

Residential access has been slower to develop but is now arguably an even larger influence on optical access deployment than Fortune 1000 business access. The copper-based access network gradually met the residential challenge with the introduction of ADSL and cable modem technologies. These technologies continued to make incremental improvements but are now approaching their limits in term of cost and bandwidth. Recent residential network architectures that use high bandwidth are key drivers in the deployment of optical fiber from the central office to neighborhood cabinets, which then extend copper VDSLx to homes.

Finally, a new driver for optical access has been the growing bandwidth demand at cellular towers. As more users adopt cellular and subscribe to new mobile data services offered via Evolution-Data Optimized (EVDO), Worldwide Interoperability for Microwave Access (WiMAX) and Long Term Evolution (LTE) networks, the bandwidth requirements of every cell site grow substantially (WiMAX Forum, 2007).

Optical systems are required to deliver the needed bandwidth and wireless carriers are asking backhaul service providers to provide Ethernet service offerings as an alternative to traditional DS1 circuit-based backhaul, which does not scale economically (Fujitsu Inc., 2006).

Emerging High-Bandwidth Broadband Applications

One residential application that is clearly gaining dominance is HDTV programming. Digital broadcasting will become effective in the United States in 2009 and many homes already subscribe to digital television cable modem or xDSL systems. As digital television becomes the norm, service providers must differentiate their offerings by

being able to offer consumers a high number of HDTV channel options. Today, each HDTV channel requires approximately 20 Mbps of bandwidth. Considering that many customers desire multiple simultaneous channels so that the family can view different programming on different television sets, it is apparent that this one application could consume most of the bandwidth on a residential access link (Han, Yue & Smith, 2006, p.3).

Traditional voice applications do not require much bandwidth, but new applications are combining voice with video and data elements. For example, voice communication is now a part of instant messaging, e-learning, virtual conferencing, and webinar[3] applications. Low latency, high quality and high bandwidth are required for these applications

Another new application that has become viable with the introduction of larger bandwidth to residential and business locations is improved security surveillance. Surveillance cameras which stream video into private networks or even the public Internet are gaining acceptance for home security, school security, business and even traffic monitoring. These emerging technologies, with high-bandwidth requirements, will help to protect lives and valuables.

Online gaming is another popular application that has grown into an extensive world market that drives bandwidth demand. New games are continually introduced with higher resolution graphics, more realistic animations, and expanded interactive play options. The user experience can also be improved by increased bandwidth, which serves to reduce latency and improve the game play response times.

Other new gadgets and devices like smart phones, MP3 players, and YouTube have increased exponentially the amount of music and movies downloaded worldwide. You can even watch TV series, soap opera and movies real time on a Web site.

For business enterprises, a number of drivers are fueling the need for added bandwidth connectivity. First, applications such as remote data backup of business-critical information to diverse physical storage locations have become a critical issue for companies of all sizes. A very high-speed uplink is required for this application.

Second, corporate expansion has fueled demand for Metropolitan Area Networking and Wide Area Networking beyond just the Fortune 5000-sized organization. There is increased demand for optical access bandwidth to many campus locations. Also many businesses expand globally and then require a private network to connect different offices in a fast, secure and reliable way. Private networking like this has given rise to Virtual Private Network (VPN) services.

A VPN is a private network that uses a public network to connect remote sites or users. Instead of using a set of dedicated, costly leased lines, a VPN uses "virtual" connections routed through the carrier's network, and even the Internet from the company's private network to the remote site or employee. This application has been widely accepted by business enterprise and it has driven a demand by employees for "access" points such as at hotels and conference sites.

WiMAX, based on the IEEE 802.16 standard, and the Third-Generation Partnership Project's Long-Term Evolution (3GPP LTE) are seen by the industry as the two leading technologies to become the fourth-generation (4G) mobile standard. The ITU is still in the process of developing the 4G specification, known as International Mobile Telecommunication – Advanced (IMT-A). Ratification of this specification is not expected before the year 2010. The new expected capabilities of these IMT-A systems will be able to support ~100 Mb/s peak data rate for high mobility access and ~1 Gb/s for fixed or nomadic low-mobility access with high quality and secure connections. Figure 5 shows the WiMAX and LTE path to IMT-A. It shows the huge performance improvement from 2G to the 3G networks and then a leapfrog to the 4G networks. Higher link speed and lower latency performance will help to dramatically improve

the user experience. Internet browsing and file downloading experiences from the 4G network will be competitive with wireline broadband like DSL, cable, or Fiber to the Home (FTTH) in the near future.

Today most wireless service providers in the U.S. lease T1 facilities for the access network backhaul from cellular towers. Recent modest increases in backhaul bandwidth requirements have been meet by simply adding an additional leased T1 facility to each cellular tower. However, as wireless service provides look to the creation of new bandwidth intensive services they are beginning to upgrade wireless base stations to support 4G technologies such as WiMAX and LTE (WiMAX Forum, 2007). Increasing the number of leased T1 facilities to these 4G tower sites is not an economically attractive option given the high number of T1 lines required and the associated recurring monthly lease expenses. Instead, an optical access solution can be deployed which provides a single high bandwidth connection facility and a much-reduced monthly leased cost per bit transported.

We are on the threshold of the new broadband era. Applications and services drive bandwidth demand in the network. Data traffic is increasing rapidly on broadband networks as more bandwidth-intensive services are launched. The user experience is determined by the end-to-end capability of the network. The increase in current network performance requirements could be tenfold in order to meet the true broadband services requirement. Optical access can offer faster speeds, lower cost per bit, and greater convenience with enhanced multimedia quality to the customers as compared to copper access.

DESCRIPTION OF ETHERNET

Ethernet was initially developed as a LAN technology. The biggest challenge in deploying Ethernet service in the Metro Area Network (MAN)

Figure 5. Path to IMT-A

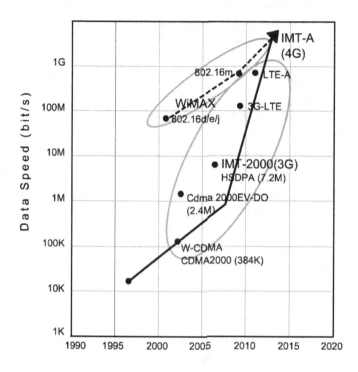

or Wide Area Network (WAN) is how to offer carrier-class features with a more reliable and manageable solution than Ethernet running in the LAN environment.

Many standards bodies, including the IEEE, ITU, Internet Engineering Task Force (IETF) and Metro Ethernet Forum (MEF), are working to define new capabilities to make Ethernet services as reliable as traditional TDM services. Ethernet has been successfully reinvented and applied in the service provider networks. Key features like QoS and Operations, Administration, and Management (OAM) have made Ethernet more carrier-grade than before. Ethernet has been used in this context as a service and a networking technology, not just an interface. The MEF has defined "Carrier Ethernet" as a ubiquitous, standardized, carrier-class service defined by five attributes that distinguish Carrier Ethernet from familiar LAN-based Ethernet. The five attributes are:

- Standardized services
- Quality of Service
- Scalability
- Service Management
- Reliability

The MEF has so far defined two types of Ethernet services that provide connectivity between two or multiple geographically dispersed locations and additional service types are being developed. The two currently defined MEF service types are E-Line and E-LAN. E-Line is a point-to-point private line service that provides Ethernet access to layer 3 services like IP VPN and Internet access. A point-to-point Ethernet Line service provides a symmetrical Ethernet connection between two subscriber locations. A point-to-point E-Line service is similar to a permanent virtual circuit (PVC) type of connection, but is less expensive and simpler to manage than a Frame Relay or ATM PVC.

Figure 6. MEF-defined E-Line & E-LAN services

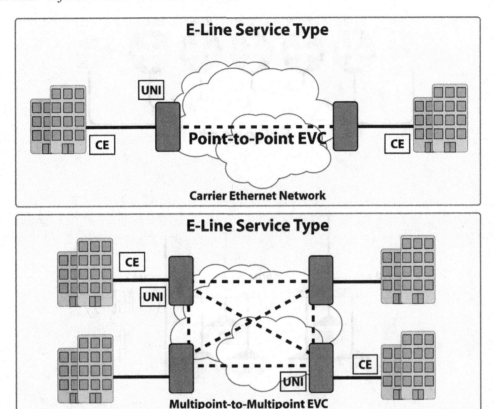

E-LAN service is a multipoint service that provides layer 2 VPN and transparent LAN services. E-LAN service provides Ethernet connectivity among multiple subscriber locations over the WAN, using Virtual LAN (VLAN) to segregate the subscriber traffic. E-LAN is much less expensive and simpler to implement and manage than on a Frame Relay and ATM network since E-LAN has a much lower equipment cost. Figure 6 illustrates the MEF definition of E-Line and E-LAN services (MEF Forum, 2008).

Enterprise users are attracted to carrier Ethernet because it is expected to provide them with a universal jack that allows bandwidth growth, addition of connections, and addition of new services without requiring a service truck roll. These benefits will potentially save significant operation costs. For example, to increase band-width to a customer today may require the physical addition of a T1 facility. This T1 deployment process may take multiple weeks to implement in a current network as the order rolls through the organization and equipment is deployed, verified, and turned-up for service.

As more and more services migrate to packet or IP-based technologies, Ethernet becomes the interface of choice for the next-generation packet-based network (Yue, 2006b). An Ethernet interface can potentially be less expensive than a SONET interface and provides a more cost-efficient way to aggregate and backhaul residential services. Ethernet interfaces are also becoming a common interface on many type of products and can now be found in a wide variety of service provider networking equipment, such as video servers, switches, service routers, core routers, WiMAX

Figure 7. Ethernet as universal jack

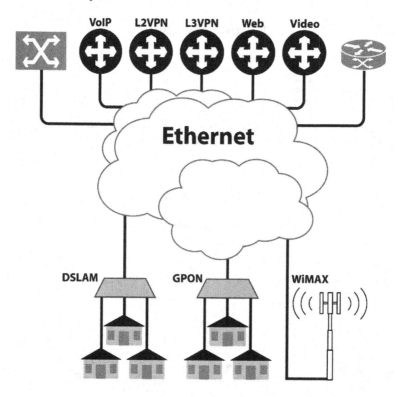

base stations, GPON OLT, DSLAMs and so on. Figure 7 illustrates Ethernet as the universal jack access interface.

However, in reality Ethernet is not yet a true replacement option. There are still problems to solve. Management of the new data network, for instance, will be a key issue. Additionally, service providers are comfortable with the current mode of circuit-based operations, and any change will require new training and education across many organizations in order to effectively deploy packet-based solutions. Also, a successful Ethernet service offering will likely require carriers to be able to deliver a consistent and high quality service experience to customers across a wide geographic area in order to capture corporate metropolitan area networking and VPN business. This will require carriers to define common service characteristics across a variety of vendors' equipment and will lead to vendor interoperability requirements as

well as requirements for equipment protection and timing/synchronization coordination.

A number of service providers have already started down the path of delivering Ethernet services. The majority of deployments have focused on E-line solutions using either Ethernet over SONET/SDH dedicated circuits or Ethernet over WDM point-to-point wavelengths (Hubbard, 2008). These E-line solutions more closely model traditional circuit-based services and can be supported in existing operational support systems (OSS).

E-LAN solutions have also been deployed and are generally best effort services using dedicated fiber optics over the last-mile connection to customer locations. These data centric service offerings differ considerably from circuit-based solutions and have often been managed outside of the traditional operation support systems of large service providers.

THE NEED FOR A SERVICE-LEVEL AGREEMENT (SLA)

A SLA is a formally negotiated agreement or contract between customers and their service provider, or between two service providers. An SLA usually specifies the levels of availability, throughput, delay, jitter, or other attributes of the service. Some SLAs even specify penalties in the case that these quality and reliability levels are violated.

Customer demand for SLAs has been increasing over the last decade. As companies have expanded their data networking needs they have often used VPN services rather than invest in their own network construction. This use of VPN services means that business critical data is carried over a network that the business does not control. An SLA allows a business to gain some measure of reporting and assurance from the service provider that the quality of the transport network will not impact their business operations.

SLAs, performance monitoring collection, fault isolation, and reporting have all been developed over time for the existing SONET/SDH and WDM transport networks. Service providers are now working with equipment vendors and various standards bodies to develop similar capabilities for the Carrier Ethernet network. Different data traffic types: voice, video and data, require different delay and jitter characteristics for best performance, which must be monitored (Yue & Gutierrez, 2003b). Also, traffic from a number of different customers may be multiplexed into the same physical port on a transport network piece of equipment. In this case, each customer's traffic must be kept isolated from others but also provide information per flow on end-to-end performance. Finally, trouble shooting on problematic flows will be necessary without impacting service to all customers. Today, many service providers substantially over-provision their network capacity so that congestion issues can be minimized, SLA performance can be guaranteed and penalties avoided. However this comes with the cost of network inefficiency and does not scale well to extensive deployment models.

This SLA reporting requirement represents a considerable issue to be addressed before carrier Ethernet will achieve full deployment.

Apart from the network QoS requirement, successful broadband deployment requires a highly reliable network infrastructure. It requires carrier-grade service, which specifically means that the service must be operational with "five nines" of reliability (a telecommunications industry term for 99.999% circuit availability, meaning the percentage of the time a network is functional and available for use by its subscribers) on a 24x7 basis, to minimize service impact to subscribers due to network outage. It requires low latency and jitter for voice and video traffic.

The high reliability for carrier-grade services includes scheduled and unscheduled maintenance events. Scheduled events include software upgrade, program/data backup, routine/scheduled maintenance. Scheduled events include hardware fault repair, software patches, problem diagnostics and some deferred maintenance repair.

Quick automated recovery and restoration of service after a fault greatly increases reliability. The quicker the recovery, the more often the system or service can go down and still meet the "five nines" criteria. System redundancy and robust upgrade procedures can provide a fast recovery time and minimize service interruption. High Mean Time Before Failure (MTBF) components that have very low probability of failure lead to maximal system reliability (McCabe, 2007, p.114).

Ethernet as a Replacement for T1 in Broadband Backhaul for Cellular Base Stations

One of the primary broadband wireless applications is to backhaul traffic from the base station (BS) into the core network. The present model for

Table 1. Wireless Backhaul Capacity

	Voice Spectrum (Mhz)	Data Spectrum (MHz)	Voice Spectral Efficiency (bit/Hz)	Data Spectral Efficiency (bit/Hz)	Total Bandwidth (Mb/s) (assume 3 sectors, peak usage: 70%)	#T1s
GSM 2G	1.2	NA	0.52	NA	1.3	1
GSM / Edge 2.75G	1.2	2.3	0.52	1	6.1	4
HSDPA 3G	NA	5	NA	2	21.0	14
LTE	NA	5	NA	3.8	39.9	26

most wireless operators is to lease T1 lines from the Local Exchange Carriers (LECs) to backhaul the wireless traffic. The current cost of leasing a T1 circuit can be expensive, especially in rural areas. Additional T1 circuits must be added, at significant cost, to provide increased backhaul bandwidth as new high bandwidth wireless services such as Internet access, gaming, and email are adopted. These new services will increasingly be available to consumers as cell towers are upgraded to high bandwidth wireless radios that use Universal Mobile Telecommunications System (UMTS)/ third generation (3G) cellular technologies, Wi-Fi, LTE (3GPP, 2007) and WiMAX (IEEE, 2005) technologies.

Wireless operators are looking for ways to reduce their operating expenses and avoid the poor scaling of simply adding additional T1 circuits. The use of optical access solutions to deliver carrier Ethernet ports for these applications will provide a lower price per transported bit. Also, these solutions will typically support easy upgrades of bandwidth from 1Mbps up to 1000Mbps via a port provisioning process performed by a technician at a central management center rather than requiring a site truck-roll.

As more wireless applications become IP-based, wireless operators will move their optical access and core infrastructure to IP based solutions (Eisenach, 2008, p.11). Table 1 shows the various backhaul capacity requirements for current and future wireless networks. As next generation broadband WiMAX and LTE networks start to roll out, the backhaul bandwidth requirements are

expected to hit at least 30Mb/s+ per cell site and then scale to hundred of megabits per second for future expansion (Han & Yue, 2006).

The data shows that when the wireless network migrates to 3G+ technologies, which require much higher backhaul capacity per cell site, the number of T1s needed in each site becomes unmanageable and is not cost-effective.

Optical carrier Ethernet becomes the choice for backhauling next generation broadband wireless traffic (Han & Yue, 2006). The physical layer will be either fiber or microwave for Ethernet transport. However, since Ethernet is still not a true replacement option during the transition, a hybrid TDM/Ethernet hardware configuration approach will be desirable as service providers transition their networks to a fuller Ethernet or packet based network. Figure 8 shows this; the service provider uses T1 interfaces to backhaul their GSM (2G) and UMTS (3G) traffic and in parallel uses Ethernet to backhaul 4G broadband traffic (like WiMAX or LTE) over the traditional SONET access network.

In another case, if the service provider is more comfortable with an Ethernet-only implementation, a second Outside Plant (OSP) cabinet containing the Ethernet network element (NE) is required at the tower. This NE provides the optical Ethernet access to the network. T1 traffic is transported via circuit emulation over Ethernet connection (MEF, 2008). (see Figure 9).

Backhaul bandwidth is a big problem to solve. Ethernet and IP will eventually be the transport protocol of choice whether the traffic is backhauled

Figure 8. Wireless Backhaul Access Network with SONET Transport

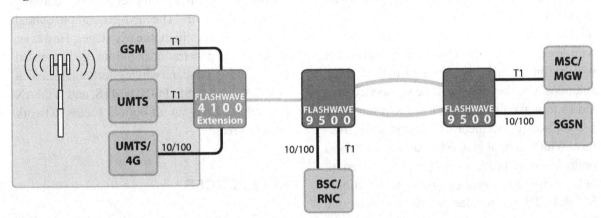

via copper or fiber or microwave facilities when carrier-class Ethernet is ready to provide a high bandwidth and reliable backhaul link to connect the radio access network to the core network.

THE FUTURE OF OPTICAL ACCESS

Optical access solutions will continue to evolve in a number of directions to better service the distinctive needs of residential, business, and cell-tower backhaul applications.

Residential networks will continue to be driven by an ever-expanding appetite for bandwidth as additional video, music, and social networking applications are conceived and adopted. Advances in Voice over IP and in the delivery of 911 services over IP will continue to drive residential focus towards a data centric environment.

Meeting this need will challenge service providers to enhance existing technology standards without complete infrastructure reinvestment. For example, continued advancements are under study for extending PON span transmission distances and increasing splitting ratios; VDSL copper loop bonding to increase bandwidth and multiple-input-multiple-output (MIMO) techniques for crosstalk cancellation; as well as new Data Over Cable Service Interface Specification (DOCSIS) 3.0 upgrades for cable providers all increase available bandwidth.

Figure 9. Wireless Backhaul Access Network with Optical Ethernet

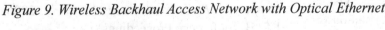

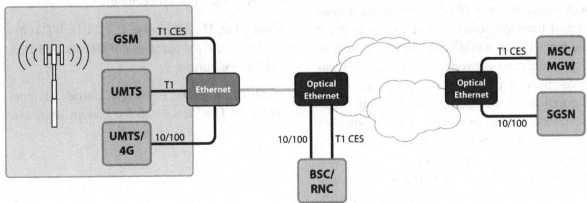

The business access and cell-tower backhaul market segments are also beginning to converge on IP and Ethernet as a higher bandwidth solution with lower overall cost. Optical access products will be the primary elements to deliver these solutions. One key issue will be for service providers to make this transition without requiring a complete replacement of existing infrastructure. With that goal in mind, a number of hybrid optical access products will be introduced to help bridge the transition from the traditional SONET/SDH dominated access world to one that can also support true carrier-class Ethernet transport. A combination of SONET/SDH, packet, and wavelength service options integrated into a single base platform will simplify operations for large carriers. A single platform, deployed in an outside plant cabinet, could be initially configured for today's SONET/SDH service delivery and be expanded with additional packet services or dedicated wavelength service in a simple "card-add" process as needed. The lengthy site survey and installation process would be eliminated for a service provider to deploy additional cabinets for each service type as the network migration progresses.

CONCLUSION

Emerging broadband services and the need for CAPEX and OPEX savings drives the vision of an optical access network. The evolution of technology towards IP and Ethernet as a converged transport solution is beginning to erase the barriers that traditionally drove the need to deploy different networks for different applications. Service providers are competing across the application spectrum and are bundling voice, data, video and even wireless services together to survive in this new environment where there is much less revenue per bit transported across a network. The need for a low-cost, operationally efficient, and performance-guaranteed network

creates many challenges for service providers and equipment vendors. The next-generation, optical access market is still in its early stages. However, the increased penetration of the new packet-based broadband wireline and wireless technologies such as WDM, EoS, DSL, PON, UMTS, and WiMAX can help to bring an all-optical access network closer to reality.

REFERENCES

American National Standards Institute (ANSI) (1991). *ANSI T1.105-1991, Digital Hierarchy - Optical Interface Rates and Formats Specifications.*

Eisenach, R. (2008). *Understanding Mobile Backhaul.* CED Magazine.

FTTH Council. (2006). *Definition of Terms.* FTTH Council.

Fujitsu Network Communications Inc. (2006). *Ethernet in the First Mile Over Point-to-Point Fiber (EFMF).* Fujitsu Network Communications Inc.

Han, S., & Yue, W. (2006). *Next-Generation Packet-Based Transport Networks Economic Study.* OFC/NFOEC 2006 Conference Proceedings.

Han, S., Yue, W., & Smith, S. (2006). *FTTx and xDSL: A Business Case Study of GPON versus Copper for Broadband Access Networks.* 2006 FTTH Conference Proceedings.

Hinderthur, H., & Friedric, L. (2003). *WDM hybrid transmission based on CWDM plus DWDM.* Lightwave Europe.

Hubbard, S. (2008, February). Carrier Ethernet Services: The View From the Enterprise. *Heavy Reading, 6*(2).

Institute of Electrical and Electronic Engineers. (*IEEE*) (2004). *IEEE 802.3ah - Ethernet in the First Mile Task Force archives.* New York: IEEE.

Institute of Electrical and Electronic Engineers (IEEE). (2005). *IEEE Standard 802.16e-2005. Mobile 802.16.* New York: IEEE.

International Telecommunications Union – Telecommunication Standardization Sector (ITU-T). (2002). *ITU-T Recommendation G.983.5: A broadband optical access system with enhanced survivability.* Geneva, Switzerland: ITU-T.

International Telecommunications Union – Telecommunication Standardization Sector (ITU-T). (2003). *ITU-T Recommendation G.984: Gigabit-capable Passive Optical Networks (GPON).* Geneva, Switzerland: ITU-T.

International Telecommunications Union – Telecommunication Standardization Sector (ITU-T). (2005a). *ITU-T Recommendation* G.983.1, Broadband optical access systems based on passive optical networks (PON). Geneva, Switzerland: ITU-T.

International Telecommunications Union - Telecommunication Standardization Sector (ITU-T). (2005b). *ITU-T Recommendation G.983.2: ONT management and control interface specification for B-PON (with amendments 1 and 2, erratum 1 and an implementer's guide).* Geneva, Switzerland: ITU-T

McCabe, J. (2007). *Network Analysis, Architecture and Design.* Amsterdan, The Netherlands, & Boston, US: Elsevier/Morgan Kaufmann Publishers.

Metro Ethernet Forum (MEF) (2008). *MEF 10. MEF Technical Specifications.*

Mocerino, J. V. (2006). Carrier-Class Ethernet Service Delivery Migrating SONET to IP & Triple Play Offerings. OFC/NFOEC Proceedings 2006 (pp.96-401).

Telcordia, (2001). *Telcordia GR 253 CORE: SONET Transport Systems: Common Generic Criteria.* Telcordia.

Third-Generation Partnership Project. (3GPP) (2007). *Technical Specification 23.402: Architecture Enhancements for Non-3GPP Accesses.* Sophia-Antipolis, France: 3GPP.

WiMAX Forum (2007). *Deployment of Mobile WiMAX Networks by Operators with Existing 2G & 3G Networks.* WiMAX Forum.

Yue, W. (2006a). *How GPON Deployment Drives the Evolution of the Packet-Based Network.* 2006 FTTH Conference Proceedings.

Yue, W. (2006b). *The Role of Pseudowires and Emerging Wireless Technologies on the Converged Packet-Based Network.* Proceedings of OFC/NFOEC, Anaheim CA, March 2006 & presented in the QWEST HPN Summit 2006.

Yue, W., & Gutierrez, D. (2003a). *Ready for Primetime: MSPPs Can Deliver Digital Video Over Existing Network.* Lightwave North America & Lightwave Europe.

Yue, W., & Gutierrez, D. (2003b). *Digital Video Transport over SONET using GFP and Virtual Concatenation.* NFOEC 2003, Conference Proceedings.

Yue, W., & Hunck, B. (2005). *Deploying Multiservice Networks Using RPR over the Existing SONET Infrastructure.* OFC/NFOEC 2005 Conference Proceedings.

Yue, W., & Mocerino, J. V. (2007). *Broadband Access Technologies for FTTx Deployment.* OFC/NFOEC 2007 Conference Proceedings.

ENDNOTES

[1]   Layer 2 switches and Layer 3 routers are computers that provide connectivity between networks or parts of networks. They transport, route and forward traffic over the network based on communications protocols such as Ethernet Media Access Control (MAC) addressing.

[2]   The Fortune 1000 is a list of the 1000 largest companies in the United States, ranked by publicly available revenue figures. This list is maintained by *Fortune* magazine. The term "Fortune 1000" is also used generically

Section 2
Modern Optical Technologies and Future Architectures for Broadband Access

Chapter 3
Active Optical Access Networks

Gerasimos C. Pagiatakis
School for Pedagogical and Technological Education (ASPETE), Greece

ABSTRACT

In this chapter, active optical access networks (AONs) are examined. AONs are a special type of optical access networks in which the sharing of optical fibers among users is implemented by means of active equipment (as opposed to passive optical networks –PONs– where sharing is achieved by using multiple passive splitters). In active optical access networks, user-side units, known as Optical Network Units (ONUs), are usually grouped in access Synchronous Digital Hierarchy (SDH) rings and fiber-interconnected to a local exchange unit, known as Optical Line Termination (OLT). In AONs (as well as in PONs) the optical fiber (originally used in the trunk network) is introduced in the access domain, namely between the customer and the local exchange. Practically, this means that the huge bandwidth provided by the optical fiber becomes directly available to the normal user. Despite the obvious financial and techno-economical issues related to the massive deployment of optical access networks, the possibilities and challenges created are enormous. This chapter examines the various units and modules composing an active optical access network and presents the basic procedures for implementing such a network.

INTRODUCTION: BACKGROUND

This chapter deals with the active optical access networks, usually referred to as AONs. In this type of optical access networks, sharing of optical fibers among users is implemented by means of active equipment, as opposed to the multiple-passive-splitter approach employed in passive optical networks –PONs– (Venieris, 2007).

Generally speaking, an optical access network can be considered as the optoelectronic infrastructure installed in the access part of the telecom network, that is the part between the subscriber and the local exchange (Figure 1). This infrastructure contains both active equipment, (installed in the customer premises, the local exchange and sometimes in

DOI: 10.4018/978-1-60566-707-2.ch003

Figure 1. An abstract view of an active optical access network

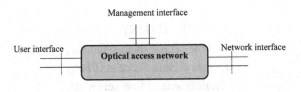

between) and optical fibers used to interconnect the above equipment. This situation should be contrasted to the conventional (copper) access network which is basically passive and merely contains the twisted-pair cable or cables installed between the subscriber and the local exchange.

From the topology point of view, the simplest optical access network would be a "direct-fiber" access network in which each customer would get his/her own pair of fibers (one fiber used for each direction of transmission). Such networks could provide very large bandwidths (nowadays a bandwidth-distance product of 500 Gbps.km is feasible), however they would be very costly due to the amount of fiber and active equipment used. Indeed, "direct-fiber" networks (also referred to as "star-topology" networks) are only deployed in cases where either the service area is small or key-account customers are to be served.

Usually, in optical access networks the "shared-fiber" approach is used which anticipates the sharing of fiber (as well as active equipment) to reduce costs. It is not until the fiber cable gets relatively close to the customer that is split into customer-dedicated pairs of fibers. In active optical access networks, this split is implemented by means of active equipment (as opposed to passive optical access networks where sharing is achieved by using multiple passive splitters).

With the explosive spread of Internet and the increasing demand in data connections and broadband services, as well as the appearance of more and more bandwidth-demanding applications, it is evident that copper-based access solutions (such as DSLs) may soon become inadequate. By exploit-

ing the huge capacity of the optical fiber, optical access networks seem to be the only solution to meet the fast-increasing bandwidth demands.

The idea of using optical fibers in the access part of the telecom network goes back to the early 1990s when the optical fiber had already established itself as the dominant transmission medium for the trunk network. Though the advent and fast spread of copper access technologies (mainly DSLs) in the early 2000s made, at least temporarily, the use of optical access technologies a less obvious choice, it appears now that these two initially competing technologies may be used either in different domains (installation of DSLs might be preferable where fast provision of services is necessary or the deployment of optical cables is costly or problematic) or in a combined manner to optimize bandwidth provided versus cost.

Generally speaking, the basic advantage of optical access networks (PONs or AONs) as compared with other access technologies (copper or wireless) is the huge bandwidth provided by the optical fiber, while their main drawback is the high cost of the active equipment and the need for massive fibre deployment in the access area. Optical access networks can offer bit rates over 1 Gbit/s which make possible the provision of services such as Fast Internet, Video on Demand (VoD) and "triple-play" services[1].

When referring to optical access networks (either active or passive), the term "Fiber In The Loop" (FITL) is generally used to imply the introduction of fiber-optics either in the whole or in some part of the access network (local loop).[2] Depending on the extent to which the fiber technology covers the local loop (that is, where the electrical-optical interface is put) various alternative and more specific terms can be used. Thus the term "Fiber To The Curb" (FTTC) refers to the arrangement where only the primary part of the access network becomes optical (the optical fiber reaches the outdoor main distribution frame). When optical fibers are deployed up to the building basement, the term "Fiber To The Building" (FTTB) is used while the situation where the fiber reaches the office itself is described as "Fiber To The Office" (FTTO). In turn, FTTB and FTTO may be considered as special cases of the "Fiber To The Home" (FTTH) or "Fiber To The Premises" (FTTP) arrangements.

In the section to follow, the general structure of AONs is presented. The following sections deal with the basic components of such networks, namely the ONU and the OLT as well as with management and deployment issues. Next the possibility of using AONs as a platform for implementing combined access solutions as well as the use of wavelength-division multiplexing (WDM) technology are examined. The chapter continues with a comparison of AONs to alternative access technologies and solutions as well as the examination of recent developments in AONs deployment world-wide. The chapter closes with some concluding remarks.

THE GENERAL STRUCTURE OF ACTIVE OPTICAL ACCESS NETWORKS

As already stated, an optical access network is, generally, the optoelectronic infrastructure between the subscriber and the local exchange (Figure 1). Through an optical access network, the customer must be provided with all available services (whether narrowband or broadband), without having to replace his/her own equipment (telephone sets, ISDN devices, modems etc.).

To achieve this:

- The optical access network (whether active or passive) must provide all the necessary user interfaces for the provision of the available telecom services (whether narrowband or broadband). Such services may be:
  ○ Usual dial-up services such as Plain

Old Telephony Service (POTS) and associated services (e.g. cardphones).
  ○ International-Service-Digital-Network (ISDN) related services whether Basic-Rate-Access (BRA) or Primary-Rate-Access (PRA)[3].
  ○ Analog leased line services[4].
  ○ Narrowband data services (of the Nx64 kbit/s type with N = 1, ..., 31)
  ○ Broadband services (currently implemented through digital subscriber loops - DSLs)
  ○ Cable television (CATV) if such service is already provided.
- The optical access network must provide the necessary interfaces to the available telecom networks, such as the telephone network and the various data networks.
- The optical access network must provide interfaces to the management network and it must be manageable locally as well as remotely.

Finally, an optical access network must be flexible and versatile (in order to be easily adaptable to the fast-changing telecommunication landscape) as well as expandable (to meet the fast-increasing demand in telecommunication services).

In its most common design, an optical access network (whether active or passive) comprises three basic units, the "Optical Network Unit" (ONU), the "Optical Line Termination" (OLT) and the transport network in between them (Figure 2). Normally, a number of ONUs (that may be installed near or inside the users' premises) are interconnected, via the transport network, to a single OLT installed inside the telephone exchange building.

The ONU is the optical-access-network termination towards customer's side (Figure 2). An ONU may be installed either inside a customer's premises (indoor ONU – FTTB/FTTO/FTTH/FTTP arrangements) or at an outdoor location (outdoor ONU – FTTC situation). Regardless

Figure 2. The general structure of an active optical access network using an SDH transport network

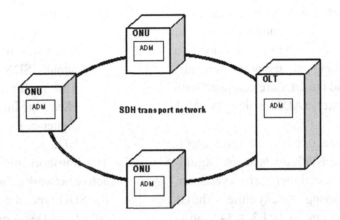

of its type, an ONU provides three (3) groups of interfaces (i) customer interfaces (towards customer equipment), (ii) transport network interfaces (towards the optical transport network) and (iii) management interfaces towards the management network to enable local as well as remote management of the ONU equipment.

The OLT is the access network termination towards the various telecom networks (Figure 3). The OLT is installed in the local exchange premises and, through the transport network, it is interconnected to the ONUs of the area served by the local exchange. An OLT provides three (3) groups of interfaces, (i) transport network interfaces (towards the optical transport network),

(ii) telecom network interfaces (towards the telephone and data networks), and (iii) management interfaces to the management network. In most access networks, the network interface between the OLT and the local exchange is of the V5.x type (V5.1 or V5.2).

The transport network contains the transmission equipment as well as the optical cable (or cables) interconnecting the ONUs to the OLT. It is exactly the transport network which characterizes an access network as passive (PON) or active (AON). In most AONs, the transport network is of the Synchronous-Digital-Hierarchy (SDH) type (Horrocks & Scarr, 1994), usually of STM-1 ($\approx$155 Mbit/s) or (more rarely) STM-4

Figure 3. Narrowband and transport (SDH-type) multiplexers of an ONU

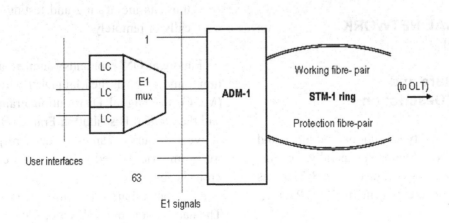

(4xSTM-1 ≈ 622 Mbit/s) capacity[5]. Such transport networks are organized in SDH rings, each one containing a number of ONUs and terminating at the network OLT. Each SDH ring contains two pairs of fibers (working pair and protection pair) while the ONUs and the OLT are equipped with Add-Drop Multiplexers (ADMs) either ADM-1 or ADM-4 [6].

In specific cases (e.g. remote ONUs or small-bandwidth applications) Plesiochronous-Digital-Hierarchy (PDH) connections (Horrocks & Scarr, 1994) may be used having capacity either at the E1 (≈ 2 Mbit/s) or (more rarely) the E3 (≈ 34 Mbit/s) level[7]. These transport connections are of the point-to-point type and are implemented by using OLTE (Optical Line Termination Equipment) devices in both the ONU and the OLT. OLTEs are capable of providing a NxE1 (usually N = 1 or 2) or E3 (≈ 34 Mbit/s) rate and they are interconnected either through a single fiber (by using different wavelengths for each direction of transmission) or a pair of fibers (one fiber per direction).

Alternative implementations may anticipate the point-to-point connections of some ONUs to a concentration node (usually called "Optical Distribution Terminal" –ODT–) which is, in turn, interconnected to the OLT through an SDH transport ring. In such arrangements, the point-to-point interconnection of the ONUs to the ODT is usually implemented by means of OLTEs (of NxE1 or E3 capacity).

THE OPTICAL NETWORK UNIT (ONU)

ONU Structure and Functional Description

Regardless of its type (indoor, outdoor) and capacity, an ONU (Figure 3) generally contains the following modules (Intracom, 1998; Siemens 1998; Vassilopoulos & Pagiatakis, 2002; Pagiatakis, 2004):

- The narrowband multiplexers. These, in turn include:
  - The Line Cards (LCs) that provide the necessary user interfaces (telephone, ISDN etc.).
  - The multiplexing unit that multiplexes the individual user signals into higher-order signals (most usually, E1).
- The transport multiplexer. Since, in most active networks, the transport network is of the SDH type, the transport multiplexer is either an ADM-1 or, more rarely, an ADM-4. Such multiplexers provide four (4) optical interfaces to be interconnected to the working and the protection fiber pair.

Apart from the above multiplexing and transmission equipment, an ONU contains:

- Power-supply equipment. Normally, an ONU is connected to the public electricity network (230 VAC in Europe) and contains an AC/DC converter as well as a power distribution unit to supply the active equipment with the necessary DC power.
- Batteries for supplying the ONU with power in case of failure of the electricity network.
- Management and supervision equipment that enables the interconnection of the ONU to the management network and, thus, its monitoring and testing either locally or remotely.

Finally, an ONU contains a number of distribution frames, such as the Main Distribution Frame (MDF), the Digital Distribution Frame (DDF) and the Optical Distribution Frame (ODF). The above distribution frames are used, respectively, for symmetric twisted pairs, coaxial cables and optical cables[8].

ONU operation can be summarized as follows: The narrowband multiplexers collect (by means

Figure 4. Line Cards with POTS, "U" or Voice-Frequency (VF) interfaces.

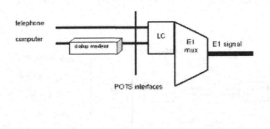

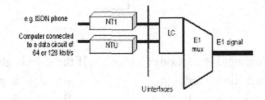

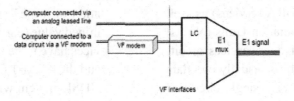

of the respective line cards) individual user signals (such as POTS, ISDN or data signals) and (by means of the multiplexing unit) multiplex them into E1 signals. E1 signals are, in turn, interconnected to the electrical interfaces of the transport ADM and are mapped (multiplexed) into the SDH aggregate signal (STM-1 or STM-4). Regarding the opposite (receiving) direction, the optical STM-1 or STM-4 aggregate signal is demultiplexed (by the ADM) into a number of E1 signals which are, in turn, demultiplexed (by the narrowband multiplexers) into the individual signals (POTS, ISDN or data) to be provided the user.

The Narrowband Multiplexers

As already mentioned, an ONU narrowband multiplexer collects the individual user signals (POTS, ISDN, data) and multiplexes them into E1 signals (standardized according to ITU-Recs G.703 and G.704). In the receiving direction, a narrowband multiplexer demultiplexes the E1 signals (received by the ONU ADM) into the individual POTS, ISDN or data signals that are finally provided to the user (see Figure 4).

The type of line cards contained into a narrowband multiplexer depends on the services to be provided to the interconnected user. The most usual types are the following:

Figure 5. Provision of E1 services (2048 Mbit/s) through an ONU. HDSL devices are used in case signal attenuation between the user and the ONU exceeds a prescribed value (e.g. 6 dB at 1 kHz)

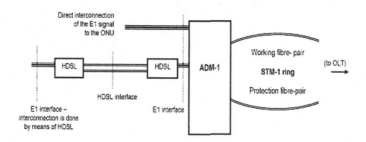

- Line cards with conventional telephony (POTS) interfaces: Such line cards are suitable for providing telephony (POTS) or cardphone services. They can be also interconnected to dial-up modems.
- Line cards with "U" interfaces: Such line cards can be connected to NT1s (for the provision of ISDN-BRA services) as well as to Network Termination Units (NTUs) for the provision of data services at bit-rates of 64 or 128 kbit/s (usually two data circuits of 64 kbit/s or a single circuit of 128 kbit/s is possible).
- Line cards with voice-frequency (VF) interfaces: Such line cards are suitable for users with analog leased-line connections as well as data services through voice-frequency modems.
- Line cards with Nx64 kbit/s interfaces (N = 3, ..., 31): Such line cards are suitable for the provision of data services at bit-rates from 192 to 1984 kbit/s.

At this point, the following remarks can be made:

- The user preserves his/her equipment (e.g. telephones, cardphones, ISDN devices, modems, NT1s etc.).
- Each line card is usually capable of providing more than one user interfaces.

- If the user is already provided with E1 services (e.g. a private telephone exchange –PBX– with ISDN–PRA facility), user equipment (in this case the PBX) is directly interconnected to an E1 interface of the ONU ADM. If the distance between the user and the ONU is such that the attenuation of the E1 signal exceeds a predetermined value (e.g. 6 dB at 1 kHz), HDSL devices have to be installed between the user and the ADM-1 (see Figure 5). Usually, an HDSL modem with one or two E1 user interfaces is installed at the customer's premises while an HDSL card is installed in the ONU)[9].

An important function of the narrowband multiplexer is the determination of the E1 interfaces between the ONU and the OLT. The following arrangements can be made:

- An open (standardized) interface (such as the V5.1) is used between the ONU and the OLT. The same interface may be used between the OLT and the telephone exchange.
- A proprietary interface may be used between the ONU and the OLT. This may be done to allow concentration in the ONU-OLT part of the network thus reducing the number of E1 signals required and, in turn,

Figure 6. Schematic of a typical subrack of a narrowband multiplexer. In this subrack, it is considered that each line card provides 15 telephone interfaces, hence two (2) such cards (30 subscribers) are interconnected to one multiplexing unit. The above subrack can serve up to 120 subscribers and produce up to four (4) E1 signals that are, in turn, interconnected to the E1 interfaces of the ADM

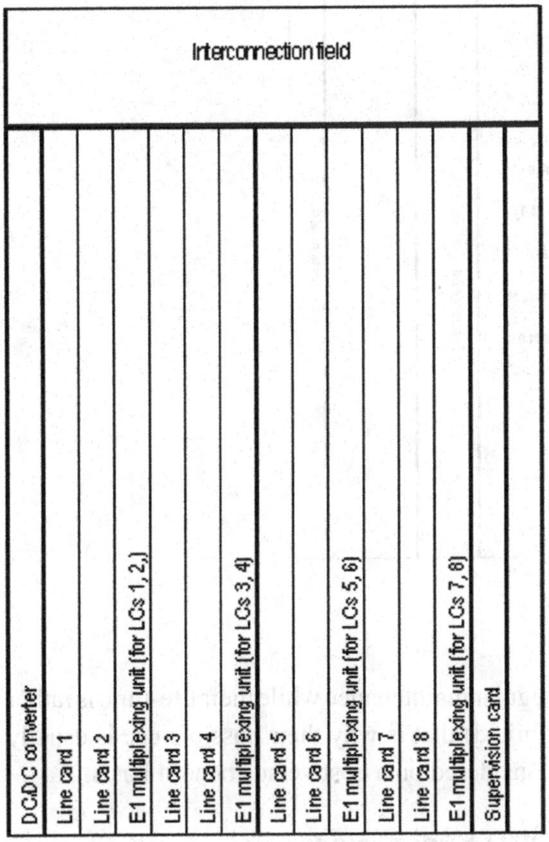

Interconnection field
DC/DC converter
Line card 1
Line card 2
E1 multiplexing unit (for LCs 1, 2,)
Line card 3
Line card 4
E1 multiplexing unit (for LCs 3, 4)
Line card 5
Line card 6
E1 multiplexing unit (for LCs 5, 6)
Line card 7
Line card 8
E1 multiplexing unit (for LCs 7, 8)
Supervision card

the number of SDH rings (thus economizing in OLT SDH equipment). Since the interface between the OLT and the telephone exchange must be an open one (V5.1 or V5.2) a "protocol converter" must be installed between the OLT and the local exchange equipment (as explained below, this converter may be the OLT cross-connect).

For the installation of narrowband multiplexers, subracks are used that are, in turn, installed into the ONU racks. Each subrack may contain one or more narrowband multiplexers (each one with the necessary line cards and the multiplexing unit) as well as power supply and possibly monitoring cards (see Figure 6).

The Add-Drop Multiplexer (ADM)

As already mentioned, ADMs maps (multiplexes) E1 signals (created by the narrowband multiplexers) or (more rarely) E3 signals into an STM-1 or STM-4 optical signal. In the receiving direction, ADMs demultiplex the received STM-1 or STM-4 optical signal into the E1 or E3 component signals which are, in turn, demultiplexed, by the narrowband multiplexers, into the individual service signals (telephony, IDSN, data).

An important advantage of the ADMs is the provision of an alternative (protection) path between the ONU and the OLT. Indeed, an ADM has two pairs of optical interfaces, the "east" and "west" pair, one of which is interconnected to the working path and the other to the protection path. In case of a failure in the working path, traffic is automatically routed to the protection path (typical switching time about 50 ms).

Physically, an ADM-1 or ADM-4 consists of a number of cards installed in a subrack. These are the tributary cards (for E1 or E3 signals) the optical card(s) interconnected to the working and the protection fiber pair as well as power supply and supervision cards (Figure 7).

ONU Power Supply

Usually, ONUs are directly power-supplied by the electrical network (230 VAC in Europe). ONUs contain an AC/DC converter (to convert the AC power to –48 or –60 VDC, depending on the telecom network power-supply voltage) as well as a power distribution unit to supply the

Figure 7. Schematic of a typical ADM-1 subrack. The two (2) 21xE1 tributary cards are interconnected to the E1 interfaces of the narrowband multiplexers and are also internally interconnected to the optical (STM-1) card. The latter has four (4) optical interfaces for the working and the protection ("east" and "west") fiber pairs.

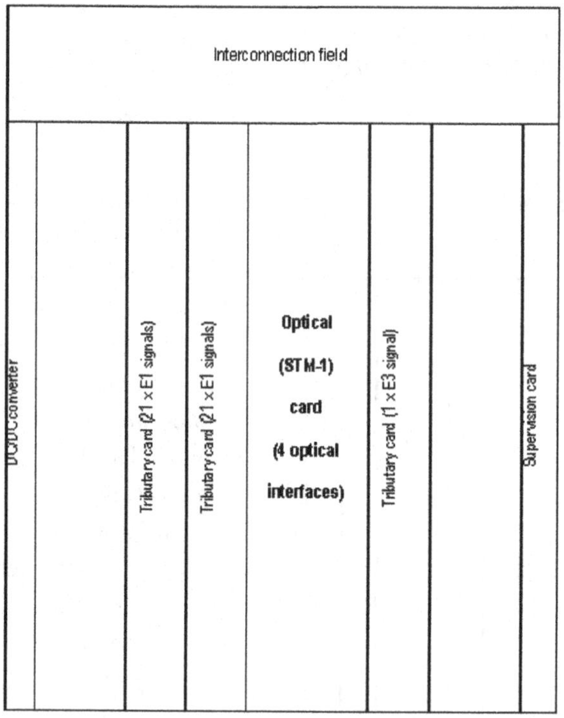

active equipment with the necessary DC power (usually ± 5 VDC).

Though ONU's power consumption depends on the number of served users and the telecom services provided, an indoor ONU requires about 1–2 kW of power and is usually connected to the electricity network typically through a 3x6 mm² cable with a 25 A fuse. Grounding is done by means of a 16 mm² cable.

An ONU may also have batteries to provide the necessary power back-up in case of failure of the power-supply networks. Depending on the number of users and the telecom traffic, batteries may power the ONU for 4 up to 8 hours. Despite the obvious advantage of providing power back-up, batteries add to the cost of the ONU (they also need

regular maintenance while their life-time is rather limited) that is why their possible use is usually considered on a cost versus benefit basis.

ONU Supervision

Each ONU has a supervision unit for the monitoring of its active and power-supply equipment. This unit may also perform diagnostic tests in the "user-ONU" (copper) part of the access network. Through the supervision unit, an ONU can be managed either locally or remotely.

The ONU Distribution Frames

As already stated, an ONU contains an MDF for copper twisted-pairs, a DDF for the interconnection of coaxial cables and an ODF for interconnecting optical fibers.

The MDF is mainly used for the interconnection of the ONU's active equipment (essentially the line cards) with either the building distribution frame (indoor ONUs) or the secondary network (outdoor ONUs). In any case, the ONU's MDF is usually divided in two (2) parts, the equipment part (to be interconnected to the line cards) and the network part (to be interconnected to either the building distribution frame or the secondary network). Obviously, for the provision of telecom services the two parts must be properly interconnected.

The MDF is, by far, the biggest distribution frame in the ONU and its size becomes critical, especially in outdoor ONUs where space is always limited. The evaluation of the necessary number of twisted-pair interfaces is based on the possible number of line-card user-interfaces. An equal number of interfaces has to be provided in the equipment part of the MDF while in the network part the number may be a little larger (say by 20%) to allow the provision of services remaining on copper (in which case the ONU's MDF may be used for the interconnection of secondary to primary copper cables). As an example, an outdoor ONU with 500-telephone-circuits capacity must have an MDF with 500 + 1.20x500 = 1100 twisted-pair interfaces[10].

ONU Types

Depending on the installation site, ONUs can be classified into two main categories:

- The outdoor ONUs (used in FTTC access networks)
- The indoor ONUs (used in FTTB/ FTTO/FTTH/FTTP networks)

The outdoor ONUs are installed in outdoor places (such as pavements) and basically replace the conventional outdoor distribution frames. In such arrangements the optical access network only covers the primary part of the subscriber network, the secondary part remaining on copper ("Fiber To The Curb" or FTTC topology).

Due to the fact that outdoor ONUs are exposed to adverse environmental conditions, the ONU equipment is housed in specially fabricated enclosures. Depending on its size and the services provided, an outdoor ONU may serve from about 100 to 750 subscribers. The typical size of an outdoor ONU (for 500 subscribers) is W x D x H = 800 x 400 x 1500 mm.

In contrast to the outdoor ONUs, indoor ONUs are installed in the customers' premises (for example, in the basement or the telecommunication room if such a room exists) and usually serve the telecom needs of the building where they are installed ("Fiber To The Building/Office/Home/Premises" or FTTB/FTTO/FTTH/FTTP arrangement).

Indoor ONUs have modular structure (they consist of individual racks) and they are expandable to up to 1000 or even 2000 subscribers (Figure 8). Depending on its size and the services provided, an indoor ONU usually contains two (2) or three (3) racks each one having dimensions W x D x H = 600 x 300 x 2200 mm (total size W x D x H = (1200–1800) x 300 x 2200 mm).

It must be noted that indoor and outdoor ONUs are basically equipped with similar equipment. The main difference is the fact that for outdoor ONUs, space and environmental protection become much more critical parameters than for indoor ONUs.

ONU Interconnection to the Optical Network

For the ONU interconnection to the optical network, a subscriber optical cable (usually up to 12 fibers) with length up to about 200 m is used. This subscriber cable is connected to the core optical

Figure 8. Schematic of a typical indoor ONU (interconnected to an SDH transport network).

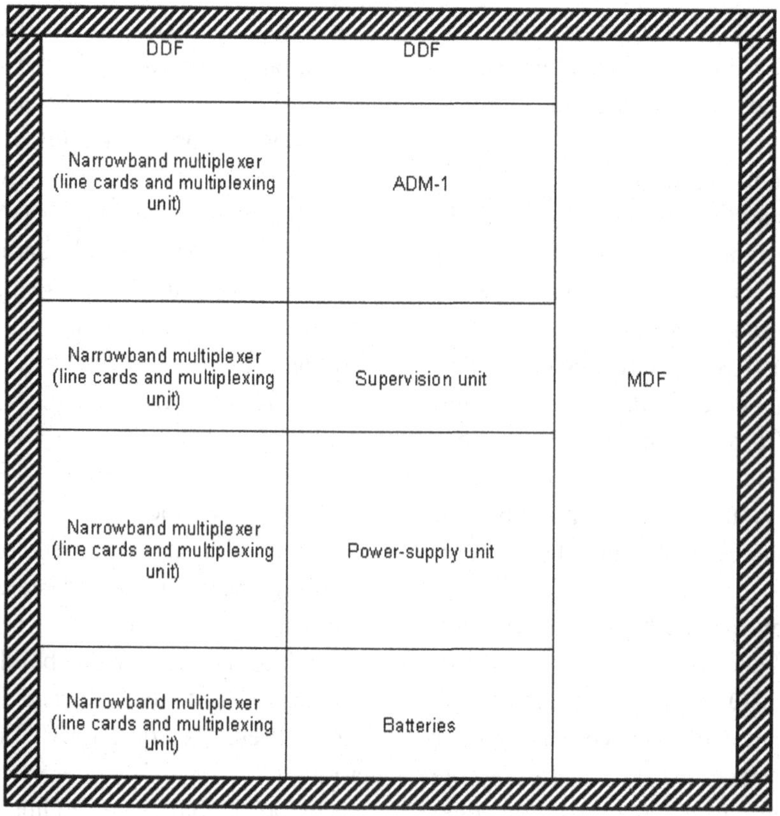

DDF
DDF
Narrowband multiplexer (line cards and multiplexing unit)
ADM-1
Narrowband multiplexer (line cards and multiplexing unit)
Supervision unit
MDF
Narrowband multiplexer (line cards and multiplexing unit)
Power-supply unit
Narrowband multiplexer (line cards and multiplexing unit)
Batteries

cable (that runs through the whole access network area) by means of an accessible interconnection box installed in a manhole.

For the interconnection of the subscriber to the core optical cable, various approaches may be adopted. A possible solution may anticipate the connection of four (4) subscriber fibers to two core fibers (in order both a working and a protection path are provided), the remaining subscriber fibers left for future use. Usually, for every two

Figure 9. Interconnection of the ONU to the subscriber and the core fiber cable (schematic).

Subscriber cable (up to 12 fibers; 4 fibers are connected to 2 fibers of the core cable
Working path
Fiber pair reserved for future use
ODF
STM-1 optical ring
to OLT
Patchcords (2 pairs)
ADM-1
Protection path

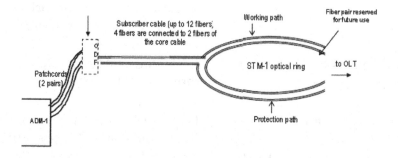

Figure 10. Block diagram of a typical OLT (though the number of the OLT ADMs is equal to the number of the access network rings, only a single ADM-1 has been drawn for the sake of simplicity).

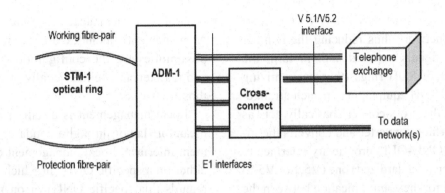

(2) interconnected core fibers, two (2) more may be also reserved for future use, in which case a 96-fiber (48-pair) core cable can provide up to 24 optical access rings (Figure 9).

The interconnection of the ONU to the subscriber optical cable can be done as follows:

- For indoor ONUs, a wall-mounted ODF (with capacity, at least equal to that of the subscriber cable) is installed near the ONU (at a distance of a few meters). At this ODF, the subscriber optical cable is terminated. Then the optical interfaces of the ONU ADM are interconnected to the ODF by means of patchcords.
- In outdoor ONUs, an ODF is installed inside the ONU. This ODF is used for the termination of the subscriber optical cable (coming from underground) and its interconnection (through patchcords) to the optical interfaces of the ADM.

THE OPTICAL LINE TERMINATION (OLT)

The Optical Line Termination (OLT) can be considered as the optical-access-network termination towards telecom network side. It is installed at the local exchange premises and it is interconnected

to all the ONUs of the optical access network serving a specific area.

The OLT basically contains the transport ADMs and possibly a cross-connect (Figure 10). The number of ADMs is equal to the number of SDH rings that form the access network served by the OLT. On the other hand, the use of the cross-connect is compulsory if the interface used between the ONU and the OLT is different than that used between the OLT and the telephone exchange (Intracom, 1998; Pagiatakis, 2004; Siemens 1998; Vassilopoulos & Pagiatakis, 2002).

OLT operation can be summarized as follows: In an access network based on an SDH-type transport network, ONUs are interconnected to the OLT through optical SDH rings. The OLT has one ADM for each one of those rings. This ADM receives the STM optical signal coming from the respective SDH ring and demultiplexes them into its tributaries (usually E1 signals). These tributaries (that consist of time-slots carrying individual user signals such as POTS, ISDN and data) are directed to an 1/0 cross-connect which "regroups" time-slots and, at the same time, converts the "internal" (between the ONU and the OLT) interface to the interface used between the OLT and the telephone exchange (usually V5.1 or V5.2)[11].

Regarding cross-connect the following remarks can be made:

- In some optical access products, a proprietary interface may be set between the ONU and the OLT. This may be done to allow concentration in the ONU-OLT part of the network thus reducing the number of E1 signals required and, in turn, the number of SDH rings (thus economizing in OLT SDH equipment). In such arrangements, the cross-connect basically acts as an interface converter and converts the internal (ONU-OLT) proprietary interface to the open (standardized) one (V5.1 or V5.2) needed for the communication between the OLT and the telephone exchange.

- If the optical access network uses an open interface between the ONU and the OLT, the use of cross-connects may be avoided. However, lack of concentration in the ONU-OLT part leads to a larger number of E1 signals and, in turn, to more rings and additional SDH equipment for the OLT.

The power-supply of the OLT is done by means of the local exchange power-supply arrangement (−48 or −60 VDC). Most usually, this arrangement anticipates the use of batteries (for power backup) as well as UPSs.

Physically, OLTs have a modular structure in the sense that they consist of individual racks and are expandable according to the needs of the specific optical access network they serve (Figure 11). The interconnection of the OLT to the core optical cable (in fact, the interconnection of the optical interfaces of the OLT ADMs to the respective optical rings) is done through an ODF while the interconnection of the OLT to the telephone exchange (and the data network nodes) is done through a DDF. Both the ODF and the DDF are usually installed in separate racks in the OLT location (Vassilopoulos & Pagiatakis, 2001).

MANAGEMENT OF ACTIVE OPTICAL ACCESS NETWORKS

Regarding optical access networks, the term "management" includes all the solutions and possibilities for the configuration, supervision and maintenance (either locally or remotely) of those networks.

Local management is usually performed by means of a laptop through a standardized management interface. Local management can be done either on a specific ONU (in which case it only regards this specific ONU) or on the OLT (in which case, the remote management of all ONUs interconnected to the OLT is possible).

Central (remote) management of an optical access network offers more possibilities and it is usually performed at the OLT site. A possible solution is the development of a Local-Area Network (LAN), usually of the Ethernet type, consisting of a server and a number of clients, the number depending on the size and capacity of the access network (Siemens, 1998).

BLOCK DIAGRAM OF AN ACTIVE OPTICAL ACCESS NETWORK

Based on sections above, a full block diagram of an active optical access network is as in Figure 12.

DEPLOYMENT OF ACTIVE OPTICAL ACCESS NETWORKS

Basic Parameters

The basic parameters for the design of an active optical access network mainly depend on the choices and decisions made by the telecom carrier. Such choices may refer to:

- The telecom services to be provided. For example, a telecom carrier may decide that

Figure 11. Schematic of a typical OLT (ODF and DDF are installed in separate racks and are not shown for the sake of simplicity).

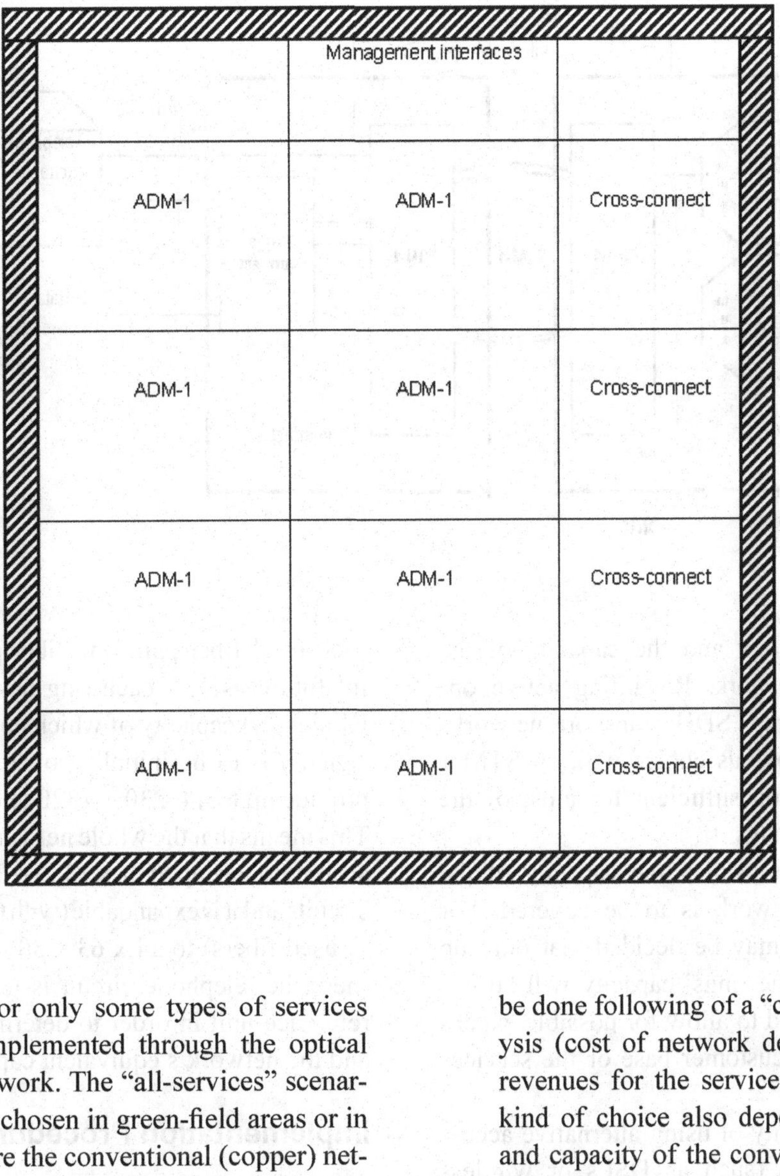

Management interfaces		
ADM-1	ADM-1	Cross-connect
ADM-1	ADM-1	Cross-connect
ADM-1	ADM-1	Cross-connect
ADM-1	ADM-1	Cross-connect

either all or only some types of services will be implemented through the optical access network. The "all-services" scenario may be chosen in green-field areas or in areas where the conventional (copper) network is either inadequate or obsolete. On the other hand, in regions where the copper network has the necessary capacity, the telecom carrier may decide that only some types of services (e.g. the data ones) will be provided through the optical network.

- The user profile. The telecom carrier may decide that either all or only a special category of users (e.g. key-account customers) will be connected. This kind of choice may

be done following of a "cost-benefit" analysis (cost of network deployment versus revenues for the services provided). This kind of choice also depends on the state and capacity of the conventional (copper) infrastructure.

- The areas where the optical access network is to be deployed. Such a choice is usually dictated by factors such as the quality and the available capacity of the existing network (in comparison with expected demand in telecom services), the cost and disturbance to be caused by installation works etc.

- The type of the ONUs to be installed (indoor, outdoor or both).

Figure 12. Block diagram of a typical optical access network (a single ONU is depicted for the sake of simplicity).

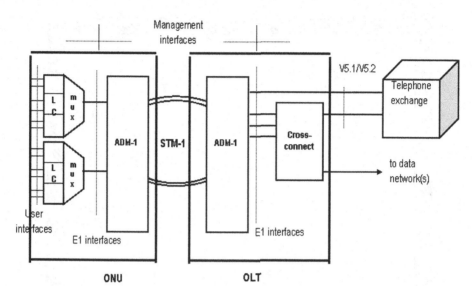

- The technology and the capacity of the transport network. Regarding active optical networks, SDH transport networks seem the obvious choice with an STM-1 capacity being sufficient for most of the ONU-OLT rings.

- The extent to which the capacity of the transport network is to be covered. For example, it may be decided that only up to 70% of the rings' capacity will be initially covered to allow for possible expansion of the customer base or the services provided.

- The possibility of using alternative access technologies (such as DSLs or wireless local loops) either independently or in a combined manner (for example, it may be decided that DSLAMs are installed inside the ONUs to reduce the copper length thus increasing the bit-rate provided to the DSL user)[12].

As an example, consider an optical access network that uses a 96-fiber cable and thus can support up to 24 SDH rings (on the basis that for each used fiber-pair, one fiber-pair is reserved for future use). If each ring has an STM-1 (63 E1 signals) capacity of which about 70% (44 E1 signals) is to be initially covered, the ring can provide up to 44 x 30 = 1320 telephone circuits. This means that the whole network (24 rings) can accommodate up to 24 x 1320 = 31680 telephone circuits and it is expandable (with the same number of used fibers) to 24 x 63 x 30 = 45360 circuits (here the telephone circuit is merely used as a reference unit in order to determine the ONU's and the network's equivalent capacity).

Implementation Procedure

General Remarks

Once some basic parameters (such as the above) have been determined, the basic steps for implementing an optical access network are the following (Vassilopoulos & Pagiatakis, 2001):

- Network planning: A detailed survey of possible customers and services is carried out and, based on this survey, the distribution of the ONUs across the access network is determined.

- Network dimensioning: The number and capacity of core and subscriber fiber cables are determined and a detailed list of equipment for the ONUs and the OLT is formed.
- Equipment purchasing.
- Cable installation: The core and subscriber fiber-cables are installed and interconnected.
- Installation of the management system (usually at the OLT site).
- Installation and commissioning of the OLT.
- Installation and commissioning of the ONUs.
- Switching of telecom services from the conventional (copper) to the optical network.

The above steps are briefly described below.

Network Planning

The customers to be interconnected are determined and the telecom services to be provided are written down in detail (e.g. number of telephone and ISDN-BRA or PRA connections, number and type of data circuits etc.). The distribution of the ONUs throughout the access network area is determined.

Network Dimensioning – Purchasing of Equipment

A detailed installation study for the core fiber cable is performed. Among others, the cable capacity (total number of fibers), the cable route as well as the fiber allocation per customer are determined.

A detailed list of the necessary equipment (both active and passive) is made. The equipment is, then, purchased from the chosen manufacturer(s).

Cable Installation

The core cable is installed. This may be the most difficult and time-consuming phase due to the problems caused by installation works in urban environments.

The subscriber cables are installed at the chosen customer premises. This includes the installation of indoor ODFs as well as the interconnection between of the subscriber cable to the core cable (through accessible underground interconnection boxes).

Installation and Commissioning of the OLT

A detailed study for the OLT installation is performed. This study determines the installation site as well as the preparation and installation works to be done (racks and power supply, multiplexing and transmission equipment, distribution frames etc.). For the OLT site, proper air-conditioning is necessary (in order the temperature and humidity be preserved within acceptable limits) as well as power back-up by batteries and UPSs.

The appropriate databases for the OLT and the telephone exchange are developed.

The OLT and the telephone exchange are interconnected through a DDF (Figure 13). This distribution frame must have enough interfaces for the interconnection of the E1 signals provided by the OLT. The necessary number of interfaces is calculated using the formula *Number of E1 interfaces = (Number of OLT ADMs) x 63 + (Number of cross-connect inputs and outputs)*.

The OLT is interconnected to the management system.

The OLT is cable interconnected to the core optical cable and thus to the access network ONUs through the ODF (figure 13).

Preliminary tests are performed to ensure that the OLT is properly working.

Figure 13. Interconnection between the transport network and the OLT (through the ODF) and the OLT and the local exchange (through a DDF). In the OLT, only a single ADM-1 has been drawn for the sake of simplicity.

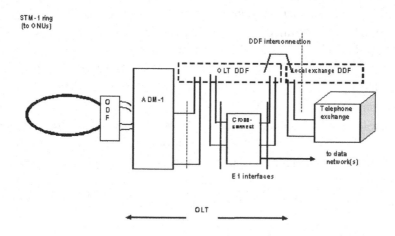

Installation and Commissioning of the ONUs

Installation sites, whether indoor or outdoor, are chosen. For indoor ONUs, the appropriate temperature is 15–25°C and the relative humidity is below 50%. Air-conditioning is recommended for all sites that do not fulfill the above requirements. Outdoor ONUs are installed in suitable outdoor locations (e.g. on pavements) ideally at the site of the conventional (copper) distribution frame to be replaced by the ONU. The ONU cabin must be protected against accidental or deliberate damage. Thermal insulation as well as ventilation of the cabin must be such that the active equipment of the ONU remains operational under adverse weather conditions (excessively high or low temperature as well as high relative humidity). Also, outdoor ONUs must be externally power-supplied and may have batteries for power back-up.

The ONUs are installed and connected to the power-supply network.

The ONUs are interconnected to the subscriber fiber cable which is, in turn connected to the core fiber cable. ONUs are put into operation and tests are performed to ensure that the ONUs are properly working and interconnected to the OLT.

Telecom services are switched from the conventional (copper) to the optical access network (Figure 14). This may cause a short interruption in the provision of telecom services. Depending on the ONU type (indoor or outdoor) the following works are usually done:

- *Indoor ONUs:* At the location of an indoor ONU, the telecom carrier has already installed an MDF (where the copper cable serving the building has been terminated) while the building has its own distribution frame for the interconnection of the indoor cabling. Obviously, for the provision of telecommunication services to the building tenants, the two distribution frames must have been interconnected. Service switching from the conventional (copper) subscriber network to the optical access network requires the disconnection of the copper cable from the telecom carrier MDF and the connection of the building distribution frame to the MDF of the ONU.
- *Outdoor ONUs:* For the provision of services through an outdoor ONU, the (copper)

Figure 14. Switching of services from the conventional (copper) subscriber network to an indoor ONU (schematic). The secondary cable and the telecom carrier MDF are removed after the switching of services to the ONU.

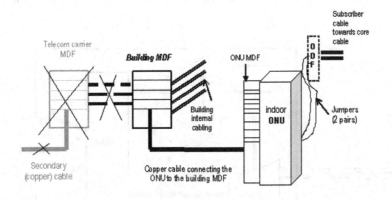

outdoor distribution frame is disconnected and the subscribers are interconnected to the ONU MDF.

ACTIVE OPTICAL ACCESS NETWORKS AS A PLATFORM FOR IMPLEMENTING COMBINED ACCESS SOLUTIONS

An important merit of the active optical access networks is the fact that they can be used as a platform for employing alternative access technologies and eventually implementing integrated access solutions.

As already stated, an ONU can be used to house DSLAMs (for the provision of DSL services) as well as fiber nodes (for access to the CATV network). Regarding the DSL case, by installing a DSLAM inside an ONU, the copper length is largely reduced (in most cases, to less than 500 m) and thus the downstream bit-rate available to the user can be largely increased (Figure 15).

Active optical access networks (usually of the FTTC type) can be used to provide VDSL2 services (at rates commonly in excess of 50 Mbit/s) by housing VDSL2 multiplexers.

Other implementations may anticipate the installation of wireless equipment in some out-

door ONUs to enable interconnection of remote customers or customers inaccessible by cable.

THE USE OF WDM TECHNOLOGIES IN ACTIVE OPTICAL ACCESS NETWORKS

WDM is a relatively new multiplexing technique that enables the simultaneous transmission of multiple digital signals, through the same fiber pair, by using different wavelengths. The WDM technique is widely used in the trunk network where (in the format of Dense WDM or DWDM) it can achieve aggregate rates of the order of Tbit/s (e.g. by multiplexing of 128 wavelengths each one carrying an STM-64 ≈ 10 Gbit/s digital signal) (EXFO, 2000).

Though basically a trunk network technology, WDM may also lend itself for use in the metro and the access domain (Kani et al., 1993; Saleh & Simmons, 1999)[13], particularly in the Coarse WDM (CWDM) format (multiplexing of up to 20 wavelengths in the 1270-1610 range). In metro and access networks, bandwidth may not be the issue (a single pair of fibers is capable of providing up to 500 Gbps.km while the distances are usually less than 40 km), however the use of

Figure 15. Possible use of the optical access network as a platform for the provision of DSL services (schematic). The DSLAM is installed inside the ONU and splits the POTS from the data signals. POTS signals are directed to the proper line cards and are multiplexed by the E1 multiplexer. Data signals are multiplexed by the DSLAM (to a NxE1 or E3 signal) which is, in turn, interconnected to the ONU ADM.

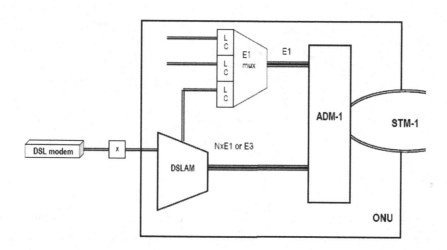

Figure 16. Possible use of WDM in the access network – the "wavelength-per-customer" scenario (schematic). The access ring is fed with four (4) wavelengths each one carrying an STM-1 signal. Optical Add-Drop Multiplexers (OADMs) are used to add and drop the proper wavelength to the respective ONU.

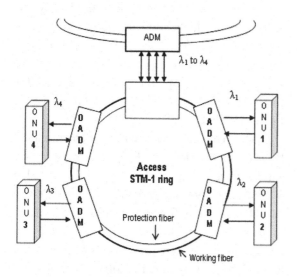

multiple wavelengths may give the possibility to new-coming service providers to hire capacity, in the form of individual wavelengths, rather than investing on infrastructures of their own.

Though the possible use of WDM in the access network is discussed as a medium or long-term solution, some scenaria have already been examined that anticipate the use of different wavelengths on either a "wavelength-per-customer" or a "wavelengths-per-service" basis (Eurescom, Project P917, 2000). Regarding the wavelength-per-customer solution, the allocation of different wavelengths to different customers (essentially to different ONUs) can be achieved by means of multi-wavelength access rings equipped with add-drop nodes that can transmit (drop) or receive (add) the particular wavelengths to/from the respective ONUs (Figure 16).

A COMPARISON OF ACTIVE OPTICAL ACCESS NETWORKS TO ALTERNATIVE ACCESS TECHNOLOGIES

Fiber access technologies can be viewed as the last step in the digitisation of the local loop which started back in the 1980s with the development of ISDN (OTE, 2000). These technologies are capable of providing access to broadband and combined services such as Fast Internet, VoD and "triple-play" (Green, 2004).

Comparing the various access network technologies (basically optical, DSL and wireless) is not a straightforward manner due to the various technical and economical factors and criteria that have to be taken into account regarding such a comparison. It is also evident that, in most cases, a comparative evaluation of the available access solution must be done on a case-by-case basis (for example, a residential area may have to be considered in an entirely different manner that an industrial or a key-account-customer one).

The main advantages of active optical access networks (as compared to DSL and wireless access technologies) are: (i) the huge bandwidth they can provide, (ii) the secure and highly reliable connections the can offer (reliability enhanced by the path-protection provided by SDH transport networks), (iii) the maturity and standardisation of SDH technology used in the transport network and (iv) their flexibility, versatility and expandability. The main disadvantages of active optical networks are: (i) their rather high cost per customer (especially in the initial phases of deployment) and (ii) the rather long time-frame for their deployment (mainly due to the necessary cable installation works). As compared to PONs, active optical access networks are a more mature and standardized technology and can generally cover larger areas however they may be more costly mainly due to the use of SDH transport networks (as opposed to the multiple splitters used by PONs).

ACTIVE OPTICAL ACCESS NETWORKS BY COUNTRY

Apart from maybe Japan, where FTTH has become the dominant technology for providing broadband services, the introduction of the optical fiber in the access network is either in the planning or in the pilot phase. The main reason for that seems to be the big investments needed for the massive deployment of fiber in access network as well as the fact that, up to a point, broadband services can be provided through the existing copper network by means of DSL technologies.

Active optical access networks have already been deployed in Sweden where there are plans to connect more than 100,000 apartments over the next coming years. In Greece, active networks already serve about 30,000 subscribers in areas of Athens, Salonica and other cities and plans have been announced to connect up to 2,000,000 subscribers by 2013. Projects of various scales (mainly trials, pilot or small-scale projects) have been or are being implemented in almost all European countries, the USA, Hong-Kong and South Korea to name a few. Plans to employ VDSL2 solutions have already announced in most European countries, the USA, Canada, Australia and New Zealand. Though the situation may vary from country to country, offered services range from the conventional telephony and data services to Voice-over-IP (VoIP), Fast Internet, Cable TV, HDTV, VoD etc. (Fiber to the premises by country. *Wikipedia, the free encyclopedia;* Very High Speed Digital Subscriber Line 2. *Wikipedia, the free encyclopedia*).

CONCLUDING REMARKS

Active optical access networks offer significant advantages over alternative access technologies, the most notable ones being huge bandwidth and high reliability, versatility and expandability. DSL technologies also seem well suited for the rapid provision of high-bit-rate services to domestic customers, however they are already reaching their limits not to mention that a large part of the conventional copper network is near the end of its expected lifetime. Of course, the extensive introduction of FITL technology (whether AON or PON) in the access domain will depend, among others, on the willingness of government organizations and private companies to undertake the massive investments needed for a large-scale deployment of fiber-optic equipment in the access area (the proper cabling of the buildings' interior and its interconnection to the outer fiber network is also a very important issue, Gauthey, 2007). However, the significant advantages of optical access networks together with the fact that a fair number of successful pilot projects have been already carried out in Europe, the USA and other parts of the world make FITL technologies not a high-end solution aiming at the far-away future but a way to provide advanced and attractive services in the short and medium term (Van den Hoven, 2008).

REFERENCES

Eurescom, Project P917, Building and Operating Broadband Access Networks (BOBAN) (2000). *Deliverable 10, DWDM technologies for access networks*. Heidelberg, Germany.

EXFO Electro-optical Engineering Inc. (2000). *Guide to WDM Technology and Testing*. Quebec City, Canada.

Gauthey, G. (2007). Regulation must play a role in the breakthrough of fiber. *Lightwave Europe, Q4/2007*, 9-10.

Green, P. E. (2004). Fiber to the Home: The Next Broadband Thing. *IEEE Communications Magazine, 42*(9), 100–107. doi:10.1109/MCOM.2004.1336726

Horrocks, J., & Scarr, R. (1994). *The Technology Guide to Telecommunications*. Surrey, UK.

Intracom, S. A. (1998). *The Fiber Access System IAS-F, Technical Description*. Athens, Greece.

Kani, J., Teshima, M., Akimoto, K., Takachio, N., Suzuki, H., & Iwatsuki, K. (2003). A WDM-based Optical Access Network for Wide-Area Gigabit Access Service. *IEEE Communications Magazine, 41*(2), S43–S48. doi:10.1109/MCOM.2003.1179497

Pagiatakis, G. C. (2004). *Fiber-Optic Telecommunications*. Salonica, Greece, Tziolas Publications (in Greek). Hellenic Organization for Telecommunications (OTE), Planning Department (2000). *Description of Digital Technologies used in the Access Network*. Athens, Greece (in Greek).

Saleh, A. A. M., & Simmons, J. M. (1999). Architectural Principles of Optical Regional and Metropolitan Access Networks. *Journal of Lightwave Technology, 17*(12), 2431–2438. doi:10.1109/50.809662

Siemens ON Training Center. (1998). *Fastlink (document AN4013E6A01)*. Munich, Germany.

Van den Hoven, G. (2008). FTTH in Europe, one year later. *Lightwave Europe, Q1/2008*, 9-10.

Vassilopoulos, C., & Pagiatakis, G. C. (2002). *Advanced Telecommunications Infrastructures and Services*. Athens, Greece, OTE Publications (in Greek).

Venieris, I. (2007). *Broadband Networks* (Chapter 8). Salonica, Greece, Tziolas Publications (in Greek). Fiber to the premises by country. *Wikipedia, the free encyclopedia*. Retrieved June 18, 2008 from http://en.wikipedia.org/wiki/Fiber_to_the_premises_by_country.

Very High Speed Digital Subscriber Line 2. *Wikipedia, the free encyclopedia*. Retrieved July 02, 2008 from http://en.wikipedia.org/wiki/Very_High_Speed_Digital_Subscriber_Line_2.

ABBREVIATION LIST

ADM: Add-Drop Multiplexer

AON: Active Optical (access) Network

CATV: CAble TeleVision

CWDM: Coarse Wavelength Division Multiplexing

DDF: Digital Distribution Frame

DSL: Digital Subscriber Line

DSLAM: Digital Subscriber Line Access Multiplexer

DWDM: Dense Wavelength Division Multiplexing

FITL: Fiber In The Loop

FTTB: Fiber To The Building

FTTC: Fiber To The Curb

FTTH: Fiber To The Home

FTTP: Fiber To The Premises

HDSL: High-bit-rate Digital Subscriber Line

HDTV: High Definition TeleVision

ISDN-BRA: Integrated Services Digital Network – Basic Rate Access

ISDN-PRA: Integrated Services Digital Network – Primary Rate Access

LC: Line Card

MDF: Main Distribution Frame

OADM: Optical Add-Drop Multiplexer

ODF: Optical Distribution Frame

ODT: Optical Distribution Terminal

OLT: Optical Line Termination

OLTE: Optical Line Termination Equipment

ONU: Optical Network Unit

PDH: Plesiochronous Digital Hierarchy

PON: Passive Optical (access) Network

POTS: Plain Old Telephone Service

SDH: Synchronous Digital Hierarchy

STM: Synchronous Transfer Module

VDSL: Very-high-speed Digital Subscriber Line 2

VF: Voice Frequency

VoIP: Voice over IP

WDM: Wavelength Division Multiplexing

ENDNOTES

[1] "Triple play" refers to the combined provision of telephony, Fast Internet and video services.

[2] Both "local loop" and "access network" refer to the part of the telecom network that lies between the subscriber and the local exchange. The term "local loop" is mainly used in connection with the conventional telephone network while the term "access network" usually refers to upgraded local loops using optical, copper or wireless technologies. In any case, a local loop (or access network) consist of the "primary" network (the part of the local loop between the local exchange and the outdoor distribution frame), the secondary network (the part between the outdoor distribution frame and the building distribution frame) and the "tertiary" network (the part between the building distribution frame and the customer equipment, e.g. the telephone set).

[3] ISDN is provided either through a Basic-Rate Access (BRA) or a Primary-Rate Access (PRA). BRA provides the equivalent of two (2) digital telephone (B) channels, each at a 64 kbit/s rate, and a signalling (D) channel at a 16 kbit/s rate (2B+D connection). In turn, PRA provides 30 digital telephone (B) channels and a signalling (D) channel, each one at a 64 kbit/s rate (30B+D connection). BRA requires the installation (by the service provider) of an NT1 device that becomes the interface between the customer equipment and the telephone network (the NT1 interface towards the telephone network is referred to as the "U" interface).

[4] Analog leased lines are direct analog connections either 2-wire or 4-wire. Analog leased lines may be used to provide a permanent connection between computers (equipped with modems). Physically, such lines may pass through local exchange premises but

they bypass telephone exchange equipment.

5   Synchronous Digital Hierarchy (SDH) anticipates the following signal levels: STM-1 (155,52 ≈ 155 Mbit/s), STM-4 (4xSTM-1 = 622,08 ≈ 622 Mbit/s), STM-16 (4xSTM-4 = 2,48832 ≈ 2,5 Gbit/s), STM-64 (4xSTM-16 = 9,9533 Gbit/s ≈ 10 Gbit/s) and STM-256 (4xSTM-64 = 39,813 Gbit/s ≈ 40 Gbit/s).

6   Add-Drop Multiplexers – level 1 (ADM-1s) map (multiplex) E1, E3 or E4 signals into an STM-1 aggregate signal. More specifically, up to 63 E1 or 3 E3 signals can be mapped into an STM-1 signal. It is also possible to multiplex a combination of E1 and E3 signals based on the rule that one (1) E3 signal "replaces" 21 E1 signals. It is also possible to map one (1) E4 signal (≈ 140 Mbit/s) into an aggregate STM-1 signal. ADMs of higher level can multiplex a multiple number of E1, E3 or E4 signals and also STM signals of lower levels. E1, E3 or E4 signals multiplexed into STM signals are called "tributaries". When the aggregate (STM) signal is optical (which is most usually the case), the ADM provide four (4) optical interfaces, two (2) for the working and two (2) for the protection fiber pair (the interfaces being sometimes referred to as "east" and "west" pair).

7   Plesiochronous Digital Hierarchy (PDH) – European version anticipates the following signal levels: E1 (2,048 ≈ 2 Mbit/s), E2 (4xE1 = 8,448 ≈ 8 Mbit/s), E3 (4xE2 = 34,368 ≈ 34 Mbit/s), E4 (4xE4 = 139,264 ≈ 140 Mbit/s) and E5 (564,992 ≈ 565 Mbit/s).

8   Distribution frames are, essentially, passive "connection matrices" that are used in order to enhance the modularity of the telecommunication networks. Main Distribution Frames (MDFs) are used for the interconnection of copper twisted pairs, Digital Distribution Frames (DDFs) are used for the interconnection of coaxial cables while Optical Distribution Frames (ODFs) are used for the interconnection of optical fibers. An ONU usually has an MDF for the interconnection of the copper subscriber network to the analog interfaces of the narrowband multiplexers (in essence, the analog interfaces of the line cards), a DDF for the interconnection of the E1 interfaces of the narrowband multiplexers to the ADM and an ODF for the interconnection of the optical interfaces of the ADM to the subscriber optical cable.

9   HDSL (High-bit rate Digital Subscriber Line) refers to a technique used to increase the maximum reach of E1 signal over normal twisted-pair cables. The most common implementation includes the installation of an HDSL modem and an HDSL card at the customer's and the local exchange premises respectively. This HDSL equipment divides the E1 signal into two sub-signals each one having a bit-rate of 1168 kbit/s and transported over a separate twisted pair. These lower-rate signals can reach a distance of about 2.7 km on common twisted pairs (of 0.4 mm diameter) without regeneration.

10   Since the exact capacity of an ONU depends on the services that provides, the telephone circuit (to be exact the 64 kbit/s bit-rate) is commonly used as a reference unit to determine ONU's capacity. For example a capacity a 500-telephone-circuit ONU is capable of providing either 500 telephone circuits or (equivalently) 250 ISDN-BRA connections or 500 64kbit/s data-circuits etc.

11   When characterising cross-connects as M/N, M denotes the hierarchy level of input/output signals and N denotes the level to which input/output signals are decomposed inside the cross-connect. Thus 1/0 cross-connect means that E1 input signals are internally decomposed into 64 kbit/s (E0) signals before being regrouped into E1 output signals.

12   A DSLAMs is a multiplexer normally in-
     stalled at the local exchange that enables the
     provision of DSL services. The interconnec-
     tion between the user-side equipment (e.g.
     the DSL modem) and the DSLAM is done
     by means of the existing copper telephone
     cables. Installing the DSLAM inside an ONU
     brings the DSLAM closer to the customer
     thus reducing the copper length and increas-
     ing the bit-rate provided.

13   The WDM technique, as applied in the trunk
     telecom network, employs the 3rd optical
     window (around 1550 nm) to allow for the
     possible use of fibre amplifiers. Due to the
     large number and the close spacing of the
     wavelengths used, this particular version of
     WDM is usually referred to as Dense WDM.
     However, in the metro and access network
     where the distances are short (hence, there
     is no need for fibre amplifiers) the full range
     between 1270 and 1610 nm is used while
     the number of possible wavelengths to be
     used is less than 20 (Coarse WDM).

Chapter 4
Wavelength Division Multiplexed Passive Optical Networks
Principles, Architectures and Technologies

Calvin C. K. Chan
The Chinese University of Hong Kong, Hong Kong

ABSTRACT

Wavelength division multiplexed passive optical network has emerged as a promising solution to support a robust and large-scale next generation optical access network. It offers high-capacity data delivery and flexible bandwidth provisioning to all subscribers, so as to meet the ever-increasing bandwidth requirements as well as the quality of service requirements of the next generation broadband access networks. The maturity and reduced cost of the WDM components available in the market are also among the major driving forces to enhance the feasibility and practicality of commercial deployment. In this chapter, the author will provide a comprehensive discussion on the basic principles and network architectures for WDM-PONs, as well as their various enabling technologies. Different feasible approaches to support the two-way transmission will be discussed. It is believed that WDM-PON is an attractive solution to realize fiber-to-the-home (FTTH) applications.

INTRODUCTION

Nowadays, access to the Internet has become an indispensable part of the daily lives in modern cities. Wide deployment of access network infrastructure has enabled the service providers to connect to all of the enterprise and residential users, thus the number of Internet users as well as the market of broadband access have experienced an explosive growth over

DOI: 10.4018/978-1-60566-707-2.ch004

the recent years. In addition to the conventional voice and broadcast video traffic, the current network is also carrying various real-time network services and interactive media-rich applications. Therefore, the demand of bandwidth in the current and next generation Internet is ever-increasing drastically. The current predominant broadband access technologies deployed, including digital subscriber line (DSL), and community antenna television (CATV), have their limitations in terms of bandwidth upgradeability, network scalability and robustness. As the network services and applications are getting more data-centric and involve more real-time interactive communications among all parties, the next generation broadband access solutions should be flexible, scalable and robust enough to meet the high bandwidth requirement and assure good quality of service for the data traffic.

Passive optical network (PON) (Effenberger et al., 2007) is a promising solution to enable high-speed broadband access. It is based on fiber-optic technology which unleashes the enormous transmission bandwidth in the optical fiber. Therefore, it solves the bandwidth bottleneck of the current copper-based access solutions and provides higher bandwidth to meet the traffic demand in the access networks. A PON is typically a point-to-multipoint optical network on which the optical line terminal (OLT) or the central office (CO) delivers services, via a long fiber feeder, to the remote node (RN), where the optical power is split and fed into multiple distribution fibers to reach many optical network units (ONUs) at the subscriber side. The infrastructure between the OLT and the ONUs does not require any electric power supply, thus can greatly ease the network management of the outside plant facilities. A single wavelength is employed at the OLT to carry the downstream traffic, mainly for service distribution, while a relatively lower bit rate upstream wavelength is also employed at the ONU to carry the requests from the subscribers back to the OLT. Both the upstream and the downstream bandwidths have to be time-shared among all ONUs, to keep the cost of the access network low and economically feasible for subscribers. In order to make PONs more economical, the full service access network (FSAN) consortium was formed by several telecommunication operators in 1995, so as to standardize the common requirements and services for PONs. The FSAN recommendations were later adopted by the International Telecommunication Union (ITU) as the ITU-T G.983 *Broadband PON* (B-PON) standards (ITU-T Recommendation G.983.1, 1998; ITU-T Recommendation G.983.2, 2000; ITU-T

Recommendation G.983.3, 1998). In a B-PON, the separation between the OLT and the ONU is 20 km and each OLT can serve up to 32 ONUs. The downstream traffic at 622-Mb/s is carried by a 1.49-μm optical carrier, while the upstream traffic at 155-Mb/s is carried by a 1.3-μm optical carrier, both of which adopt time-division multiple access (TDMA) for bandwidth sharing among all ONUs. The 1.55-μm wavelength window is reserved for analog video overlay. In 2003, ITU has released the ITU-T G.984 recommendations for the next generation PON, called *Gigabit-capable PON* (G-PON) (ITU-T Recommendation G.984.1, 2003; ITU-T Recommendation G.984.2, 2003; ITU-T Recommendation G.984.3, 2003), in which the transmission speeds in downstream and upstream directions are increased to 2.5 Gb/s and 1.25 Gb/s / 2.5 Gb/s, respectively. It has also adopted the framing mechanism based on generic framing procedure (GFP). In 2001, IEEE 802.3 standard group also started the 802.3ah working group (IEEE 802.3ah EFM), to standardize the transport of Ethernet frames on PONs (EPON) (Kramer & Pesavento, 2002), due to the popularity of Ethernet in both metro and access arenas. It specifies a symmetric transmission speed of 1 Gb/s for both downstream and upstream traffic, with 16 ONUs per OLT. In recent years, the advent of the 10-Gb/s Ethernet technology is also enhancing the transmission speed of PONs to the 10-Gb/s regime.

In general, all of these PON standards mentioned above are based on simple power splitting at the RN and adopt TDMA to share the transmission bandwidth among all ONUs. The number of ONUs supported is limited by the power budget as well as the bandwidth sharing. Moreover, media access control (MAC) protocol is needed to coordinate the transmission of the upstream data packets from all ONUs so as to avoid any possible collision at the RN. The work in (Kramer, Mukherjee, & Pesavento, 2002) is a good example of a dynamic bandwidth allocation protocol designed for EPON. Furthermore, the transmission distance between each ONU and the OLT may vary, which lead to variation in the signal power as well as the phase alignment of the received upstream data frames at the OLT. Therefore, ranging procedure has been proposed to achieve the upstream traffic synchronization. In addition, burst mode receivers are adopted at the OLT to adapt their gain according to the received peak intensity as well as to recover the clock and phase of the received upstream data frames.

With the ever-increasing bandwidth demand in broadband access and the recent availability of low cost optical components, it is becoming feasible and practical to upgrade the PONs by employing wavelength division multiplexing (WDM) technique such that each ONU is served by a dedicated set of wavelengths to communicate with the OLT. Therefore, WDM passive optical networks (WDM-PON) (Banerjee et *al.*, 2005) has recently attracted much attention from both research community as well as service providers. In a WDM-PON, each ONU can enjoy a dedicated bandwidth, which is scalable according to the need of the individual ONU. The ranging problem in conventional PONs is eliminated since all upstream wavelengths are multiplexed at the RN without any signal collision. These further enhance the system capacity and network flexibility. Symmetric two-way communication is supported and enables the optical access networks not only for service distribution, but also for data networking. Therefore, WDM-PON is a promising solution to support a robust and large-scale next generation optical access network.

In this chapter, we will provide a comprehensive discussion on the basic principles, network architectures, and various enabling technologies. We will describe the basic principles of WDM-PON as well as the WDM architecture and technologies suitable for WDM-PON. Various feasible schemes and enabling technologies to achieve the downstream and the upstream transmissions will be reviewed.

PRINCIPLES OF WDM-PON

Figure 1 shows the network architecture of a WDM-PON in a tree-topology. In a WDM-PON, each ONU will be served by a dedicated set of wavelengths to communicate with the OLT. Its architecture is similar to that of a PON except that the power splitter at the RN is replaced by a WDM multiplexer. The OLT collects the information from the outside networks and provisions the broadband services to all subscribers. The WDM source at the OLT generates the downstream wavelength channels for all ONUs and they are carried on a fiber feeder to the RN where the individual wavelengths are routed to their destined ONUs, via the WDM multiplexer and the respective distribution fibers. Each ONU, which is usually resided at the basement of the buildings or houses, contains a pair of optical transceivers. The optical receiver detects the destined downstream wavelength and retrieves the received data. Sometimes, the received data may be further distributed to more subscribers in the same access area, via other media, such as digital subscriber line or mobile connections, etc. The optical transmitter at the ONU modulates the upstream data on a designated upstream wavelength before being transmitted back to the OLT, via the RN. All the upstream wavelengths are then detected at the OLT, via the WDM receiver. In general, the connection between the OLT and an ONU is realized by the dedicated set of wavelengths on a point-to-point basis, thus no sharing of bandwidth among the ONUs is needed while enhanced privacy is achieved. Also, there is no point-to-multipoint MAC protocol required. The data rates for the wavelengths may be different, hence the bandwidth for each ONU is scalable according to the need and different varieties of services can be supported.

Figure 1. A typical WDM-PON

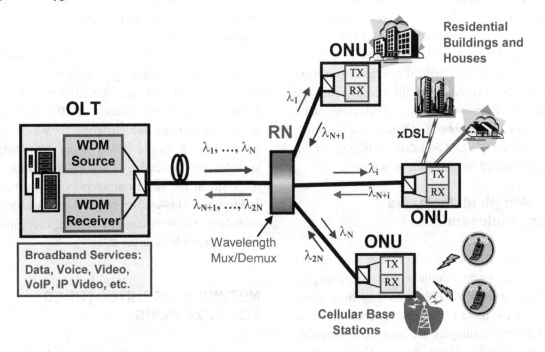

WDM TECHNOLOGIES FOR WDM-PONS

In order to support multiple wavelengths as well as the feature of wavelength routing in WDM-PONs, WDM based devices and technologies are very crucial. In this section, we will discuss some of the WDM technologies for WDM-PONs.

Wavelength Grids

In WDM-PONs, the channels are represented by distinct wavelengths over the transmission window of fiber. The wavelengths are defined in equally-spaced wavelength grids, standardized by ITU (ITU-T Recommendation G.692, 1998). The most commonly used channel spacing for dense-WDM (DWDM) is 100 GHz or 200 GHz. There are also standards for 50-GHz or even narrower channel spacing. However, with such narrow channel spacing, the DWDM wavelengths have to be well-stabilized, by means of temperature stabilization or wavelength locking techniques,

to avoid any possible wavelength drift due to environmental changes, especially when the ONUs are deployed in the field. Any wavelength drift will lead to severe crosstalk to the neighboring channels. Moreover, the drifted wavelength will also get deviated from the filter passband of the in-line WDM components, such as the wavelength multiplexers/demultiplexers, optical filters, etc. So, the signal power will be largely filtered off and the signal will also be severely distorted. In order to relax such stringent requirement, coarse-WDM (CWDM) (ITU-T Recommendation G.695, 2003) has been standardized to provide 18 wavelengths, spaced by 20 nm, over the fiber transmission window from 1271 nm to 1611 nm (ITU-T Recommendation G.694.2, 2002). Such wide channel spacing inevitably allows large wavelength drift, thus the wavelength transmitters do not need to be incorporated with a temperature control circuit. This greatly reduces the cost and the power consumption of the transmitters. However, relatively higher loss at the shorter wavelengths limits the power budget and the limited number of available

CWDM wavelengths restricts scalability. Besides, as some of the CWDM wavelengths are chosen around the high water absorption spectrum (around 1400 nm) of the conventional optical fiber, a new kind of optical fiber with the water absorption peak removed should be employed to avoid the excessive absorption. As the CWDM components require a wide passband (~13 nm), CWDM multiplexers and CWDM add-drop multiplexers are mostly realized by thin-film technology.

Wavelength Multiplexers / Demultiplexers

Most of the technologies to realize wavelength multiplexers or demultiplexers are based on planar lightwave circuit (PLC) as they have good potential for monolithic integration with other components. Examples are arrayed waveguide gratings (AWG) (Smit & Van Dam, 1996), thin-film filters or multilayer interference filters (Gerken & Miller, 2003), echelle grating, diffraction grating (Soole et al., 1994), etc. Apart from PLC technologies, wavelength demultiplexers can also be implemented using fiber Bragg gratings (FBGs) (Bilodeau et al., 1995). In order to ensure the devices made by these technologies to be practical and deployable in optical access networks, two critical issues, namely the polarization sensitivity and the thermal sensitivity, have to be carefully considered. Polarization dependency leads to severe polarization dependent loss while thermal dependency leads to mismatch between the laser wavelength and the multiplexer passband, which translates to high loss in signal power and also possibly filtering-induced distortion. Recently, various techniques have been reported to achieve polarization insensitive (Spiekman, L. H. et al., 1996; Takahashi, Hibino, Ohmori, & Kawachi, 1993) or athermal (Inoue et al., 1997; Keil et al., 2001; Lo & Kuo, 2003; Ooba, Hibino, Inoue, & Sugita, 2000; Pedersen, Demeester, & Smit, 1996) wavelength multiplexers. Thus, they are suitable to be deployed in the field.

In particular, AWG is a promising device to support wavelength multi-/demultiplexing in large channel count at the RN. It routes each specific wavelength to a unique output port. It is based on silica-on-silicon technology with low propagation loss (<0.05 dB/cm) and high fiber coupling efficiency. An important feature of AWG is the free spectral range (FSR) which defines the wavelength periodicity. A cyclic AWG supports multiple wavelengths, spaced by an integer multiple of the FSR, at the same output port, as illustrated in Figure 2. This cyclic property enables special wavelength assignment and routing mechanism on the same AWG at the RN.

NETWORK ARCHITECTURES FOR WDM-PONS

The tree-shaped network architecture, as shown in Figure 1, is the most common approach to realize a WDM-PON, as it is the most similar to the conventional PONs. Hence, any migration of the PON to WDM-PON can be achieved by simply incorporating wavelength-dependent devices at the OLT, the RN and the ONUs. Nevertheless, there are several other feasible network architectures proposed for WDM-PONs over the recent years.

The earliest proposal of WDM-PON in tree topology was the passive photonic loop (PPL) (Wagner, Kobrinski, Robe, Lemberg, & Smoot, 1988; Wagner & Lemberg, 1989), as shown in Figure 3, in which the OLT was carrying the services, via a long feeder, to a remote node where both the downstream and the upstream wavelengths were routed by a single passive wavelength multiplexer to and from their destined subscribers or ONUs, respectively. Therefore, in each connection between the OLT and an ONU, only one common piece of feeder fiber and one piece of distribution fiber were required to carry both the downstream and the upstream wavelengths. Alternatively, a pair of feeder fiber and distribution fibers might

Figure 2. Schematic diagram illustrating the operation of a wavelength router: (left) Interconnectivity scheme (a_i denotes the signal at input port a with frequency i) and (right) frequency response. (©2008, IEEE. Used with permission.)

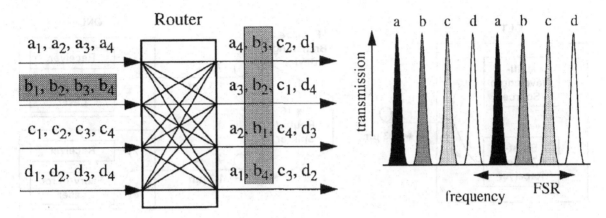

be employed to separate the downstream and the upstream wavelengths. This approach required a pair of wavelength multiplexers at the RN, one to demultiplex the downstream wavelengths and the other to multiplex the upstream wavelengths.

In (Feldman, Harstead, Jiang, Wood, & Zirngibl, 1998; Giles et *al.*, 1996), an interesting WDM-PON architecture, named as composite PON (CPON), was proposed. It employed WDM in the 1550-nm band for the downstream traffic and a single wavelength in the 1300-nm band to support the upstream traffic, shared via TDMA, as shown in Figure 4. At the RN, there required a WDM demultiplexer to route the downstream

wavelengths and one power splitter to combine the upstream signals from all transmitting ONUs. Examples of integrated device combining the WDM router and power splitter were demonstrated in (Inoue, Himeno, Moriwaki, & Kawachi, 1995; Zirngibl, Doerr, & Joyner, 1998). In addition, a burst-mode receiver was required at the OLT to receive the upstream signal and synchronize the data frames from different transmitting ONUs.

In (Mayer, Martinelli, Pattavina, & Salvadori, 2000), a multi-stage WDM-PON architecture was proposed. Several AWGs with smaller port-count were arranged in multiple stages and by making use of the cyclic wavelength routing property of

Figure 3. Passive photonic loop (PPL) architecture employing wavelength multi/demultiplexing and routing at the remote node. (©2008, IEEE. Used with permission.)

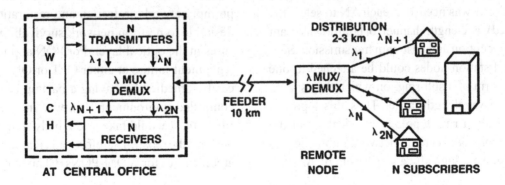

Figure 4. A single-fiber CPON. The downstream transmission uses dense WDM in the 1.5-μm wavelength window, and upstream uses power-combining in the 1.3-μm window. (©2008, IEEE. Used with permission.)

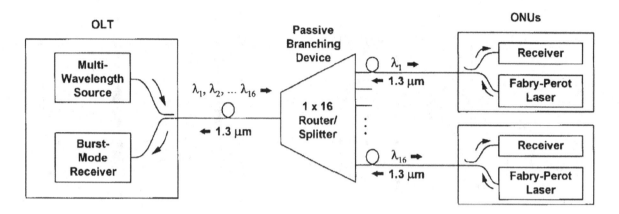

the AWG, reuse of a given wavelength for more than one ONUs was possible. Figure 5 shows an example of a multi-stage WDM-PONs with 32 ONUs, using two laser sources, each generating the same set of 16 wavelengths. This approach offered good scalability to support more ONUs by simply employing more wavelengths or having more stages of AWGs.

Apart from these architectures in tree-topology, WDM-PONs can also be realized using ring topology (An et *al.*, 2004; Iannone at *al.*, 2000). In the transparent WDM network featuring shared virtual rings (Iannone at al., 2000), a network hub node and multiple access nodes (ANs) were connected in series in form of a ring, as shown in Figure 6. The network hub node sent out multiple wavelength channels, each of which was destined for one particular AN. A wavelength add-drop multiplexer was needed at each AN to select the assigned wavelength channel for both downstream signal reception and upstream transmission. Several end station nodes could be attached to one AN in form of a sub-ring, on which they shared the same wavelength carrier. Thus, this approach was simply a ring topology physically, but the connections between the network hub node and the ANs were logically a star topology. Another

example is a network prototype, called SUCCESS (An et *al.*, 2004), in which the end stations were attached to the ANs in tree or star topology. So, it was simply multiple conventional PONs attached to a WDM ring. This approach employed dynamic wavelength allocation and a scheduling algorithm to provide flexible bandwidth sharing across multiple attached PONs. This could enhance the network scalability.

Network survivability is also a crucial issue in network management of WDM-PONs. Any component or fiber failure would lead to huge loss of data or even business. Subscribers are now requesting high-availability services and connections. Thus, the networks should provide resilience against failures, for instance, in case of possible catastrophic events such as fire or flooding. In order to facilitate effective and prompt network protection and restoration, it is desirable to perform network survivability measures in the optical layer. For PON applications, equipment failure at either OLT or ONU can be easily remedied by having a backup unit in the controlled environment. However, in case of any fiber cut, it would take a relatively long time to perform the repair. Therefore, survivable network architectures for WDM-PONs (Chan, 2007) have

Figure 5. An example of a multi-stage AWG-based WDM-PON using 16 wavelength values to support 32 ONUs. (©2008, IEEE, OSA. Used with permission.)

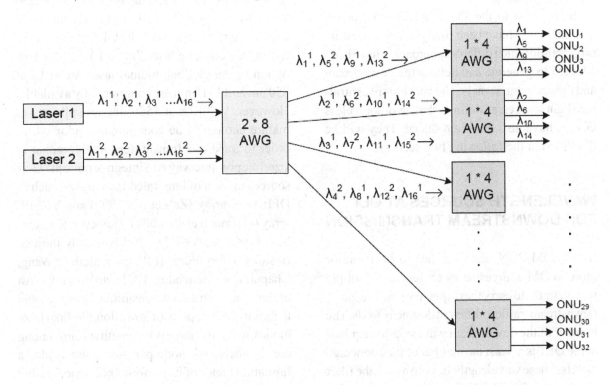

Figure 6. A transparent WDM network featuring shared virtual rings. (©2008, IEEE. Used with permission.)

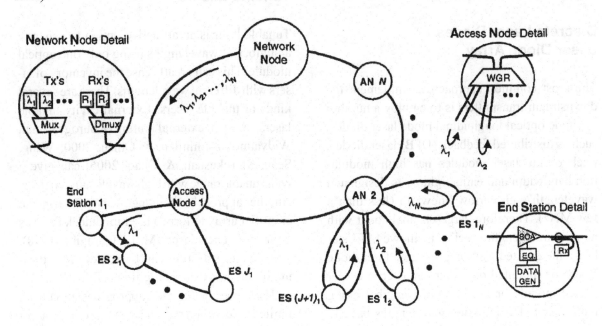

recently been introduced to protect the networks against any fiber cut.

In addition to the above WDM-PON architectures, there have been many other interesting proposals in which different approaches have been adopted to generate and deliver the downstream and the upstream signals. The concepts of centralized light sources, as well as colorless or sourceless ONUs, have also been introduced. They will be discussed in the following two sections.

WAVELENGTH SOURCES AT OLT FOR DOWNSTREAM TRANSMISSION

In a WDM-PON, the OLT has to host one or more WDM sources so as to generate multiple wavelengths to carry the respective point-to-point downstream traffic to different destined ONUs. The number of the downstream wavelengths required at the OLT is at least the number of the connected ONUs. These wavelengths have to match the filter passbands of the WDM multiplexer employed at the RN so as to route the individual downstream wavelengths to their respective ONUs. There are several categories of the WDM sources to support downstream transmission:

Discrete Laser Diodes or Laser Diode Array

The most common approach to implement the downstream transmitters is to employ a number of single optical longitudinal mode laser diodes, such as distributed feedback (DFB) laser diodes. Each of the laser modules has high modulation bandwidth and emits light at its designated wavelength with narrow linewidth (less than a few MHz). Therefore, large channel count with narrow wavelength spacing is allowed. To stabilize the emitted wavelengths, each laser diode has to be mounted on a thermal electric cooler (TEC) module or requires wavelength locking mechanism. The emission wavelengths have to

be chosen to match with the filter passbands of the WDM multiplexer at the OLT as well as that at the RN. However, wavelength inventory and management is required at the OLT. Vertical cavity surface-emitting laser (VCSEL) is a low-cost option for single-longitudinal laser. VCSELs at 850 nm and 1310 nm are commercially available. However, VCSELs at 1550 nm are still not yet mature enough to be commercially adopted. In order to make the transmitter more compact in size, one possible way is to integrate multiple laser sources in form of integrated laser arrays, such as DFB laser array (Zah et al., 1992) and VCSEL array (Hofmann et al., 2008). However, the number of laser diodes integrated is usually limited. Besides, Fabry-Pérot (FP) laser diodes (Wang, Chapuran, & Menendez, 1991) is also a low-cost option as a downstream transmitter. However, due to its intrinsic property of multi-longitudinal laser modes, it requires large wavelength spacing among the channels and mode partition noise is also a limitation factor of its performance. One feasible method to improve its performance is by means of injection locking, which will be discussed in the later subsection.

Tunable Lasers

Tunable laser is an attractive option to generate a number of wavelengths using only one optical module. This can greatly ease the inventory of lasers with different wavelengths. There are several kinds of tunable lasers. External-cavity tunable lasers based on external grating (Kouroger, Imai, Widyatmoko, Shimizu, & Ohtsu, 2000; Wang, Seah, Murukeshan, & Chao, 2006) can give a wide tunable range with high wavelength accuracy. Another approach is to integrate an AWG and an amplifier array to form a laser module, known as multi-frequency laser (MFL) (Zirngibl, 1998). Wavelength selection can be achieved by turning on the individual amplifiers (Zirngibl et al., 1994). However, these two approaches suffer from limited modulation rate and the issue of long-term stability, due to their long cavity lengths. On the other hand, DFB and distributed Bragg reflector (DBR) lasers can achieve tunable wavelength output by means of thermal control or driving current control (Hong, Kim, & Makino, 1998). Wavelength tuning based on thermal control has very slow response time, while driving current control can improve the wavelength tuning time to the order of nanoseconds. By adopting multi-section laser design, such as superstructure-grating DBR laser (Öberg et al., 1995) and grating-assisted directional coupler with rear sampled grating reflector (GCSR) laser (Rigole et al., 1995), etc., wavelength tuning can be more flexibly achieved. However, they are susceptible to mode-hopping, which degrades the transmission performance. In general, tunable laser can ease the inventory management of lasers, but precise wavelength tuning control is required. It has also found application as a backup laser source for protection in WDM systems (Delorme, 2000).

Broadband Light Source for Spectral Slicing

Instead of using wavelength-specific optical sources, broadband light source (BLS) can be employed to generate multiple individual wavelengths by means of spectral slicing technique (Zirngibl, Doerr, & Stulz, 1996) via narrowband optical filters, such as Fabry-Perot filter or AWG. There are several feasible options for the BLS, namely, superluminescent light emitting diode (LED) (Glance et al., 1996; Han, Son, Lim, Choi, & Chung, 2004; Iannone, Frigo, & Darcie, 1995; Liou, Koren, Burrows, Zyskind, & Dreyer, 1997; Liou et al., 1998; Reeve, Hunwicks, Methley, Bickers, & Hornung, 1988; Wagner, & Chapuran, 1990), amplified spontaneous emission (ASE) from Erbium-doped fiber amplifier (EDFA) (Jung, Shin, Lee, & Chung, 1998; Lee, Chung, & DiGiovanni, 1993) and reflective semiconductor optical amplifier (RSOA) (Park et al., 2007). In (Lee, Kim, Han, & Lee, 2006), the high-power

EDFA ASE light was passed through a piece of nonlinear dispersion-shifted fiber to further broaden the incoherent spectrum to 130 nm, by means of continuous-wave (CW) supercontinuum generation. The generated broadband spectrum was spectrally sliced into more than 100 channels by use of a Lyot-Sagnac filter with a wavelength spacing of 1 nm and a channel bandwidth of 0.5 nm. In general, the main drawbacks of using spectral slicing of broadband incoherent sources to generate the downstream wavelengths were the limited transmission distance as well as limited bit-rate (< 1 Gb/s), due to the presence of various kinds of noises, such as incoherent intensity noise, mode-partition noise and optical beat noise. In addition, the crosstalk effect among the spectrally sliced wavelengths also imposed system performance limitations (Murtaza, & Senior, 1996; Jang, Lee, & Chung, 1999; Feldman, 1997). In (Han, et al., 2004), forward error correction (FEC) technique was adopted to improve the system performance of the spectrally-sliced LED signal in a WDM-PON.

Spectral slicing technique can also be applied to coherent sources, such as FP laser diode (Holloway, Keating, & Sampson, 1997; Woodward, Iannone, Reichmann, & Frigo, 1998) or mode-locked femosecond laser (Mikulla et al., 1999), to general multiple wavelengths. Due to the coherence nature of the light source, they do not suffer from beat noise among the spectral components, thus can enhance the transmission capacity. In (Nuss, Knox, & Koren, 1996), a chirped-pulse femtosecond laser was demonstrated by spectral slicing a 100-fs optical pulse, which had a spectral width of 4.4 THz. 44 WDM channels, spaced by 100 GHz. By temporal spreading of the femosecond pulse, by means of fiber dispersion, before performing data encoding and spectral slicing, bit-interleaved WDM/TDM modulation format for a WDM-PON (Stark et al., 1997) was achieved. Similar bit-interleaved WDM/TDM format can also be achieved by spectral slicing of amplified LED sources (Liou et al., 1997; Liou et al., 1998),

Figure 7. A WDM access system using the loop-back spectrally sliced LED source, time delayed, and multiplexed WDM transmitter. (©2008, IEEE. Used with permission.)

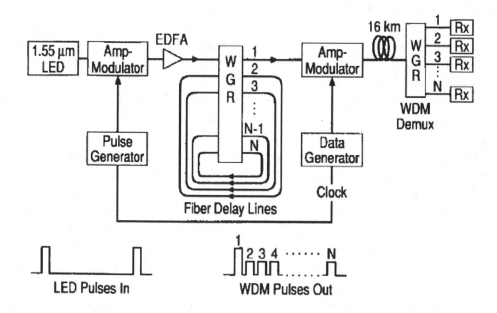

using a sequentially-pulsed MFL (Giles, Zirngibl, & Joyner, 1997), a rapidly tunable laser (Kinoshita, Okayasu, & Shibata, 1997), or shared multiple DFB lasers (Chae, & Oh, 1998). Figure 7 shows an example of a 24-channel WDM-PON in WDM/TDM format, using the loop-back spectrally sliced LED source, time delayed, and multiplexed WDM transmitter (Liou et al., 1998).

Injection-Locking / Wavelength-Seeding of Fabry-Pérot Laser Diodes or Reflective Semiconductor Optical Amplifiers

As discussed in previous section, both FP laser diodes and RSOAs are feasible and economical BLSs for spectral slicing. However, their performances are limited due to the possible mode partition noise among the multi-longitudinal modes present in the FP lasers, as well as the incoherent ASE noise of the RSOAs. These can be alleviated by mean of injection locking or seeding of these light sources. When a FP laser diode is injection-locked by an external wavelength, it operates in a quasi-single-longitudinal mode with much higher (> 20 dB) side-mode-suppression ratio. Thus, mode partition noise is largely suppressed. Similarly, when a RSOA is wavelength-seeded, its intensity noise is much reduced by the amplitude squeezing effect of the RSOA. In general, the injection wavelength can be a coherent wavelength from a DFB laser (Lu et al., 2007) or a spectrally sliced one from an incoherent BLS (Healey et al., 2001; Lee, Choi, Mun, Moon, & Lee, 2005; Payoux, Chanclou, & Brenot, 2006). Alternatively, self-seeding technique can be adopted to improve signal quality by reflecting a spectrally sliced portion of its output back to its laser cavity. All of these techniques can greatly enhance the system performance. However, the data modulation rate is usually limited by the intrinsic modulation bandwidth of the device. Recently injection-locked FP laser diodes (Lee et al., 2005; Lu et al., 2007) or wavelength seeded RSOAs (Healey et al., 2001; Payoux et al., 2006) have been employed at the OLT as the downstream transmitters. Similarly, they have also been widely deployed as the upstream transmitters at the ONUs. More examples and system performances will be further discussed in the next section.

TRANSMITTERS AT ONUS FOR UPSTREAM TRANSMISSION

In a WDM-PON, each ONU can send data back to the OLT via the upstream wavelengths. Therefore, each ONU may have to be incorporated with a light source to support the upstream transmission. The upstream wavelength generated has to match the respective filter passband of the WDM router at the RN. Also, the wavelength monitoring or stabilization may be needed as any wavelength drift will lead to excessive power loss and distortion due to filtering effect at the RN. These increase the system complexity and cost to maintain a wavelength-specific laser source at each ONU. Therefore, the concept of WDM-PONs with *centralized light sources* at the OLT has been proposed such that all the optical sources for upstream transmission are also resided at the OLT. No laser source is installed at the ONU and thus the cost of the ONUs can be kept low. Alternatively, another idea of *colorless ONUs* has also been proposed such that the ONUs are not installed with any wavelength-specific devices. This greatly eases the inventory, maintenance and management of the wavelengths at the subscriber side and also facilitates volume production of the identical ONU modules. All these different approaches in support of upstream transmission in WDM-PONs, will be briefly discussed, as follows.

ONUs with Wavelength-Specific Laser Sources

At each ONU in a WDM-PON, the common approach is to install a wavelength specific coherent light source, such as DFB laser diode, FP laser diode or VCSEL. The emission wavelengths of all the upstream transmitters have to be carefully chosen to match with the filter passbands of the WDM multiplexer at the RN and the OLT along the upstream path. Therefore, the wavelengths have to be stabilized or wavelength locked, and wavelength inventory and management are required.

Colorless ONUs with Non-Wavelength-Specific Optical Sources

Spectral Slicing of Broadband Light Source

The upstream transmitter can be realized by installing a BLS, such as LED (Glance et al., 1996; Han, et al., 2004; Iannone et al., 1995; Jung et al., 1998; Zirngibl et al., 1995), etc. at each ONU and the upstream wavelengths are generated by means of spectral slicing at the WDM multiplexer at the RN. Therefore, no wavelength matching between the upstream light source and the filter passbands of WDM multiplexer at the RN is needed. Due to the incoherence nature of the BLS, the system performance is usually limited. Figure 8 and Figure 9 show two examples of WDM-PONs employing spectrally-sliced LED as the upstream transmitters at the ONUs.

Injection-Locked Fabry-Pérot Laser Diode

At each ONU, a directly modulated FP laser diode is installed as the low-cost upstream transmitter. It is operated in quasi-single-longitudinal mode by means of injection locking technique. There are several options to generate the external injection wavelength to the FP laser diode. In (Kim, Kang, & Lee, 2000; Lee et al., 2005; Park et al., 2004; Shin et al., 2003; Shin et al., 2006), the injection wavelengths were spectrally sliced from a BLS, based on EDFA's ASE, resided at the OLT. The FP laser had a usable wavelength range of about 40 nm. To increase the injection efficiency and

Figure 8. A WDM-PON with bi-directional spectral slicing of LED source at both the OLT and the ONU. (©2008, IEEE. Used with permission.)

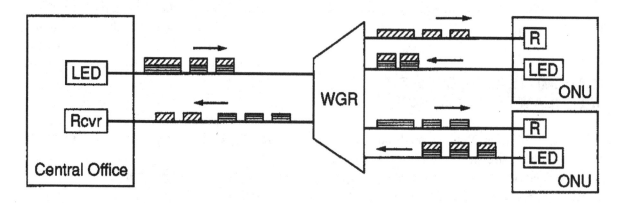

the output power, the front facet reflectivity was reduced to 1% by applying anti-reflection coating. The transmission bit-rate was 155-Mb/s and could be operated over a wide temperature range (Shin et *al.*, 2003). The same technique was also extended to realize the downstream transmitters at the OLT (Lee et *al.*, 2005; Park et *al.*, 2004; Shin et *al.*, 2006). As shown in Figure 10, two BLSs at different bands (C- and L-band) were employed at the OLT for injection of the spectrally-sliced light

into the FP laser diodes located at both the OLT and the ONUs. In (Hann, Kim, & Park, 2005), the FP laser diode at the ONU was self-injection-locked by reflecting back a portion of its own power, via a FBG at the RN, into its own laser cavity. No BLS was needed. The transmission bit-rate was 1.25 Gb/s. In (Lee, Kim, Han, & Lee, 2006; Lee, Lee, Han, Lee, & Kim, 2007), a CW supercontinuum (SC) source was employed at the OLT as the BLS, which was spectrally sliced before being injected

Figure 9. A WDM-PON with a spectrally sliced of LED source as the upstream transmitter at the ONU and a cyclic AWG as the WDM router at the RN and the central office (CO). (©2008, IEEE. Used with permission.)

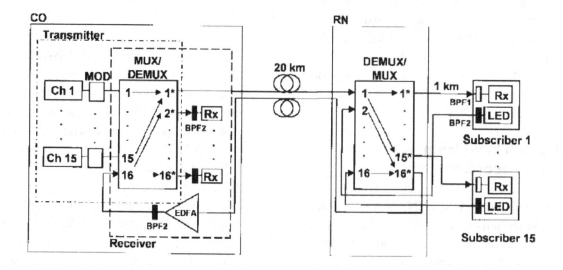

Figure 10. A WDM-PON employing wavelength-locked Fabry–Pérot laser diodes at both the OLT and the ONUs. (©2008, IEEE. Used with permission.)

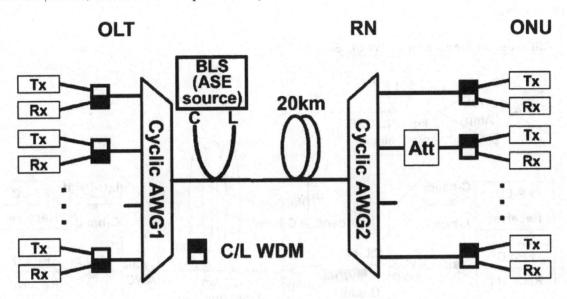

to the FP laser diodes at the ONUs, as well as those at the OLT, as shown in Figure 11. The CW SC source has a spectral width of 130 nm. The demonstrated transmission bit-rates for both the C-band downstream and the L-band upstream wavelengths are at 622 Mb/s. Similarly, in (Wen, & Chae, 2006), a pulsed form of the SC source was employed as the BLS for injection locking the FP laser diodes at the ONUs.

In (Xu et *al.*, 2007), a CW seeding wavelength generated from an array of DFB laser diodes at the OLT were distributed to the ONUs for injection locking of the FP laser diodes at the ONUs. Due to the highly coherence nature of the seeding light, 10-Gb/s transmission rate for the upstream transmission was demonstrated. In (Chan, Chan, Tong, Tong & Chen, 2002; Hung, Chan, Chen, & Tong, 2003), the downstream data-carrying wavelength was employed as the injection signal to the FP laser diode, which was directly intensity modulated with the upstream data, at each ONU. The downstream modulation formats were chosen as non-return-to-zero (NRZ) with reduced extinction ratio (Chan et *al.*, 2002), differential

phase-shift keying (DPSK) with constant intensity envelope (Hung et *al.*, 2003) (see Figure 12), or inverse-return-to-zero (IRZ) (Tse, Lu, Chen, & Chan, 2007). The relatively low intensity fluctuation in the downstream signal did not induce much interference to the superimposed upstream data.

Injection-Locked VCSEL

In (Wong, Zhao, Chang-Hasnain, Hofmann, & Amann, 2006), an injection-locked single-mode VCSEL was employed as a low-cost upstream transmitter at the ONU. Similar to (Chan et *al.*, 2002; Hung et *al.*, 2003; Tse et *al.*, 2007), the 2.5-Gb/s modulated downstream wavelength was used to injection-lock the VCSEL, which was directly modulated with 2.5-Gb/s upstream data. However, due to the relatively short laser cavity, the number of available cavity modes was small. For colorless operation, identical tunable VCSELs with wide wavelength tuning range (1530 nm to 1620 nm) (Yuen et *al.*, 2001) might be placed at the ONUs.

Figure 11. A WDM-PON employing a CW supercontinuum source as the broadband light source for injection locking of the Fabry-Pérot lasers at both the OLT and the ONUs. (©2008, IEEE. Used with permission.)

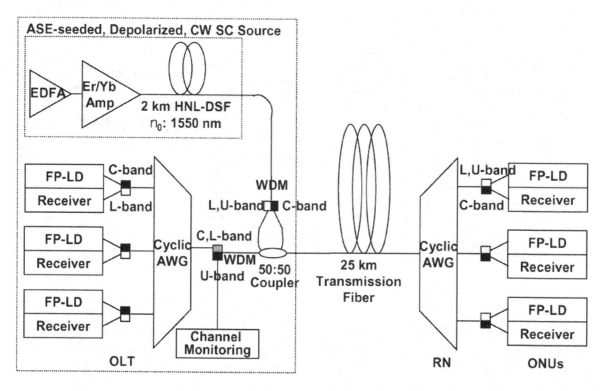

Figure 12. A WDM-PON employing downstream DPSK data and upstream OOK data with the FP laser diode at the ONU being injected-locked by the downstream wavelength. (©2008, IEEE. Used with permission.)

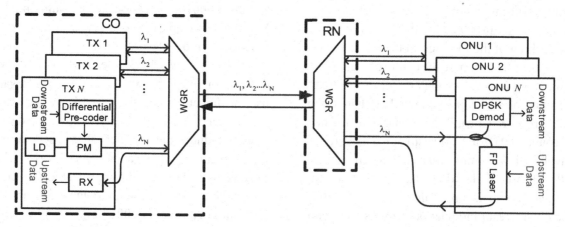

Wavelength-Seeded Reflective Semiconductor Optical Amplifier

At each ONU, a directly-modulated single-port RSOA (Feuer et *al.*, 1996) is installed as the low-cost upstream transmitter. As it exhibits a broad-band ASE spectrum, the upstream wavelengths are generated by means of wavelength seeding. There are several options to realize wavelength seeding of RSOA. In (Briand, Payoux, Chanclou, & Joindot, 2007; Borghesani, Lealman, Poustie, Smith, & Wyatt, 2007; Healey et *al.*, 2001; Payoux, Chanclou, Moignard, & Brenot, 2005; Payoux et *al.*, 2006), the seeding wavelengths for the RSOAs at the ONUs were generated via spectral slicing of a BLS, such as LED (Payoux et *al.*, 2006; Fayoux et *al.*, 2005; Briand et *al.*, 2007) or EDFA's ASE (Borghesani et *al.*, 2007), resided at the OLT. As shown in Figure 13, the LED-based wavelength-seeded RSOA was directly modulated by 1.25-Gb/s (Fayoux et *al.*, 2005) or 2.5-Gb/s (Borghesani et *al.*, 2007) upstream data and could operate at high temperature (80°C) (Borghesani et *al.*, 2007). The limited system performance due to the incoherent nature of the seeding signal could also be alleviated by means of FEC (Briand et *al.*, 2007). In (Park, Kim, & Park, 2006) and (Yeh, Chien & Chi, 2008), the seeding wavelength was a coherent light from a DFB laser or a self-seeded FP laser, respectively. Their respective upstream transmission bit rates were 1.25-Gb/s and 2.5-Gb/s, respectively. In (Lee et *al.*, 2005; Lee et *al.*, 2006; Lin, Lee, & Liu, 2008; Takesue, & Sugie, 2003; Yu, Kim, & Kim, 2007) gain-saturated RSOAs seeded by the modulated downstream wavelengths were employed as the upstream transmitters at the ONUs. However, the extinction ratio of the downstream NRZ data had to be kept low in order to facilitate the suppression of intensity fluctuation in the downstream wavelength, via gain saturation, before the upstream data was directly modulated onto it. Figure 14 illustrates the principle of the data suppression on the downstream wavelength by means of gain saturation in the SOA. The sup-pression of the "1" level of the residual downstream component could be further enhanced by means of the negative wavelength detuning method (Lee et *al.*, 2006; Yu et *al.*, 2007) for selective spectral filtering. On the other hand, the reduced extinction ratio of the downstream NRZ data might suffer from system penalty. However, it could be alleviated by using a FP etalon before detection (Lin et *al.*, 2008), so as to improve its extinction ratio and in turn the performance of the downstream data reception at the ONU.

In (Kang, Kim, Choi, Lee, &. Han, 2007; Kwon, Won, & Han, 2006; Wong, Lee, & Anderson, 2006), the RSOA at the ONU was self-seeded by reflecting a spectrally sliced portion of its own output power back to its laser cavity. In (Xu et *al.*, 2007), a WDM-PON was proposed with the downstream subcarrier multiplexed (SCM) data modulated on each downstream wavelength. A shared interferometric filter was employed at the RN so as to simultaneously separate the carrier power portions and the SCM sidebands on all downstream wavelengths, as shown in Figure 15. The extracted carrier power portion served as the seeding signal to the directly modulated RSOA, while the SCM sidebands were received as the downstream data, at the respective ONU. As the upstream data was modulated at the baseband spectrum, it would avoid the possible interference from the backscattered downstream SCM data on the same feeder fiber. A similar approach using downstream optical single-sideband (SSB) data was reported in (Presi, Proietti, D'Errico, Contestabile & Ciaramella, 2008), in which the carrier portions of all downstream wavelengths were extracted by means of a periodic optical notch filter. In (Cho et *al.*, 2008; Jang, Lee, Seol, Jung, & Kim, 2007), SCM technique was also adopted to carry both the downstream and the upstream data on the same the downstream wavelength, without much interference from each other. At the OLT, the downstream data was modulated either at the passband (Arellano, Bock, Prat, & Langer, 2006; Cho et *al.*, 2008; Jang et *al.*, 2007)

Figure 13. A WDM-PON with a single superluminescent LED for spectrally sliced seeding wavelength for the RSOA-based OLT and ONUs. (©2008, IEEE. Used with permission.)

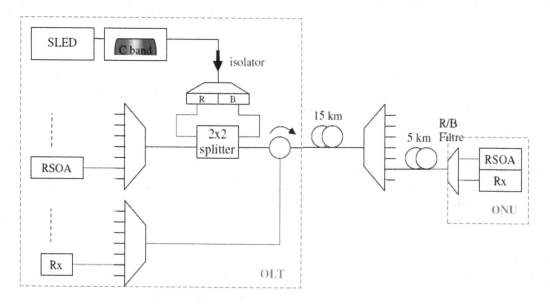

or at the baseband (Kang, & Han, 2006) of the downstream wavelength, before being injected to the RSOA at the ONU, where the upstream data was directly modulated in the baseband or the passband spectrum, respectively. Figure 16 shows the example as in (Jang et *al.*, 2007).

In (Prat, Arellano, Polo, & Bock, 2005), the downstream data and the upstream carrier were

Figure 14. Principle of the data suppression on the downstream wavelength using gain-saturated SOA. (©2008, IEEE. Used with permission.)

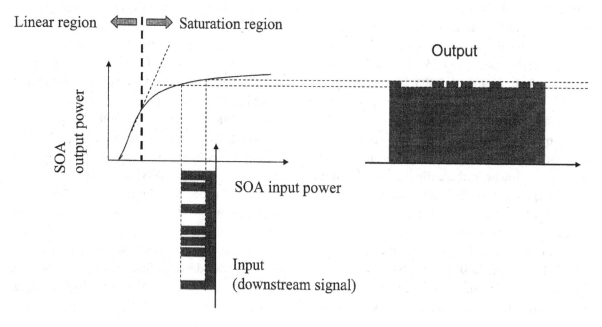

Figure 15. A RSOA-based WDM-PON with a single shared interferometric filter for carrier-reuse upstream transmission. (©2008, IEEE. Used with permission.)

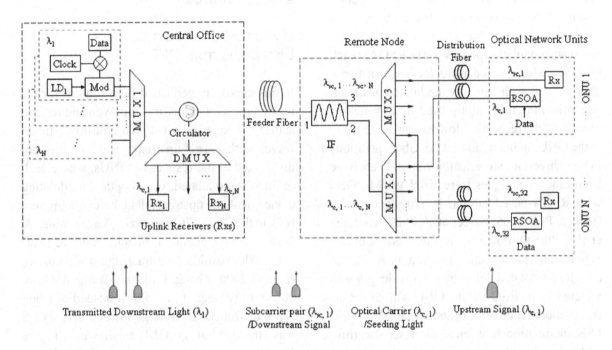

Figure 16. A bidirectional RSOA based WDM-PON utilizing a SCM signal down-link and a baseband signal for up-link. (©2008, IEEE. Used with permission.)

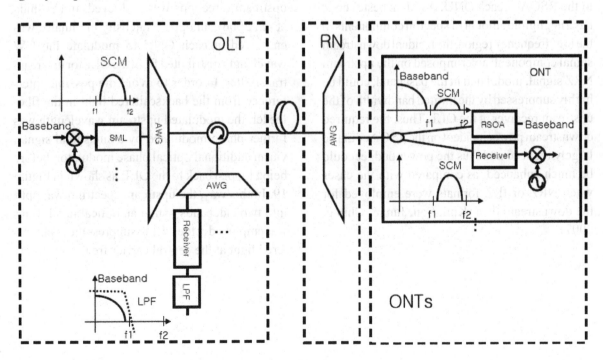

sent from the OLT time multiplexed in a single burst, as shown in Figure 17. The first burst section was the downstream data and then, after a guard band, unmodulated optical carrier was sent for upstream modulation purposes. The RSOA at each ONU served as a photodetector for the downstream data, as well as the directly modulated upstream transmitter in a time-multiplexed mode, synchronized with the respective downstream transmitter at the ONU, in burst mode. 1.25-Gb/s operations in both directions were demonstrated. There have been other techniques to realized WDM-PONs with RSOA-based ONUs. In (Arellano, Polo, Bock, & Prat, 2005; Garcés et *al.*, 2007; Martínez et *al.*, 2008), the downstream data was modulated in optical frequency-shifted keying (FSK) format and the modulated downstream wavelength was injected to the RSOA at the ONU with direct intensity modulation of the upstream data. As optical FSK modulation broadened the laser spectrum and reduced light coherence, Rayleigh scattering crosstalk was reduced. In (Kim, Jun, Takushima, Son, & Chung, 2007), Manchester coding format was chosen for the downstream data and each modulated downstream wavelength was injected to the RSOA at each ONU. As Manchester code has a negligible amount of spectral components in the low-frequency region, the residual downstream signal component superimposed on the upstream NRZ signal, modulated in the baseband, could be highly suppressed by the limited bandwidth of the upstream receiver at the OLT. Thus, the required downstream power incident on the RSOA could be largely reduced and thus the power budget could be much enhanced, as compared with the cases when NRZ or IRZ formats were employed for the downstream data (Kim, Son, Jun, & Chung, 2007).

Colorless ONUs without Optical Source

Dedicated Upstream Carriers Distributed from OLT

In this approach, dedicated light sources for the upstream wavelengths are employed and resided at the OLT, together with that for the downstream wavelengths. The upstream wavelengths are distributed to the respective ONUs, where each of them is modulated, via an optical modulating device, with the upstream data, before being sent back to the OLT (Altwegg, Azizi, Vogel, Wang, & Wyler, 1994; Nakamura, Suzuki, Kani, & Iwatsuki, 2006; Yoshida, Kimura, Kimura, Kumozaki, & Imai, 2006; Zhang, Lin, Huo, Wang, & Chan, 2006). In (Zhang et *al.*, 2006), instead of using CW light from DFB lasers, a supercontinuum BLS was employed at the OLT, as shown in Figure 18. The upstream wavelengths in C-band were spectrally sliced, via the AWG at the RN, before being modulated with the upstream data, via an optical intensity modulator, at each ONU. 10-Gb/s operations in both the downstream and the upstream directions were achieved. In (Yoshida et *al.*, 2006), an optical intensity modulator was employed at each ONU to modulate the CW wavelength distributed from the OLT for upstream transmitter. In order to avoid the possible interference from the backscattered light in the fiber feeder, the modulated upstream wavelength was further phase-modulated by a sinusoidal signal, via an additional optical phase modulator, before being looped back to the OLT, as shown in Figure 19. In this way, the upstream spectrum was split into two sidebands, while an optical notch filter was employed at the OLT to suppress the backscattered light at the central carrier frequency.

Figure 17. A WDM-PON with the RSOA served as both modulator and photodetector at each ONU. (©2008, IEEE. Used with permission.)

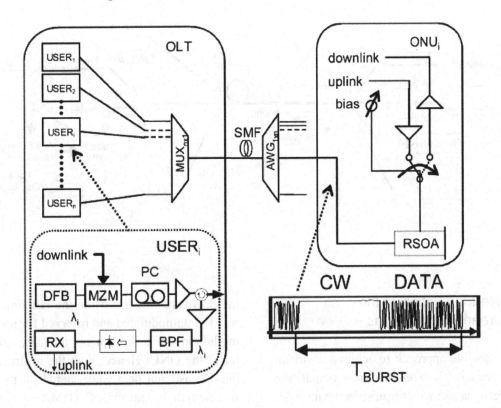

Figure 18. A WDM PON utilizing a centralized supercontinuum broadband light source at the OLT and colorless ONUs. PCF: photonic crystal fiber. (©2008, IEEE. Used with permission.)

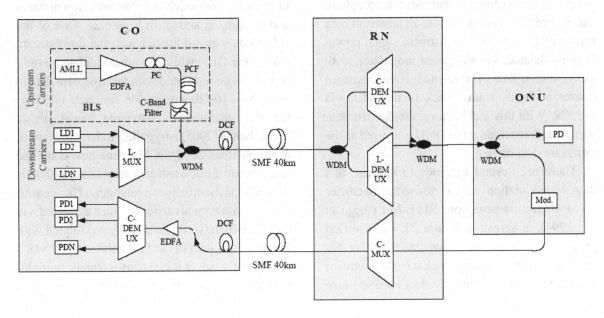

Figure 19. A single-fiber WDM PON with optical loopback method using phase modulation. (©2008, IEEE. Used with permission.)

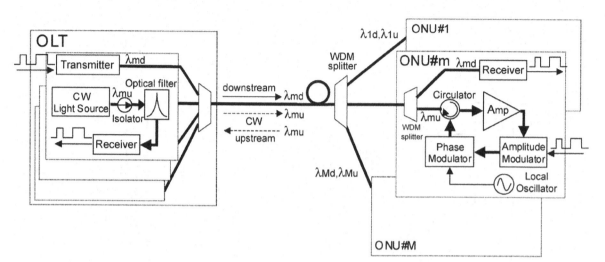

Direct Re-Modulation of the Downstream Wavelengths

Another feasible approach to achieve upstream data transmission is to re-use or re-modulate the downstream carrier with the upstream data at the ONUs, thus no light source is required at the ONUs. In WDM-PONs using centralized light sources at the OLT, the downstream signal is delivered to the ONU, where it is partially split and fed into an optical receiver for downstream data reception. The rest of the signal is fed into an upstream data transmitter where the downstream signal power is re-modulated, via an optical modulator, with the upstream data. The re-modulated upstream carrier is finally routed back to the OLT, via the RN. With this architecture, the downstream wavelength received at the ONU is re-used as the upstream data carrier.

There are several methods to achieve this data re-modulation on the downstream carrier for upstream transmission. RITE-Net (Frigo et al., 1994), as shown in Figure 20, was the first proposal to re-use the downstream carrier for upstream transmission. On each downstream wavelength, half of the time was used to carry the downstream data, while the remaining half was left unmodulated and reserved for upstream modulation, via the optical intensity modulator at the ONU. Hence, both the upstream and the downstream data were shared on the same wavelength by means of TDMA or dynamic bandwidth allocation. However, this required synchronization for proper upstream modulation. In (Akanbi, Yu, & Chang, 2006), optical carrier suppression and separation technique was adopted to generate two sidebands for each downstream wavelength, as shown in Figure 21. One of the sidebands was modulated with the downstream data at the OLT, while the other sideband would be intensity-modulated with the upstream data at the ONU. 10-Gb/s operation for each sideband was demonstrated. In (Attygalle, Nadarajah, & Nirmalathas, 2005; Attygalle, Anderson, Hewitt, & Nirmalathas, 2006), SCM technique was applied to modulate the downstream data at the passbands of each downstream wavelength. The central downstream optical carrier was then extracted, via a FBG, which would be directly modulated with the upstream data in the baseband at the ONU.

Recently, several constant-intensity modulation schemes, including FSK (Deng, Chan, Chen,

---

Figure 20. RITE-Net: A WDM-PON with wavelength reuse for upstream transmission. (©2008, IEEE. Used with permission.)

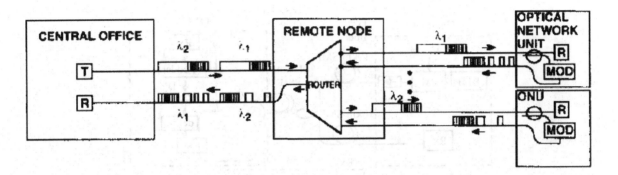

& Tong, 2003; Prat, Polo, Bock, Arellano, & Olmos, 2005), IRZ (Deng, Chan, & Chen, 2007; Chung et al., 2006), Manchester coding (Chung et al., 2006; Chung, Kim, & Kim, 2008; Kim, Park, Park, Yoon, & Kim, 2006), PSK-Manchester (Li, Dong, Wang, & Lu, 2005), DPSK (Hung, Chan, Chen, & Lin, 2003), etc. were adopted for the downstream data, such that the upstream data could be readily superimposed onto the received downstream wavelength, via simple intensity modulation at the ONUs. They had the common property that there would be minimal possible

Figure 21. A WDM-PON with the upstream and downstream channels generated by optical carrier suppression and separation technique. The downstream carriers (A) before, and (B) after the optical carrier suppression are shown in the insets. (©2008, IEEE. Used with permission.)

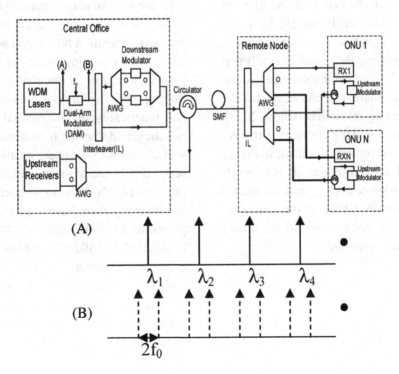

Figure 22. A WDM-PON using DPSK modulation format for both the downstream and the upstream data. (©2008, IEEE. Used with permission.)

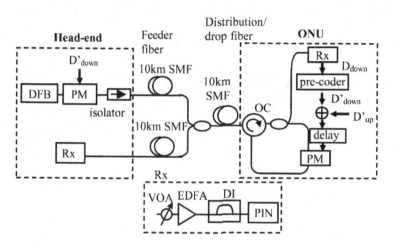

interference between the downstream and the upstream data, thus these schemes could support transmission at higher bit-rates, say 2.5 Gb/s or 10 Gb/s per wavelength. In (Xu, & Tsang, 2008; Xu, & Tsang, 2008; Zhao, Chen, & Chan, 2007), DPSK was proposed as the modulation format for the upstream data, while the downstream data was modulated in either NRZ with reduced extinction ratio (Xu, & Tsang, 2008; Zhao et al., 2007) or IRZ (Xu, & Tsang, 2008). These schemes required an optical phase modulator at each ONU for upstream data modulation. It was shown that both the dispersion tolerance and the requirement for modulation synchronization at the ONUs could be largely relaxed. In (Chow, 2008), both the downstream and the upstream data were modulated in DPSK format, as shown in Figure 22. At each ONU, a novel scheme (Pun, Chan, & Chen, 2005) was adopted to re-write the upstream phase information on the DPSK-modulated downstream wavelengths. 10-Gb/s operations in both directions were demonstrated.

SUMMARY

In summary, various WDM-PON architectures and their approaches to support two-way traffic have been reviewed. WDM-PON has emerged as a promising solution to realize fiber-to-the-home (FTTH), as it can offer high-capacity data delivery and flexible bandwidth provisioning, so as to meet the ever-increasing bandwidth requirements as well as the quality of service requirement of the next generation broadband access networks. The maturity and lower cost of the WDM components available in the market are also among the major driving forces to enhance the feasibility and practicality of commercial deployment. It is anticipated that the broadband access market will continue to grow drastically, thus network scalability is one of the crucial issues to upgrade the WDM-PONs so as to support more subscribers. Recently, there have been several interesting proposals to realize hybrid WDM/TDM PONs (Bock, Prat, & Walker, 2005; Hsueh, Rogge, Shaw, Kazovsky, & Yamamoto, 2004; Shin et al., 2005; Talli, & Townsend, 2006). A hybrid WDM/TDM PON is typically an integration of a WDM-PON in the first stage and multiple conventional PONs in the second stage, as shown in Figure 23. At

Figure 23. A typical WDM/TDM-PON. (©2008, IEEE. Used with permission.)

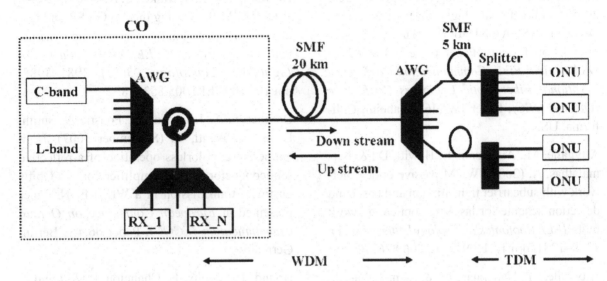

each ONU of the WDM-PON in the first stage, a conventional PON is attached such that the received downstream signal power from the first stage is further split and fed into several distribution fibers, so as to connect a larger number of subscribers. This requires a careful design of the power budget as well as the time-division multiple access protocol for the second-stage conventional PONs (Bock, & Prat, 2005; Kim, Gutierrez, An, & Kazovsky, 2005).

On the other hand, as the broadband services are getting more data centric, access networks have been evolving from conventional distribution networks to high-capacity peer-to-peer networks with guaranteed service level agreement (SLA) and quality of service (QoS) requirements. Thus the bandwidth provisioning should be more flexible and more networking functions should be supported by the network architectures and its higher layer protocols. The downstream and upstream traffic are becoming more symmetric. Therefore, other advanced techniques, such as optical code-division-multiple-access (OCDMA) over WDM-PON (Kitayama, Wang, & Wada, 2006), etc. are emerging so as to realize a full-service, scalable and robust next generation broadband access network.

REFERENCES

Akanbi, O., Yu, J., & Chang, G. K. (2006). A new scheme for bidirectional WDM-PON using upstream and downstream channels generated by optical carrier suppression and separation technique. *IEEE Photonics Technology Letters, 18*(2), 340–342. doi:10.1109/LPT.2005.861975

Altwegg, L., Azizi, A., Vogel, P., Wang, Y., & Wyler, F. (1994). LOCNET: A fiber in the loop system with no light source at the subscriber end. *IEEE/OSA . Journal of Lightwave Technology, 12*(3), 535–540. doi:10.1109/50.285337

An, F., Kim, K. S., Gutierrez, D., Yam, S., Hu, E., Shrikhande, K., & Kazovsky, L. G. (2004). SUCCESS: a next-generation hybrid WDM/TDM optical access network architecture. *IEEE/OSA . Journal of Lightwave Technology, 22*(11), 2557–2569. doi:10.1109/JLT.2004.836768

Arellano, C., Bock, C., Prat, J., & Langer, K. D. (March, 2006). RSOA-based optical network units for WDM-PON. *IEEE/OSA Optical Fiber Communication Conference / National Fiber Optic Engineers Conference (OFC/NFOEC)*, Paper OTuC1, Anaheim, California, USA.

Arellano, C., Polo, V., Bock, C., & Prat, J. (March, 2005). Bidirectional single fiber transmission based on a RSOA ONU for FTTH using FSK-IM modulation formats. Paper presented at *IEEE/OSA Optical Fiber Communication Conference / National Fiber Optic Engineers Conference (OFC/NFOEC)*, Paper JWA46, Anaheim, California, USA.

Attygalle, M., Anderson, T., Hewitt, D., & Nirmalathas, A. (2006). WDM passive optical network with subcarrier transmission and baseband detection scheme for laser-free optical network units. *IEEE Photonics Technology Letters, 18*(11), 1279–1281. doi:10.1109/LPT.2006.876770

Attygalle, M., Nadarajah, N., & Nirmalathas, A. (2005). Wavelength reused upstream transmission scheme for WDM passive optical networks. *IEE Electronics Letters, 41*(18), 1025–1027. doi:10.1049/el:20052468

Banerjee, A., Park, Y., Clarke, F., Song, H., Yang, S., & Kramer, G. (2005). A review of wavelength-division multiplexed passive optical network (WDM-PON) technologies for broadband access. *OSA Journal of Optical Networking, 4*(11), 737–758. doi:10.1364/JON.4.000737

Bilodeau, F., Johnson, D. C., Theriault, S., Malo, B., Albert, J., & Hill, K. O. (1995). An all-fiber dense wavelength-division multiplexer/demultiplexer using photoimprinted Bragg gratings. *IEEE Photonics Technology Letters, 7*(4), 388–390. doi:10.1109/68.376811

Bock, C., & Prat, J. (2005). Scalable WDMA/TDMA protocol for passive optical networks that avoids upstream synchronization and features dynamic bandwidth allocation. *OSA Journal of Optical Networking, 4*(4), 226–236. doi:10.1364/JON.4.000226

Bock, C., Prat, J., & Walker, S. D. (2005). Hybrid WDM/TDM PON using the AWG FSR and featuring centralized light generation and dynamic bandwidth allocation. *IEEE/OSA . Journal of Lightwave Technology, 23*(12), 3981–3988. doi:10.1109/JLT.2005.853138

Borghesani, A., Lealman, I. F., Poustie, A., Smith, D. W., & Wyatt, R. (September 2007). High temperature, colorless operation of a reflective semiconductor optical amplifier for 2.5-Gbit/s upstream transmission in a WDM-PON. Paper presented at *European Conference on Optical Communication (ECOC)*, Paper 06.4.1, Berlin, Germany.

Briand, J., Payoux, F., Chanclou, P., & Joindot, M. (2007). Forward error correction in WDM PON using spectrum slicing. *Optical Switching and Networking, 4*(2), 131–136. doi:10.1016/j.osn.2006.10.005

Chae, C. J., & Oh, N. H. (1998). WDM/TDM PON system employing a wavelength-selective filter and a continuous-wave shared light source. *IEEE Photonics Technology Letters, 10*(9), 1325–1327. doi:10.1109/68.705631

Chan, C. K. (2007). Protection architectures for passive optical networks. In C. Lam (Ed.), *Passive Optical Networks, Principles and Practice*, (pp.243-264). USA: Academic Press, Elsevier Inc.

Chan, L. Y., Chan, C. K., Tong, D. T. K., Tong, F., & Chen, L. K. (2002). Upstream traffic transmitter using injection-locked Fabry–Pérot laser diode as modulator for WDM access networks. *IEE Electronics Letters, 38*(1), 43–45. doi:10.1049/el:20020015

Cho, K. Y., Murakami, A., Lee, Y. J., Agata, A., Takushima, Y., & Chung, Y. C. (February, 2008). Demonstration of RSOA-based WDM PON operating at symmetric rate of 1.25 Gb/s with high reflection tolerance. Paper presented at *IEEE/OSA Optical Fiber Communication Conference / National Fiber Optic Engineers Conference (OFC/NFOEC)*, Paper OTuH4, San Diego, California, USA.

Chow, C. W. (2008). Wavelength remodulation using DPSK down-and-upstream with high extinction ratio for 10-Gb/s DWDM-passive optical networks. *IEEE Photonics Technology Letters*, *20*(1), 12–14. doi:10.1109/LPT.2007.911009

Chung, H. S., Kim, B. K., & Kim, K. J. (2008). Effects of upstream bit rate on a wavelength-remodulated WDM-PON based on Manchester or inverse-return-to-zero coding. *ETRI Journal*, *30*(2), 255–260.

Chung, H. S., Kim, B. K., Park, H., Chang, S. H., Chu, M. J., & Kim, K. J. (October, 2006). Effects of inverse-RZ and Manchester code a wavelength re-used WDM-PON. Paper presented at *IEEE Lasers and Electro-Optics Annual Meeting (LEOS)*, Paper TuP3, Montréal, Québec, Canada.

Delorme, F. (2000). 1.5-mm tunable DBR lasers for WDM multiplex spare function. *IEEE Photonics Technology Letters*, *12*(6), 621–623. doi:10.1109/68.849063

Deng, N., Chan, C. K., & Chen, L. K. (2007). A centralized-light-source WDM access network utilizing inverse-RZ downstream signal with upstream data remodulation. *Optical Fiber Technology*, *13*(1), 18–21. doi:10.1016/j.yofte.2006.03.006

Deng, N., Chan, C. K., Chen, L. K., & Tong, F. (2003). Data remodulation on downstream OFSK signal for upstream transmission in WDM passive optical network. *IEE Electronics Letters*, *39*(24), 1741–1743. doi:10.1049/el:20031092

Effenberger, F., Cleary, D., Haran, O., Kramer, G., Li, R. D., Oron, M., & Pfeiffer, T. (2007). An introduction to PON technologies. *IEEE Communications Magazine*, *45*(3), S17–S25. doi:10.1109/MCOM.2007.344582

Feldman, R. D. (1997). Crosstalk and loss in wavelength division multiplexed systems employing spectral slicing. *IEEE/OSA . Journal of Lightwave Technology*, *15*(10), 1823–1831. doi:10.1109/50.633564

Feldman, R. D., Harstead, E. E., Jiang, S., Wood, T. H., & Zirngibl, M. (1998). An evaluation of architectures incorporating wavelength division multiplexing for broad-band fiber access. *IEEE/OSA . Journal of Lightwave Technology*, *16*(9), 1546–1558. doi:10.1109/50.712236

Feuer, M., Wiesenfeld, J., Perino, J., Burrus, C., Raybon, G., Shunk, S., & Dutta, N. (1996). Single-port laser-amplifier modulators for local access. *IEEE Photonics Technology Letters*, *8*(9), 1175–1177. doi:10.1109/68.531827

Frigo, N. J., Iannone, P. P., Magill, P. D., Darcie, T. E., Downs, M. M., & Desai, B. N. (1994). A wavelength-division multiplexed passive optical network with cost-shared components. *IEEE Photonics Technology Letters*, *6*(11), 1365–1367. doi:10.1109/68.334841

Garcés, I., Aguado, J. C., Martínez, J. J., López, A., Villafranca, A., & Losada, M. A. (2007). Analysis of narrow-FSK downstream modulation in colourless WDM PONs. *IEE Electronics Letters*, *43*(8), 471–472. doi:10.1049/el:20073908

Gerken, M., & Miller, D. A. B. (2003). Wavelength demultiplexer using the spatial dispersion of multilayer thin-film structures. *IEEE Photonics Technology Letters*, *15*(8), 1097–1099. doi:10.1109/LPT.2003.815318

Giles, C. R., Feldman, R. D., Wood, T. H., Zirngibl, M., Raybon, G., & Strasser, T. (1996). Access PON using downstream 1550-nm WDM routing and upstream 1300-nm SCMA combining through a fiber-grating router. *IEEE Photonics Technology Letters, 8*(11), 1549–1552. doi:10.1109/68.541579

Giles, C. R., Zirngibl, M., & Joyner, C. (1997). 1152-subscriber WDM access PON architecture using a sequentially pulsed multifrequency laser. *IEEE Photonics Technology Letters, 9*(9), 1283–1284. doi:10.1109/68.618505

Glance, B., Liou, K. Y., Koren, U., Burrows, E. C., Raybon, G., & Burrus, C. A. (1996). A single-fiber WDM local access network based on amplified LED transceivers. *IEEE Photonics Technology Letters, 8*(9), 1241–1242. doi:10.1109/68.531849

Han, K. H., Son, E. S., Lim, K. W., Choi, H. Y., & Chung, Y. C. (2004). Bi-directional WDM PON using light-emitting diodes spectrum-sliced with cyclic arrayed-waveguide grating. *IEEE Photonics Technology Letters, 16*(10), 2380–2382. doi:10.1109/LPT.2004.833865

Hann, S., Kim, T. Y., & Park, C. S. (September, 2005). Direct-modulated upstream signal transmission using a self-injection locked F-P LD for WDM-PON. Paper presented at *European Conference on Optical Communications (ECOC)*, Paper We3.3.3, Glasgow, United Kingdom.

Healey, P., Townsend, P., Ford, C., Johnston, L., Townley, P., & Lealman, I. (2001). Spectral slicing WDM-PON using wavelength-seeded reflective SOAs. *IEE Electronics Letters, 37*(19), 1181–1182. doi:10.1049/el:20010786

Hofmann, W., Wong, E., Böhm, G., Ortsiefer, M., Zhu, N. H., & Amann, M. C. (2008). 1.55-mm VCSEL arrays for high-bandwidth WDM-PONs. *IEEE Photonics Technology Letters, 20*(4), 291–293. doi:10.1109/LPT.2007.915631

Holloway, W. T., Keating, A. J., & Sampson, D. D. (1997). Multiwavelength source for spectrum-sliced WDM access networks and LANs. *IEEE Photonics Technology Letters, 9*(7), 1014–1016. doi:10.1109/68.593384

Hong, J., Kim, H., & Makino, T. (1998). Enhanced wavelength tuning range in two-section complex-coupled DFB lasers by alternating gain and loss coupling. *IEEE/OSA . Journal of Lightwave Technology, 16*(7), 1323–1328. doi:10.1109/50.701412

Hsueh, Y. L., Rogge, M. S., Shaw, W. T., Kazovsky, L. G., & Yamamoto, S. (2004). SUCCESS-DWA: Highly scalable and cost effective optical access network architecture. *IEEE Optical Communication Magazine, 42*(8), S24–S30. doi:10.1109/MCOM.2004.1321383

Hung, W., Chan, C. K., Chen, L. K., & Lin, C. L. (September, 2003). System characterization of a robust re-modulation scheme with DPSK downstream traffic in a WDM access network. Paper presented at *European Conference on Optical Communications (ECOC)*, Paper We3.4.5, Rimini, Italy.

Hung, W., Chan, C. K., Chen, L. K., & Tong, F. (2003). An optical network unit for WDM access networks with downstream DPSK and upstream remodulated OOK data using injection-locked FP laser. *IEEE Photonics Technology Letters, 15*(10), 1476–1478. doi:10.1109/LPT.2003.818055

Iannone, P. P., Frigo, N. J., & Darcie, T. E. (March 1995). WDM passive-optical-network architecture with bidirectional optical spectral slicing. Paper presented at *Optical Fiber Communication Conference (OFC)*, Paper TuK2, Anaheim, California, USA.

Iannone, P. P., Reichmann, K. C., Smiljanic, A., Frigo, N. J., Gnauck, A. H., Spiekman, L. H., & Derosier, R. M. (2000). A transparent WDM network featuring shared virtual rings. *IEEE/OSA . Journal of Lightwave Technology, 18*(12), 1955–1963. doi:10.1109/50.908802

IEEE. 802.3ah EFM. Ethernet in the First Mile Task Force, http://www.ieee802.org/3/efm/.

Inoue, Y., Himeno, A., Moriwaki, K., & Kawachi, M. (1995). Silica-based arrayed-waveguide grating circuit as optical splitter/router. *IEE Electronics Letters, 31*(9), 726–727. doi:10.1049/el:19950497

Inoue, Y., Kaneko, A., Hanawa, F., Takahashi, H., Hattori, K., & Sumida, S. (1997). Athermal silica-based arrayed-waveguide grating multiplexer. *IEE Electronics Letters, 33*(23), 1945–1946. doi:10.1049/el:19971317

ITU-T Recommendation G.692. (1998). Optical interfaces for multichannel systems with optical amplifiers.

ITU-T Recommendation G.983.1. (1998). Broadband optical access systems based on Passive Optical Networks (PON).

ITU-T Recommendation G.983.3. (1998). A broadband optical access systems with increased service capability by wavelength allocation.

ITU-T Recommendation G.983.2. (2000). ONT management and control interface specifications for ATM PON.

ITU-T Recommendation G.694.2. (2002). Spectral Grids for WDM Applications: CWDM Wavelength Grid.

ITU-T Recommendation G.695. (2003). Optical interfaces for coarse wavelength division multiplexing applications.

ITU-T Recommendation G.984.1. (2003). Gigabit-capable Passive Optical Networks G-PON): General characteristics.

ITU-T Recommendation G.984.2. (2003). Gigabit-capable Passive Optical Networks G-PON): Physical Media Dependent (PMD) layer specification.

ITU-T Recommendation G.984.3. (2003). Gigabit-capable Passive Optical Networks G-PON): Transmission convergence layer specification.

Jang, S., Lee, C. S., Seol, D. M., Jung, E. S., & Kim, B. W. (March, 2007). A bidirectional RSOA based WDM-PON utilizing a SCM signal down-link and a baseband signal for up-link. Paper presented at *IEEE/OSA Optical Fiber Communication Conference / National Fiber Optic Engineers Conference (OFC/NFOEC)*, Paper JThA78, Anaheim, California, USA.

Jang, Y. S., Lee, C. H., & Chung, Y. C. (1999). Effects of crosstalk in WDM system using spectrum-sliced light sources. *IEEE Photonics Technology Letters, 11*(6), 715–717. doi:10.1109/68.766795

Jung, D. K., Shin, S. K., Lee, C. H., & Chung, Y. C. (1998). Wavelength-division-multiplexed passive optical network based on spectrum-slicing techniques. *IEEE Photonics Technology Letters, 10*(9), 1334–1336. doi:10.1109/68.705634

Kaneko, A., Kamei, S., Inoue, Y., Takahashi, H., & Sugita, A. (2000). Athermal silica-based arrayed-waveguide grating (AWG) multi/demultiplexers with new low loss groove design. *IEE Electronics Letters, 36*(4), 318–319. doi:10.1049/el:20000261

Kang, J. M., & Han, S. K. (2006). A novel hybrid WDM/SCM-PON sharing wavelength for up- and down-link using reflective semiconductor optical amplifier. *IEEE Photonics Technology Letters, 18*(3), 502–504. doi:10.1109/LPT.2005.863632

Kang, J. M., Kim, T. Y., Choi, I. H., Lee, S. H., & Han, S. K. (2007). Self-seeded reflective semiconductor optical amplifier based optical transmitter for upstream WDM-PON link. *IET OptoElectronics, 1*(2), 77–81. doi:10.1049/iet-opt:20050116

Keil, N., Yao, H. H., Zawadzki, C., Bauer, J., Bauer, M., Dreyer, C., & Schneider, J. (2001). Athermal all-polymer arrayed-waveguide grating multiplexer. *IEE Electronics Letters, 37*(9), 579–580. doi:10.1049/el:20010406

Kim, B. K., Park, H., Park, S. J., Yoon, B. Y., & Kim, B. T. (September, 2006). WDM passive optical networks with symmetric up/down data rates using Manchester coding based re-modulation. Paper presented at *European Conference on Optical Communication (ECOC)*, Paper Tu4.5.5, Cannes, France.

Kim, H. D., Kang, S. G., & Lee, C. H. (2000). A low-cost WDM source with an ASE injected Fabry-Pérot semiconductor laser. *IEEE Photonics Technology Letters, 12*(8), 1067–1069. doi:10.1109/68.868010

Kim, K. S., Gutierrez, D., An, F. T., & Kazovsky, L. G. (2005). Design and performance analysis of scheduling algorithms for WDM-PON under SUCCESS-HPON architecture. *IEEE/OSA . Journal of Lightwave Technology, 23*(11), 3716–3731. doi:10.1109/JLT.2005.857729

Kim, S. Y., Jun, S. B., Takushima, Y., Son, E. S., & Chung, Y. C. (2007). Enhanced performance of RSOA-based WDM PON by using Manchester coding. *OSA Journal of Optical Networking, 6*(6), 624–630. doi:10.1364/JON.6.000624

Kim, S. Y., Son, E. S., Jun, S. B., & Chung, Y. C. (March, 2007). Effects of downstream modulation formats on the performance of bidirectional WDM-PON using RSOA. Paper presented at *IEEE/OSA Optical Fiber Communication Conference / National Fiber Optic Engineers Conference (OFC/NFOEC)*, Paper OWD3, Anaheim, California, USA.

Kinoshita, T., Okayasu, M., & Shibata, N. (July 1997). Stable operation condition of optical WDM PDS system employing a rapidly tunable laser diode and wavelength router. Paper presented at *Optoelectronics and Communications Conference (OECC)*, Paper 10A1–5. 368–369. Seoul, Korea.

Kitayama, K., Wang, X., & Wada, N. (2006). OCDMA over WDM PON-solution path to gigabit-symmetric FTTH. *IEEE/OSA . Journal of Lightwave Technology, 24*(4), 1654–1662. doi:10.1109/JLT.2006.871030

Kouroger, M., Imai, K., Widyatmoko, B., Shimizu, T., & Ohtsu, M. (2000). Continuous tuning of an electrically tunable external cavity semiconductor laser. *OSA Optics Letters, 25*(16), 1165–1167. doi:10.1364/OL.25.001165

Kramer, G., Mukherjee, B., & Pesavento, G. (2002). IPACT a dynamic protocol for an Ethernet PON (EPON). *IEEE Communications Magazine, 40*(2), 74–80. doi:10.1109/35.983911

Kramer, G., & Pesavento, G. (2002). Ethernet passive optical network (EPON): building a next-generation optical access network. *IEEE Communications Magazine, 40*(2), 66–73. doi:10.1109/35.983910

Kwon, H. C., Won, Y. Y., & Han, S. K. (2006). A self-seeded reflective SOA-based optical network unit for optical beat interference robust WDM/SCM-PON link. *IEEE Photonics Technology Letters, 18*(17), 1852–1854. doi:10.1109/LPT.2006.881212

Lee, J. H., Kim, C. H., Han, Y. G., & Lee, S. B. (2006). Broadband, high power, erbium fibre ASE-based CW supercontinuum source for spectrum-sliced WDM PON applications. *IEE Electronics Letters, 42*(9), 67–68. doi:10.1049/el:20060713

Lee, J. H., Kim, C. H., Han, Y. G., & Lee, S. B. (2006). WDM PON upstream transmission at 1.25 Gb/s using Fabry–Pérot laser diodes injected with spectrum-sliced, depolarized, CW supercontinuum source. *IEEE Photonics Technology Letters, 18*(20), 2108–2110. doi:10.1109/LPT.2006.883288

Lee, J. H., Lee, K., Han, Y. G., Lee, S. B., & Kim, C. H. (2007). Single, depolarized, CW supercontinuum-based wavelength-division-multiplexed passive optical network architecture with C-Band OLT, L-Band ONU, and U-Band monitoring. *IEEE/OSA . Journal of Lightwave Technology, 25*(10), 2891–2897. doi:10.1109/JLT.2007.903637

Lee, J. S., Chung, Y. C., & DiGiovanni, D. J. (1993). Spectrum-sliced fiber amplifier light source for multi-channel WDM applications. *IEEE Photonics Technology Letters, 5*(12), 1458–1461. doi:10.1109/68.262573

Lee, S. M., Choi, K. M., Mun, S. G., Moon, J. H., & Lee, C. H. (2005). Dense WDM-PON based on wavelength-locked Fabry–Pérot laser diodes. *IEEE Photonics Technology Letters, 17*(7), 1579–1581. doi:10.1109/LPT.2005.848558

Lee, W., Cho, S. H., Park, M. Y., Lee, J. H., Kim, C., Jeong, G., & Kim, B. W. (2006). Wavelength filter detuning for improved carrier reuse in loop-back WDM-PON. *IEE Electronics Letters, 42*(10), 596–597. doi:10.1049/el:20060289

Lee, W., Park, M. Y., Cho, S. H., Lee, J., Kim, C., Jeong, G., & Kim, B. W. (2005). Bidirectional WDM-PON based on gain-saturated reflective semiconductor optical amplifiers. *IEEE Photonics Technology Letters, 17*(11), 2289–2462. doi:10.1109/LPT.2005.858153

Li, Z. H., Dong, Y., Wang, Y., & Lu, C. (2005). A novel PSK-Manchester modulation format in 10-Gb/s passive optical network system with high tolerance to beat interference noise . *IEEE Photonics Technology Letters, 17*(5), 1118–1120. doi:10.1109/LPT.2005.845663

Lin, S. C., Lee, S. L., & Liu, C. K. (2008). Simple approach for bidirectional performance enhancement on WDM-PONs with direct modulation lasers and RSOAs. *OSA Optics Express, 16*(6), 3636–3643. doi:10.1364/OE.16.003636

Liou, K. Y., Koren, U., Burrows, E. C., Zyskind, J. L., & Dreyer, K. (1997). A WDM access system architecture based on spectral slicing of an amplified LED and delay-line multiplexing and encoding of eight wavelength channels for 64 subscribers. *IEEE Photonics Technology Letters, 9*(4), 517–519. doi:10.1109/68.559407

Liou, K. Y., Koren, U., Dreyer, K., Burrows, E. C., Zyskind, J. L., & Sulhoff, J. W. (1998). A 24-channel WDM transmitter for access networks using a loop-back spectrally-sliced light-emitting diode. *IEEE Photonics Technology Letters, 10*(2), 270–272. doi:10.1109/68.655381

Lo, Y. L., & Kuo, C. P. (2003). Packaging a fiber Bragg grating with metal coating for an athermal design. *IEEE/OSA . Journal of Lightwave Technology, 21*(5), 1377–1383. doi:10.1109/JLT.2003.810925

Lu, H. H., Ma, H. L., Chuang, Y. W., Chi, Y. C., Liao, C. W., & Peng, H. C. (2007). Employing injection-locked Fabry–Pérot laser diodes to improve bidirectional WDM–PON performances. *Optics Communications, 270*(2), 211–216. doi:10.1016/j.optcom.2006.09.054

Martínez, J. J., Gregorio, J. I. G., Lucia, A. L., Velasco, V. A., Aguado, J. C., & Binué, M. Á. L. (2008). Novel WDM-PON architecture based on a spectrally efficient IM-FSK scheme using DMLs and RSOAs. *IEEE/OSA .Journal of Lightwave Technology, 26*(3), 350–356. doi:10.1109/JLT.2007.909864

Mayer, G., Martinelli, M., Pattavina, A., & Salvadori, E. (2000). Design and cost performance of the multistage WDM PON access networks. *IEEE/OSA . Journal of Lightwave Technology, 18*(2), 125–143. doi:10.1109/50.822785

Mikulla, B., Leng, L., Sears, S., Collings, B. C., Arend, M., & Bergman, K. (1999). Broad-band high-repetition-rate source for spectrally sliced WDM. *IEEE Photonics Technology Letters, 11*(4), 418–420. doi:10.1109/68.752534

Murtaza, G., & Senior, J. M. (1996). WDM cross-talk analysis for systems employing spectrally-sliced LED sources. *IEEE Photonics Technology Letters, 8*(3), 440–442. doi:10.1109/68.481143

Nakamura, H., Suzuki, H., Kani, J., & Iwatsuki, K. (2006). Reliable wide-area wavelength division multiplexing passive optical network accommodating gigabit Ethernet and 10-Gb Ethernet services. *IEEE/OSA . Journal of Lightwave Technology, 24*(5), 2045–2051. doi:10.1109/JLT.2006.871057

Nuss, M. C., Knox, W. H., & Koren, U. (1996). Scalable 32 channel chirped-pulse WDM source. *IEE Electronics Letters, 32*(14), 1311–1312. doi:10.1049/el:19960854

Öberg, M., Rigole, P. J., Nilsson, S., Klinga, T., Backbom, L., & Streubel, K. (1995). Complete single mode wavelength coverage over 40 nm with a super structure grating DBR Laser. *IEEE/OSA . Journal of Lightwave Technology, 13*(9), 1892–1898. doi:10.1109/50.464740

Ooba, N., Hibino, Y., Inoue, Y., & Sugita, A. (2000). Athermal silica-based arrayed-waveguide grating multiplexer using bimetal plate temperature compensator. *IEE Electronics Letters, 36*(21), 1800–1801. doi:10.1049/el:20001267

Park, S. B., Jung, D. K., Shin, D. J., Shin, H. S., Yun, I. K., & Lee, J. S. (2007). Colorless operation of WDM-PON employing uncooled spectrum-sliced reflective semiconductor optical amplifiers. *IEEE Photonics Technology Letters, 19*(4), 248–250. doi:10.1109/LPT.2007.891197

Park, S. J., Kim, G. Y., & Park, T. S. (2006). WDM-PON system based on the laser light injected reflective semiconductor optical amplifier. *Optical Fiber Technology, 12*(2), 162–169. doi:10.1016/j.yofte.2005.07.006

Park, S. J., Lee, C. H., Jeoung, K. T., Park, H. J., Ahn, J. G., & Song, K. H. (2004). Fiber-to-the-home services based on wavelength division multiplexing passive optical network. *IEEE/OSA . Journal of Lightwave Technology, 22*(11), 2582–2591. doi:10.1109/JLT.2004.834504

Payoux, F., Chanclou, P., & Brenot, R. (September 2006). WDM PON with a single SLED seeding colorless RSOA-based OLT and ONUs. Paper presented at *European Conference on Optical Communication (ECOC)*, Paper Tu4.5.1, Cannes, France.

Payoux, F., Chanclou, P., Moignard, M., & Brenot, R. (September 2005). Gigabit optical access using WDM PON based on spectrum slicing and reflective SOA. Paper presented at *European Conference on Optical Communications (ECOC)*, Paper We3.3.5, Glasgow, United Kingdom.

Prat, J., Arellano, C., Polo, V., & Bock, C. (2005). Optical network unit based on a bidirectional reflective semiconductor optical amplifier for fiber-to-the-home networks. *IEEE Photonics Technology Letters, 17*(1), 250–252. doi:10.1109/LPT.2004.837487

Prat, J., Polo, V., Bock, C., Arellano, C., & Olmos, J. J. (2005). Full-duplex single fiber transmission using FSK downstream and IM remote upstream modulations for fiber-to-the-home. *IEEE Photonics Technology Letters, 17*(3), 702–704. doi:10.1109/LPT.2004.840930

Presi, M., Proietti, R., D'Errico, A., Contestabile, G., & Ciaramella, E. (February, 2008). A full-duplex symmetric WDM-PON featuring OSSB downlink modulation with optical down-conversion. Paper presented at *IEEE/OSA Optical Fiber Communication Conference/National Fiber Optic Engineers Conference (OFC/NFOEC)*, Paper OThT4, San Diego, California, USA.

Pun, S. S., Chan, C. K., & Chen, L. K. (2005). Demonstration of a novel optical transmitter for high-speed differential phase-shift-keying / inverse return-to-zero (DPSK/Inv-RZ) orthogonally modulated signals. *IEEE Photonics Technology Letters, 17*(12), 2763–2765. doi:10.1109/LPT.2005.859412

Reeve, M. H., Hunwicks, A. R., Methley, S. G., Bickers, L., & Hornung, S. (1988). LED spectral slicing for single-mode local loop application. *IEE Electronics Letters, 24*(7), 389–390. doi:10.1049/el:19880263

Rigole, P. J., Nilsson, S., Backbom, L., Klinga, T., Wallin, J., & Stalnacke, B. (1995). 114-nm wavelength tuning range of a vertical grating assisted co-directional coupler laser with a super structure grating distributed Bragg reflector. *IEEE Photonics Technology Letters, 7*(7), 697–699. doi:10.1109/68.393177

Shin, D. J., Jung, D. K., Lee, J. K., Lee, J. H., Choi, Y. H., & Bang, Y. C. (2003). 155 Mbit/s transmission using ASE-injected Fabry–Pérot laser diode in WDM-PON over 70oC temperature range. *IEE Electronics Letters, 39*(18), 1331–1332. doi:10.1049/el:20030850

Shin, D. J., Jung, D. K., Shin, H. S., Kwon, J. W., Hwang, S., Oh, Y., & Shim, C. (2005). Hybrid WDM/TDM-PON with wavelength-selection-free transmitters. *IEEE/OSA . Journal of Lightwave Technology, 23*(1), 187–195. doi:10.1109/JLT.2004.840031

Shin, D. J., Keh, Y. C., Kwon, J. W., Lee, E. H., Lee, J. K., & Park, M. K. (2006). Low-cost WDM-PON with colorless bidirectional transceivers. *IEEE/OSA . Journal of Lightwave Technology, 24*(1), 158–165. doi:10.1109/JLT.2005.861122

Smit, M. K., & Van Dam, C. (1996). PHASAR-based WDM-devices: Principles, design and applications. *IEEE Journal on Selected Topics in Quantum Electronics, 2*(2), 236–250. doi:10.1109/2944.577370

Soole, J. B. D., Bhat, R., LeBlanc, H. P., Andreadakis, N. C., Grabbe, P., Caneau, C., Koza, & M. A. (1994). Wavelength precision of monolithic InP grating multiplexer/demultiplexers. *IEE Electronics Letters, 30*(8), 664-666.

Spiekman, L. H., Amersfoort, M. R., De Vreede, A. H., van Ham, F. P. G. M., Kuntze, A., & Pedersen, J. W. (1996). Design and realization of polarization independent phased array wavelength demultiplexers using different array orders for TE and TM. *IEEE/OSA . Journal of Lightwave Technology, 14*(6), 991–995. doi:10.1109/50.511599

Stark, J. B., Nuss, M. C., Knox, W. H., Cundiff, S. T., Boivin, L., & Dreyer, K. (1997). Cascaded WDM passive optical network with a highly shared source. *IEEE Photonics Technology Letters, 9*(8), 1170–1172. doi:10.1109/68.605539

Takahashi, H., Hibino, Y., Ohmori, Y., & Kawachi, M. (1993). Polarization-insensitive arrayed-waveguide wavelength multiplexer with birefringence compensating film. *IEEE Photonics Technology Letters, 5*(6), 707–709. doi:10.1109/68.219718

Takesue, H., & Sugie, T. (2003). Wavelength channel data rewrite using saturated SOA modulator for WDM networks with centralized light sources. *IEEE/OSA . Journal of Lightwave Technology, 21*(11), 2546–2556. doi:10.1109/JLT.2003.819532

Talli, G., & Townsend, P. D. (2006). Hybrid DWDM-TDM long-reach PON for next-generation optical access. *IEEE/OSA . Journal of Lightwave Technology, 24*(7), 2827–2834. doi:10.1109/JLT.2006.875952

Tse, Y. T., Lu, G. W., Chen, L. K., & Chan, C. K. (November, 2007). Upstream OOK remodulation scheme using injection-locked FP laser with downstream inverse-RZ data in WDM passive optical network. Paper presented at *Asia Pacific Optical Communications Communication Conference (APOC)*, Paper 6784-69, Wuhan, PRC.

Wagner, S., Kobrinski, H., Robe, T. J., Lemberg, H. L., & Smoot, L. S. (1988). Experimental demonstration of a passive optical subscriber loop. *IEE Electronics Letters, 24*(6), 344–346. doi:10.1049/el:19880234

Wagner, S., & Lemberg, H. (1989). Technology and system issues for a WDM-based fiber loop architecture. *IEEE/OSA . Journal of Lightwave Technology, 7*(11), 1759–1768. doi:10.1109/50.45899

Wagner, S. S., & Chapuran, T. E. (1990). Broadband high-density WDM transmission using superluminescent diodes. *IEE Electronics Letters, 26*(11), 696–697. doi:10.1049/el:19900454

Wang, L. A., Chapuran, T. E., & Menendez, R. C. (1991). Medium-density WDM system with Fabry-Perot laser diodes for subscriber loop applications. *IEEE Photonics Technology Letters, 3*(6), 554–556. doi:10.1109/68.91033

Wang, P., Seah, L. K., Murukeshan, V. M., & Chao, Z. X. (2006). Electronically tunable external-cavity laser diode using a liquid crystal deflector. *IEEE Photonics Technology Letters, 18*(15), 1612–1614. doi:10.1109/LPT.2006.879509

Wen, Y. J., & Chae, C. J. (2006). WDM-PON upstream transmission using Fabry–Pérot laser diodes externally injected by polarization-insensitive spectrum-sliced supercontinuum pulses. *Optics Communications, 260*(2), 691–695. doi:10.1016/j.optcom.2005.11.029

Wong, E., Lee, K. L., & Anderson, T. (2006). Low-cost WDM passive optical network with directly-modulated self-seeding reflective SOA. *IEE Electronics Letters, 42*(5), 299–301. doi:10.1049/el:20060097

Wong, E., Zhao, X., Chang-Hasnain, C., Hofmann, W., & Amann, M. C. (2006). Optically injection-locked 1.55-mm VCSELs as upstream transmitters in WDM-PONs. *IEEE Photonics Technology Letters, 18*(22), 2371–2373. doi:10.1109/LPT.2006.885292

Woodward, S. L., Iannone, P. P., Reichmann, K. C., & Frigo, N. J. (1998). A spectrally sliced PON employing Fabry–Perot lasers. *IEEE Photonics Technology Letters, 10*(9), 1337–1339. doi:10.1109/68.705635

Xu, L., & Tsang, H. K. (2008). Colorless WDM-PON optical network unit (ONU) based on integrated nonreciprocal optical phase modulator and optical loop mirror. *IEEE Photonics Technology Letters, 20*(10), 863–865. doi:10.1109/LPT.2008.921851

Xu, L., & Tsang, H. K. (2008). WDM-PON using differential-phase-shift-keying remodulation of dark return-to-zero downstream channel for upstream. *IEEE Photonics Technology Letters, 20*(10), 833–835. doi:10.1109/LPT.2008.919598

Xu, Z., Wen, Y. J., Zhong, W., Attygalle, M., Cheng, X., & Wang, Y. (2007). WDM-PON architectures with a single shared interferometric filter for carrier-reuse upstream transmission. *IEEE/OSA . Journal of Lightwave Technology, 25*(12), 3669–3677. doi:10.1109/JLT.2007.909341

Xu, Z. W., Wen, Y. J., Zhong, W. D., Chae, C. J., Cheng, X. F., & Wang, Y. X. (2007). High-speed WDM-PON using CW injection-locked Fabry-Pérot laser diodes. *OSA Optics Express, 15*(6), 2954–2962.

Yeh, C. H., Chien, H. C., & Chi, S. (February 2008). Cost-effective colorless RSOA-based WDM-PON with 2.5-Gbit/s uplink signal. Paper presented at *IEEE/OSA Optical Fiber Communication Conference / National Fiber Optic Engineers Conference (OFC/NFOEC)*, Paper JWA95, San Diego, California.

Yoshida, T., Kimura, S., Kimura, H., Kumozaki, K., & Imai, T. (2006). A new single-fiber 10-Gb/s optical loopback method using phase modulation for WDM optical access networks. *IEEE/OSA . Journal of Lightwave Technology, 24*(2), 786–796. doi:10.1109/JLT.2005.862441

Yu, J., Kim, N., & Kim, B. W. (2007). Remodulation schemes with reflective SOA for colorless DWDM PON. *OSA Journal of Optical Networking, 6*(8), 1041–1054. doi:10.1364/JON.6.001041

Yuen, W., Li, G. S., Nabiev, R. F., Jansen, M., Davis, D., & Chang-Hasnain, C. J. (October 2001). Electrically-pumped directly-modulated tunable VCSEL for metro DWDM applications. Paper presented at *Gallium Arsenide Integrated Circuit (GaAs IC) Symposium*, Paper TuA1.2, Baltimore, Maryland, USA.

Zah, C. E., Favire, F. J., Pathak, B., Bhat, R., Caneau, C., & Lin, P. S. D. (1992). Monolithic integration of a multi-wavelength compressive-strained multi-quantum-well distributed feedback laser array with a star coupler and optical amplifiers. *IEE Electronics Letters, 28*(25), 2361–2362.

Zhang, B., Lin, C. L., Huo, L., Wang, Z. X., & Chan, C. K. (March, 2006). A simple high-speed WDM PON utilizing a centralized supercontinuum broadband light source for colorless ONUs. Paper presented at *IEEE/OSA Optical Fiber Communication Conference / National Fiber Optic Engineers Conference (OFC/NFOEC)*, Paper OTuC6, Anaheim, California, USA.

Zhao, J., Chen, L. K., & Chan, C. K. (March, 2007). Novel re-modulation scheme to achieve colorless high-speed WDM-PON with enhanced tolerance to chromatic dispersion and re-modulation misalignment. Paper presented at *IEEE/OSA Optical Fiber Communication Conference / National Fiber Optic Engineers Conference (OFC/NFOEC)*, Paper OWD2, Anaheim, California, USA.

Zirngibl, M. (1998). Multifrequency lasers and applications in WDM networks. *IEEE Communications Magazine, 36*(12), 39–41. doi:10.1109/35.735875

Zirngibl, M., Doerr, C. R., & Joyner, C. H. (1998). Demonstration of a splitter/router based on a chirped waveguide grating router. *IEEE Photonics Technology Letters, 10*(1), 87–89. doi:10.1109/68.651116

Zirngibl, M., Doerr, C. R., & Stulz, L. W. (1996). Study of spectral slicing for local access applications. *IEEE Photonics Technology Letters, 8*(5), 721–723. doi:10.1109/68.491607

Zirngibl, M., Joyner, C. H., Stulz, L. W., Dragone, C., Presby, H. M., & Kaminow, I. P. (1995). LARNet, a local access router network. *IEEE Photonics Technology Letters, 7*(2), 215–217. doi:10.1109/68.345927

Zirngibl, M., Joyner, C. H., Stulz, L. W., Koren, U., Chien, M. D., Young, M. G., & Miller, B. I. (1994). Digitally tunable laser based on the integration of a waveguide grating multiplexer and an optical amplifier. *IEEE Photonics Technology Letters, 6*(4), 516–518. doi:10.1109/68.281813

Chapter 5
Broadband Optical Access using Centralized Carrier Distribution

Chi-Wai Chow
National Chiao Tung University, Taiwan

ABSTRACT

Passive optical network (PON) is considered as an attractive fiber-to-the-home (FTTH) technology. Wavelength division multiplexed (WDM) PON improves the utilization of fiber bandwidth through the use of wavelength domain. A cost-effective solution in WDM PON would use the same components in each optical networking unit (ONU), which should thus be independent of the wavelength assigned by the network. Optical carriers are distributed from the head-end office to different ONUs to produce the upstream signals. Various solutions of colorless ONUs will be discussed. Although the carrier distributed WDM PONs have many attractive features, a key issue that needs to be addressed is how best to control the impairments that arise from optical beat noise induced by Rayleigh backscattering (RB). Different RB components will be analyzed and RB mitigation schemes will be presented. Finally, some novel PONs including signal remodulation PONs, long reach PONs and wireless/wired PONs will be highlighted.

INTRODUCTION

It is now well known that there are various technologies for broadband access providing high-speed Internet access and triple-play services including data, voice, and video. Most well-established broadband access platforms now are digital subscriber line (DSL)-based and the Cable Modem-based. It is generally agreed that fiber-to-the-home (FTTH)

provides the bandwidth and flexibility in upgrades when considering high-speed broadband access, especially with data rate of 1 Gbit/s or above. Passive optical network (PON) is considered as an attractive FTTH technology since it is highly cost-effective. A PON is a point-to-multipoint network architecture in which passive optical power splitters are used to enable a single optical fiber to serve multiple users. A PON consists of a service provider's central office or head-end office and a number of optical networking units (ONUs) near

DOI: 10.4018/978-1-60566-707-2.ch005

end users. A PON is cost-effective since generally, there is no active component between the head-end office and the ONUs. Its configuration also reduces the amount of fiber and head-end office equipment required compared with point-to-point architectures. In conventional PON, downstream signals are broadcast to each ONU sharing a fiber. Upstream signals are combined using a multiple access protocol, invariably time division multiple access (TDMA). The optical line terminal (OLT) in the head-end office will "range" the ONUs in order to provide time slot assignments for upstream communication.

The first generation of Gigabit PONs (GPONs) has been standardized. They typically offer 1.244 to 2.488 Gbit/s, shared among 32 customers via passive optical splitters using a TDMA protocol. Whilst these PONs offer significant bandwidth increases compared to the copper-based approaches, they may not provide the best ultimate solution for network operators seeking to significantly reduce the cost of delivering future broadband services in order to sustain profit margins (Payne, D. B., & Davey, R. P., 2002). Hence research attention has recently turned to wavelength division multiplexed (WDM) PONs, or hybrid WDM-TDM PONs. (Talli, G., & Townsend, P. D., 2006)

WDM PON is a type of PON that uses multiple optical wavelengths to increase the bandwidth available to end users and to improve the utilization fiber bandwidth. WDM PON is capable to provide more bandwidth over longer distances by increasing the link loss budget of each wavelength, making it less sensitive to the optical losses incurred at each optical splitter when compared with conventional TDM-based PON. There is no standard for WDM PON now. By some definitions WDM PON uses a dedicated wavelength for each ONU. This means WDM PON can enable a number of ONUs located at customer premises, each working at different wavelengths, to share the same optical amplifiers and backhaul fiber in the network. However, one great challenge in this WDM PON is the transmitter (Tx) at the ONU,

located at the customer premise, which must have a wavelength that is precisely aligned with a specifically allocated WDM grid wavelength. A cost-effective solution would employ the same components in each ONU, which should thus be independent of the wavelength (colorless) assigned by the network. Optical carriers are distributed from the head-end office to different ONUs to produce the upstream signals. The advantages of this scheme are that the cost of wavelength referencing and control is shared among many users rather than being borne by individual users and no multi-wavelength source inventory is required for the end users. Besides, only a single optical laser source is necessary for all the ONUs in a TDM PON if hybrid WDM-TDM PON architecture is used (Talli, G., et al., 2002).

The organization of the chapter is as follows: in section II, various colorless ONUs for the WDM PON will be discussed. In section III, the Rayleigh backscattering (RB) components generated in the carrier distributed PON will be analyzed. Then, in section IV, several RB mitigation schemes will be presented. Some novels PON architectures will be highlighted in section V. Finally, a conclusion will be presented in section VI.

COLORLESS ONUS

Various colorless ONUs have been proposed for the WDM PON. The simplest scheme is to use tunable laser at the Tx. The wavelength is tuned at the installation or for a reconfiguration. The wavelength tuning speed is not critical. However, tunable lasers are too expensive up to now for ONU, since its cost is not shared. A cost-effective tunable ONU based on "set-and-forget" architecture is currently under development within the European Union-funded project PIEMAN (Photonic Integrated Extended Metro and Access Network) (Townsend, P. D., Talli, G., Chow, C. W., MacHale, E. M., Antony, C., Davey, R., De Ridder, T., Qiu, X. Z., Ossieur, P., Krimmel, H.

Figure 1. Schematic for WDM PON based on the spectrum-sliced RSOA. AWG: arrayed waveguide grating, BPF: band-pass filter.

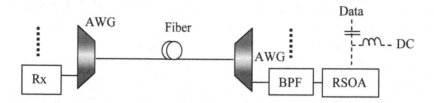

G., Smith, D. W., Lealman, I., Poustie, A., Randel, S., & Rohde, H., 2007). The tuning element is based on a high performance thin film filter, which has good thermal stability and is capable of being manufactured in high volume. Tuning of the laser is achieved by rotation of the thin film filter and it is performed by a miniature pizeo-electric ceramic motor. An interesting alternative is to use a broadband light source with the spectrum-slicing technique. Spectrum-sliced superluminescent diodes, reflective semiconductor optical amplifiers (RSOAs) and erbium-doped fiber amplifiers (EDFAs) (Chapuran, T. E., Wagner, S. S., Menendez, R. C., Tohme, H. E., & Wang, L. A., 1991; Park, S. B., Jung, D. K., Shin, D. J., Shin, H. S., Yun, I. K., Lee, J. S., Oh, Y. K., & Oh, Y. J., 2007; McCoy, A. D., Horak, P., Thomsen, B. C., Ibsen, M., Mokhtar, M. R., & Richardson, D. J., 2004) are interesting sources, since the center wavelength and width can be chosen and they are relatively low cost. Figure 1 shows the schematic of WDM PON based on the spectrum-sliced RSOA as described in (Park, S. B., et al., 2007). The RSOA was directly modulated at 155 Mbit/s data rate using bias-T. The modulated signal spectrum-sliced at the arrayed waveguide grating (AWG) was transmitted over a 25-km single mode fiber (SMF). The band-pass filter (BPF) between RSOA and AWG was used to suppress the side-modes due to the cyclic characteristic of AWG. Bit-error rate (BER) better than 10^{-10} was achieved over the temperature range from 0 °C to 60 °C. The power margins were more than 5.3 dB. Colorless operations was demonstrated

in outdoor applications over the temperature range from 20 °C to 80 °C. These schemes are also the most inexpensive when compared with other schemes introduced later in this section. This is because some light sources, such as the superluminescent diode, are relatively low cost, and temperature control is usually not required. However, the modulation speeds of these schemes are less than 1 Gbit/s and transmission distances are limited, due to dispersion and slicing loss. Besides, spectrum-slicing with incoherent light introduces excess intensity noise (EIN) generated by the spontaneous-spontaneous beat noise that falls within the receiver (Rx) pass-band.

For higher data rates transmission, injection locking or seeding techniques are preferred. These candidates include amplified spontaneous emission (ASE)-injected Fabry-Perot laser diodes (FPLDs) (Shin, D. J., Keh, Y. C., Kwon, J. W., Lee, E. H., Lee, J. K., Park, M. K., Park, J. W., Oh, Y. K., Kim, S. W., Yun, I. K., Shin, H. C., Heo, D. Lee, J. S., Shin, H. S., Kim, H. S., Park, S. B., Jung, D. K., Hwang, S., Oh, Y. J., Jang, D. H., & Shim, C. S., 2006; Shin, D. J., Jung, D. K., Shin, H. S., Kwon, J. W., Hwang, S., Oh, Y., & Shim, C., 2006; Mun, S. G., Moon, J. H., Lee, H. K., Kim, J. Y., & Lee, C. H., 2008) self-injected FPLDs (Hann, S., Kim, T. Y., & Park, C. S., 2005), ASE-injected RSOAs (Healey, P., Townsend, P., Ford, C., Johnston, L., Townley, P., Lealman, I., Rivers, L., Perrin, S., & Moore, R., 2001; Shin, H. S., Jung, D. K., Kim, H. S., Shin, D. J., Park, S. B., Hwang, S. T., Oh, Y. J., & Shim, C. S., 2005; Shin, H. S., Jung, D. K., Shin, D. J., Park,

Figure 2. Schematic for WDM PON based on the injection locked FPLD. AWG: arrayed waveguide grating, BLS: broadband light source.

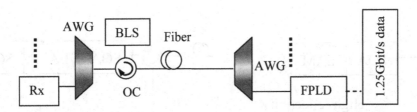

S. B., Lee, J. S., Yun, I. K., Kim, S. W., Oh, Y. J., & Shim, C. S., 2006; Payoux, F., Chanclou, P., & Brenot, R., 2006) continuous wave (CW)-injected RSOAs (Feuer, M. D., Wiesenfeld, J. M., Perino, J. S., Bums, C. A., Raybon, G., Shunk, S. C., & Dutta, N. K., 1996; Arellano, C., Bock, C., Prat, J. & Langer, K. D., 2006) self-injected RSOAs (Lee, K. L., & Wong, E., 2006), and injection locked vertical cavity surface emitting lasers (VCSELs) (Wong, E., Zhao, X., Chang-Hasnain, C. J., Hofmann, W., & Amann, M. C., 2006). The cost could be higher than the schemes of spectrum-slicing. In the ASE injecting architectures, the broadband light source (BLS) in the head-end office generates a broadband ASE to feed the Tx with spectrum-sliced ASE. Then, the injected FPLD or RSOA operates as a single-mode light source with negligible crosstalks to other WDM channels in the WDM PON. Figure 2 shows the schematic of a WDM PON based on injection locked FPLD (Mun, S. G., Moon, J. H., Lee, H. K., Kim, J. Y., & Lee, C. H., 2008). It consisted of a broadband light source (BLS) for a seed light, AWGs, and directly modulated FPLDs. The BLS was realized by using ASE from EDFA pumped by 1480 pump laser. The BLS was then spectrum-sliced and launched to injection lock the FPLD which was modulated at 1.25 Gbit/s. It is important that the injected ASE is unpolarized and incoherent to mitigate instabilities due to polarization and back reflection (Fujiwara, M., Suzuki, H., Yoshimoto, N., 2006). The noise characteristics of the ASE-injected FPLD and ASE-injected RSOA have been reported [21, 22]. Ref. (Mun, S. G., et al.,

2008) investigates the effect of AWG passbands and FPLD reflectivities in self-injection locking, showing wider AWG passband (flat-top passband with 3-dB bandwidth of 0.61 nm) and lower front-end facet reflectivity (0.1%) can provide better transmission performance. One of the key concerns of using broadband spectrum-sliced light injected FPLDs is how to increase both output power and spectral bandwidth of the BLS. Recently, BLS of using mutually injected FPLDs (Choi, K. M., & Lee, C. H., 2005; Ji, H. C., Yamashita, I., & Kitayama, K. I., 2008) and CW supercontinuum (SC) source (providing 130 nm incoherent, depolarized light source, with a total output power of ~500 mW) (Lee, J. H., Kim, C. H., Han, Y. G., & Lee, S. B., 2006; Lee, J. H. Lee, K., & Kim., C. H., 2007) are demonstrated. To further reduce the cost of ONU, it would be desirable if the ONU could operate uncooled over a wide temperature range. High temperature (up to 80°C), colorless (over the C-band) and high-speed (2.5 Gbit/s) operation of a wavelength seeded RSOA as the upstream Tx has been reported (Borghesani, A., Lealman, I. F., Poustie, A., Smith, D. W., & Wyatt, R., 2007).

For modulation speed at 10 Gbit/s or above, the electro-absorption modulator (EAM) is the best suited component. It has large modulation bandwidth, low polarization dependent loss and wide wavelength operation range. EAM is relatively expensive nowadays and has high insertion loss, hence, usually one or two SOAs are added to compensate the losses inside the ONU (Payoux, F., Chanclou, & P., Genay, N., 2007). Monolithic

Figure 3. Schematics of RONU designs using (a) REAM-SOA and (b) SOA-EAM-SOA. OC: optical circulator

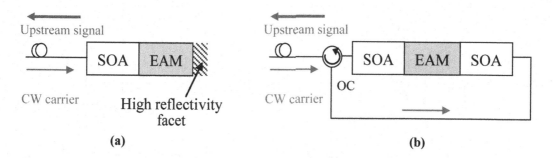

integration of SOAs and EAM on the same chip to avoid coupling losses and to reduce packaging cost has been developed (MacHale, E. K., Talli, G., Townsend, P. D., Borghesani, A., Lealman, I., Moodie, D. G., & Smith, D. W., 2008). These schemes are the most expensive when compared with other schemes described previously in this section. Figures 3(a) and (b) show the configurations of the reflective-EAM-SOA (REAM-SOA) and the SOA-EAM-SOA respectively. For the REAM-SOA described in (MacHale, E. K., et al., 2008), it used a double-pass design, comprising a mode-expanded SOA, with an anti-reflection coating and an EAM with a high-reflection coating on the back facet. Both SOA and EAM were buried heterostructure multi-quantum well (MQW) devices with a common blocking layer. The EAM used a MQW active section that was butt coupled to the tensile strained InGaAs MQW SOA active section. The REAM-SOA chip had passive alignment features for low cost volume assembly. In the experimental demonstration, optical signal at wavelength 1534.25 nm was launched into the REAM-SOA which was modulated at 10 Gbit/s non-return-to-zero (NRZ) data (3 Vpp). The SOA section was driven at 130 mA. The device was capable of supporting 10 Gbit/s upstream operation in a 128-way split 100 km reach PON.

ISSUES OF CENTRALIZED CARRIER DISTRIBUTION

As discussed in the introduction, the carrier distributed WDM PONs provide many attractive features, however, when a single drop fiber is used to reach the customer, the carrier distributed from the head-end office and the upstream signal must share the same path, giving rise to interferometric beat noise caused by RB (Talli, G., Cotter, D., & Townsend, P. D., 2006) and localized Fresnel back-reflections (Fujiwara, M., Kani, J., Suzuki, H., & Iwatsuki, K., 2006). The maximum split ratio achievable in each TDM PON is limited by the levels of RB and back reflection present in the system. While the level of back reflection can in principle be controlled by setting appropriate return loss specifications for the various optical components used in the network, the RB is an intrinsic phenomenon in fiber propagation and its level is fixed by the fiber type and configuration used. RB is generated by the distributed reflections caused by the random index fluctuation along the silica optical fiber (Gysel, P., & Staubli, R. K., 1990). This process is considered as one of the major impairments in bidirectional systems (Staubli, R. K., & Gysel, P., 1991) and Raman amplification (Jiang, S., Bristiel, B., Jaouen, Y., Gallion, P., & Pincemin, E., 2007). RB light causes interferometric noise fluctuations at the input of the silica optical fiber which results in a noise floor of ~32 dB below the original input power [25 km standard single mode fiber (SSMF) is used]. When optical amplification is present, the additional noise power due to ASE will further degrade the received signal. The RB noise is partially polarized in nature, with a colored power spectral density (PSD) proportional to the PSD of the generating input signal (Talli, G., Chow, C. W., & P. D. Townsend, 2008). This is different from the ASE, which is typically assumed to be a white noise.

The ratio of the total backscattered power P_{RB}, to the input power P_0, in a length of SSMF L, is given by

$$\frac{P_{RB}}{P_0} = \kappa(1 - e^{-2\alpha L}), \qquad (1)$$

where α is the loss coefficient per unit length. For the SSMF $\alpha = 0.2$ dB/km $= 4.6 \times 10^{-2}$ km^{-1} at a wavelength of 1550 nm. The Rayleigh factor κ is defined as

$$\kappa = \frac{\gamma S}{2\alpha}, \qquad (2)$$

where γ is the fractional power loss per unit length due to RB and S is the fraction of this power that is recaptured by the fiber. A measured value of $\kappa = 6.7 \times 10^{-4}$ of SSMF at wavelength of 1550 nm.

Figure 4 shows the two dominant contributions to the RB in the carrier distributed PONs, which interfere with the upstream signal at the Rx. The first contribution, Carrier-RB, is generated by the backscatter of the CW carrier being delivered to the reflective ONU (RONU). The second contribution, Signal-RB, is generated by the modulated upstream signal at the output of the RONU. Backscattered light from this upstream signal re-enters the RONU, where it is re-modulated and reflected towards the Rx. The RB noise is partially polarized in nature, with a colored PSD proportional to the PSD of the input signal. Hence, the spectra of Carrier-RB and the CW carrier are the same, while the Signal-RB is modulated twice by the RONU and has a broader spectrum. The relative impact of the two components depends on the exact network configuration and hence, for a full understanding, separate analysis of each effect is needed.

SOLUTIONS TO MITIGATE RAYLEIGH BACKSCATTERING

Dual-Feeder Fiber Architecture

Due to the unique features of the RB, by using proper network architecture design and using advanced modulation formats, RB induced signal degradation can be mitigated. A simple approach to mitigate RB in carrier distributed PONs could be simply performed by reducing the backscattered power that reaches the Rx at head-end office. If we consider a carrier distributed PON with two stages of splitters such as the one in Figure 5, where for clarity the AWGs are not shown, we can see clearly that most of the Carrier-RB power is generated by the optical carrier in the fiber before the first splitter, conventionally called feeder fiber. This is mainly due to the loss introduced by the first splitter and also because the feeder fiber usually accounts for most of the access length. The architecture discussed here to mitigate the backscattering impairments uses two separate feeder fibers before the splitters, one to deliver the optical carrier and the other to transmit the modulated upstream signal to the Rx (MacHale, E. K., Talli, G., & Townsend, P. D., 2006; Talli, G., Chow, C. W., MacHale, E. K., & Townsend, P. D., 2007). In this architecture, the Carrier-RB generated before the first splitter cannot reach the Rx and thus it cannot interfere with the upstream signal. It is worth to mention that although the Carrier-RB generated in the distribution and drop sections, and the Signal-RB still propagates

Figure 4. Schematic of WDM PON using centralized light source. AWG: arrayed waveguide grating, OC: optical circulator, Rx: receiver, Carrier-RB and Signal-RB: carrier and signal generated Rayleigh backscattering (Chow et al., 2007)

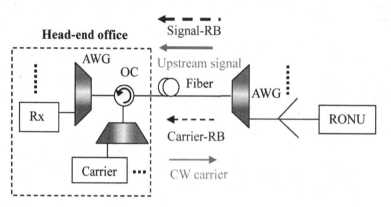

to the Rx, the overall power is greatly reduced compared with a system with a single feeder fiber. A theoretical and experimental description can be found in (Talli, G., et al., 2007). It is also important to note that this scheme maintains the benefit of a single drop and distribution fiber in the path to each ONU, which could reduce the cost of fiber and fiber connections in the network. The circulator, normally used to inject the carrier in single-feeder schemes, is also not required, since the optical splitter employed to combine the two feeder fibers performs this function.

Spectral Reshaped Modulation Formats

Advanced modulation formats have been proposed and demonstrated to mitigate RB noise in carrier distributed PONs. Phase-shift-keying (PSK)-Manchester coding is proposed to reduce interferometric beat noise (Li, Z., Dong, Y., Wang, Y., & Lu, C., 2005). In this scheme, the envelope of the PSK-Manchester signal is constant, while the wavelength of the carrier experiences up-chirp at rising edge of the modulation data and down-chirp at the falling edge. An optical notch filter is used to isolate the CW carrier component to reduce the carrier beat noise. Approaches using light source scrambling in terms of amplitude

Figure 5. Schematic of a carrier distributed PON using dual-feeder fiber architecture (Talli et al., 2007).

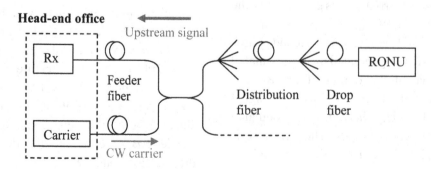

(Pepeljugoski, P. K., & Lau, K. Y., 1992), phase (Monroy, I. T., Tangdiongga, E., Jonker, R., & de Waardt, H., 2000), polarization (Heismann, F., Gray, D. A., Lee, B. H., & Smith, R. W., 1994) or frequency (Prat, J., Polo, V., Bock, C., Arellano, C., & Olmos, J. J. V., 2005) have been proposed for RB mitigation. Another technique which has been shown to be effective in this respect, uses a phase modulator (PM) at the RONU to spectrally broaden the upstream signal (Yoshida, T., Kimura, S., Kimura, H., Kumozaki, K., & Imai, T., 2006; Chow, C. W., Talli, G., & Townsend, P. D., 2007) in order to reduce the spectral overlap of the signal and the RB, and hence, reduce the beat noise falling within the Rx bandwidth. In this scheme, the phase modulation index is set so as to suppress the center wavelength, i.e., to make the amplitude of the zeroth order Bessel function equal to zero. The smallest modulation index that achieves the required suppression is ~2.4. Although the schemes proposed in (Yoshida, T., et al., 2006; Chow, C. W., et al., 2007) can efficiently suppress the noise caused by the Carrier-RB, they are much less effective in reducing the Signal-RB noise. Hence, Signal-RB mitigation using a gain-saturated SOA in the RONU has been demonstrated (MacHale, E. K., Talli, G., Chow, C. W., & Townsend, P. D., 2007). Dithering the bias current of a RSOA can also reduce coherent crosstalk, as reported in (Urban, P. J., Koonen, A. M. J., Khoe, G. D., & de Waardt, H., 2007). Optical carrier suppression and separation technique (OCSS) (Akanbi, O., Yu, J., & Chang, G. K., 2005) is also proposed to generate two wavelength channels using a single laser source at the head-end office. Then, one wavelength is externally modulated for downstream transmission while the other CW wavelength is delivered and modulated at the ONU to provide the upstream data transmission. Using this scheme, bidirectional operation over a single fiber can be achieved with reduced power penalties. RB mitigation can also be achieved by wavelength shifting (Prat, J., Omella, M., & Polo, V., 2007) or wavelength conversion (Shea, D. P., & Mitchell, J. E., 2006). These RB mitigation schemes usually complicate the ONU design.

Recently, a modulation scheme, namely carrier suppressed subcarrier amplitude modulated phase shift keying (CSS-AMPSK) (Chow, C. W., Talli, G, Ellis, A. D., & Townsend, P. D., 2008), is shown to be highly effective at mitigating noise generated by Carrier-RB and Signal-RB. For efficient Raleigh noise mitigation it is important to minimize the spectral overlap between the upstream signal and both types of RB, ensuring that the majority of the frequency components of the resultant electrical beat noise fall outside the Rx bandwidth. The CSS-AMPSK is generated by driving a Mach-Zehnder modulator (MZM) with an AMPSK drive signal, which is generated by modulating a sinusoidal clock signal at w GHz, with a duobinary signal using a microwave mixer. By biasing the MZM for minimum transmission, the CW carrier wavelength is suppressed, and two copies of the AMPSK drive signal are translated to sub-carriers at $\pm w$ GHz with respect to the optical carrier frequency. The fundamental frequency shift associated with subcarrier drive of a MZM biased for minimum transmission ensures a frequency translation each time a signal passes through the modulator, ensuring that the upstream signal is shifted away from the backscattered CW signal, and the Signal-RB components are further shifted away from the upstream signal. Hence, the CSS-AMPSK is expected to have strong tolerance to both RB components. Figure 6 shows the CSS-AMPSK modulator configuration consisting of a double balanced mixer with an electrical clock (CK) connected to the local oscillator (LO) input. The NRZ baseband data is effectively duobinary encoded by an encoder at the intermediate frequency (IF) input of the mixer. The mixer performs an RF up-conversion of the duobinary data. At the mixer RF output, an electrical duobinary signal on a carrier is generated. There is no particular synchronization requirement between the electrical sine wave and

Figure 6. RONU design of CSS-AMPSK modulation (Chow et al., 2008).

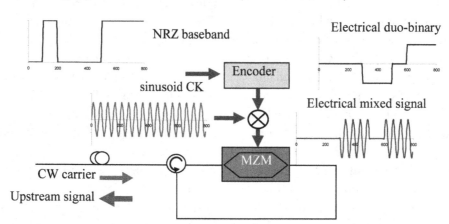

baseband data. The proof of concept experiments used commercially available LiNbO$_3$ MZM, but alterative schemes using polarization insensitive semiconductor modulators could be more practical (Leclerc, O., Brindel, P., Rouvillain, D., Pincemin, E., Dany, B., Desurvire, E., Duchet, C., Boucherez, E., Bouchoule, S., 1999).

Experimental setups have been proposed to analyze the RB performance quantitatively. Figures 7(a) and 7(b) show the experimental setups used to study the Carrier-RB and Signal-RB, respectively, which simulate the impairments of a real PON by generating two interfering signals in different arms of an interferometer. The modulator configurations (MODs) are operating in transmission mode without the optical circulator or reflective facet. By locating the modulators at different points in the setup it is possible to emulate the Carrier-RB and Signal-RB components that would be expected in a real carrier distributed network employing RONUs. For Carrier-RB analysis [Figure 7(a)], MOD$_1$ was used to generate the data signal and the Carrier-RB was generated by the unmodulated carrier propagating through 25 km of SSMF. On the other hand, for Signal-RB analysis [Figure 7(b)], the remodulated backscattering was generated by firstly modulating the optical signal with MOD$_2$. The remodulation of the backscattered signal generated by the 25 km

SSMF was then emulated by MOD$_3$. At the Rx, this interferes with the upstream signal, generated by a CW carrier delivered via the lower arm and simultaneously modulated by MOD$_3$. In both experiments a variable optical attenuator (VOA) was used to vary the signal power and generate different optical signal to Rayleigh noise ratios (OSRNRs).

A theoretical derivation of the RB properties and of the interferometric noise that it generates can be found in (Staubli, R. K., et al., 1991). The actual spectrum of the high speed modulated signal needs to be considered in order to obtain an accurate prediction of the system performance. The noise generated by the beating of signal and RB

$$\sigma_{s,b}^2 = \Re^2 \int_{-\infty}^{+\infty} \left| H_e(f) \right|^2 k \left[S_s(f) * S_b(f) + S_b(f) * S_s(f) \right] df,$$

(3)

where $S_s(f)$ and $S_b(f)$ are the PSD of upstream and backscattering signal respectively. They may be reshaped by the optical filter inside the Rx and the AWGs. * denotes deterministic cross-correlation, $\Re$ is the photodetector responsivity, $He(f)$ is the photodetector normalized frequency response, such that $He(0) = 1$. The noise generated by the RB beating with itself:

Figure 7. Experimental setup to emulate (a) Carrier-RB and (b) Signal-RB. MOD: modulator, VOA: variable optical attenuator, PC: polarization controller, AWG: array waveguide grating (Chow et al., 2008).

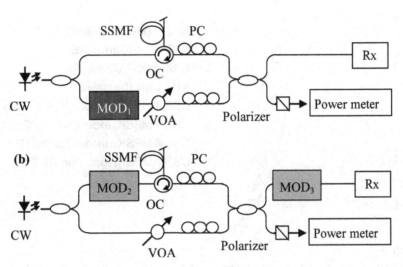

$$\sigma_{b,b}^2 = \Re^2 \int_{-\infty}^{+\infty} \left| H_e(f) \right|^2 \frac{1+p^2}{2} S_b(f) * S_b(f) df,$$

(4)

where k and p are the polarization coefficients, with $p=2k-1$. The polarization coefficient k equals to 1 for completely polarized backscattering ($p=1$) aligned with the signal polarization and $k=0.5$ for a completely depolarized backscattered field ($p=0$). In practice, roughly 1/3 of the RB light is polarized. RB analysis in terms of extinction ratios (Wong, E., Zhao, X., Chang-Hasnain, C. J., Hofmann, W., & Amann, M. C., 2007), signal bandwidths (Marki, C. F., Alic, N., Gross, M., Papen, G., Esener, S., & Radic, S., 2007) and optical filters positions in the network (Arellano, C., Polo, V., Prat, J., 2007) have also been discussed.

Figure 8 shows the experimental and modeled power penalties at a BER of 10^{-9} for the blue-side CSS-AMPSK (selected by using 30 GHz offset-AWG) and NRZ signals (for comparison) at 10 Gbit/s as a function of OSRNRs. For NRZ, even relatively low levels of RB can significantly degrade the BER due to the complete spectral overlap between upstream signal and RB components. For example, at the 1-dB power penalty window, the OSRNR of NRZ is about 25 dB. This means that the signal power should be 25 dB greater than the RB noise in order to have less than 1 dB power penalty. The CSS-AMPSK improves the required OSRNR (at 1 dB penalty) by 12 and 7 dB in the Carrier-RB and Signal-RB cases respectively. Results from the model are in good agreement with the experiments. The mitigation of the Carrier-RB is due to two effects. Firstly, the compact spectrum of CSS-AMPSK has less spectral overlap with the CW carrier, reducing the total beat noise power in the Rx. Secondly, the optical power of the Carrier-RB is concentrated at the CW carrier frequency, which is more strongly attenuated by the offset-AWG than the blue-side CSS-AMPSK signal, whose power is concentrated at +10 GHz. Furthermore, for Signal-RB, when the backscattered CSS-AMPSK is re-modulated in the RONU, the additional frequency shifts will transform odd harmonic components of the drive signal (±10 GHz) to even harmonics (0 and ±20 GHz). The spectral overlap of the upstream signal and the Signal-RB is small due to the

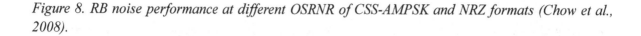

Figure 8. RB noise performance at different OSRNR of CSS-AMPSK and NRZ formats (Chow et al., 2008).

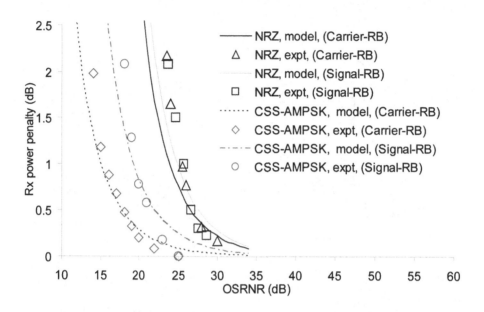

compact spectrum of the AMPSK modulation, but bigger than the Carrier-RB case. The optical filtering effect is also less effective since the +20 GHz component is within the AWG bandwidth. Hence experimental and modeling results show a slightly worse performance in the Signal-RB case. It is worth to mention that although the proposed scheme can mitigate the RB noise, it complicates the RONU design. Besides, offset-AWG is used and it attenuates the upstream signal.

SIGNAL REMODULATION CARRIER DISTRIBUTED PONS AND OTHER NOVEL PONS

As described in previous section, in the carrier distributed PON, optical carrier is distributed from the head-end office to different ONUs to produce the upstream signals. This is cost-effective since wavelength referencing and control are performed in the head-end office, and the same components can be used in each ONU. Remodulation of down-

stream signal to generate upstream signal may further reduce the cost by wavelength reuse. In this section, several signal remodulation carrier distributed PONs will be highlighted, and some novel PONs, long reach (LR)-PONs and wireless/ wired PONs will be discussed. Several remodulation schemes have been proposed, including using both downstream and upstream on-off keying (OOK) (Chan, L. Y., Chan, C. K., Tong, D. T. K., Tong, F., & Chen, L. K., 2002); downstream differential phase-shift-keying (DPSK) and upstream OOK (Hung, W., Chan, C. K., Chen, L. K., Tong, F., 2003); downstream inverse return-to-zero (IRZ) and upstream OOK (Lu, G. W., Deng, N., Chan, C. K., & Chen, L. K., 2005); downstream low extinction-ratio OOK and upstream DPSK (Zhao, J., Chen, L. K., Chan, C. K., 2007). Relatively high residual CW background between IRZ pulses and the finite extinction-ratio in the OOK is required in the IRZ-OOK and low extinction-ratio OOK-DPSK remodulation schemes, respectively, in order to provide high enough optical power for the integrity of the upstream signals. These may reduce the Rx sensitivity in the downstream signal. Because of this, downstream DPSK and upstream orthogonal DPSK/intensity modulation (IM) scheme (Tian, Y., Su, Y., Yi, L., Leng, L., Tian, X., He, H., & Xu, X., 2006), and downstream and upstream DPSK (Chow, C. W., 2008) have been proposed with reduced power penalty introduced to the downstream signal.

As bandwidths grow, the traditional approach of separating access and metro networks will become prohibitively expensive: (i) from a capex viewpoint, due to the large number of network elements and interfaces to interconnect them and (ii) from an opex viewpoint, due to network design complexity, large number of network elements, large footprint and high electrical power consumption. Integrating the metro and access networks is particularly suitable for highly populated areas, where the cost of land for head-end office is high. Hence, LR-PONs have been proposed (Shea, D. P., Ellis, A. D., Payne, D. B., Davey, R. P., & Mitchell, J. E., 2003). It usually has four new features when compared with traditional PON: high data rate in both upstream and downstream signals (> 1 Gbit/s); reach extension to > 100 km; a high split ratio (> 100); and using WDM (Shea, D. P., et al. 2003; Nesset, D., Davey, R. P., Shea, D., Kirkpatrick, P., Shang, S. Q., Lobel, M., & Christensen, B., 2005; Shea, D. P., & Mitchell, J. E., 2007; Talli, G., Chow, C. W., MacHale, E. K., & Townsend, P. D., 2006; Ruffin, A. B., Downie, J. D., & Hurley, J., 2008; Jia, Z., Yu, J., Yeo, Y. K., Wang, T., & Chang, G. K., 2006). 10 Gbit/s bidirectional LR-PON in 1024-way split, 110 km reach, using Super forward error correction (FEC) and electronic dispersion compensation (EDC) has been demonstrated (Nesset, D., et al., 2005). 1024-way split, 100 km, 10 Gbit/s LR-PON using Super FEC and EDC in upstream and high saturated power optical amplifier in the upstream is reported (Shea, D. P., et al., 2007). A 256-way split, 116 km LR hybrid WDM-TDM PON employing RONUs by combining a dual-feeder fiber scheme with spectrally-broadened upstream signal

has also been demonstrated recently, without using FEC or EDC (Talli, G., et al., 2006). LR-PON with 128-way split, 100 km reach without in-field amplification using 10 Gbit/s duobinary signals and ultra-low loss optical fiber is reported (Ruffin, A. B., et al., 2008). Recently, a LR-PON using orthogonal frequency division multiplexing (OFDM) was demonstrated, achieving a split-ratio of 256 (Chow, C. W., Yeh, C. H., et al., 2008). Figure 9 shows its performance at back-to-back and LR transmission. Inset: Constellation diagrams of (a) downstream and (b) upstream signals after 100 km SSMF transmission without dispersion compensation. In these LR-PONs, in order to extend the reach, optical amplifiers are usually needed to compensate the signal losses during the transmission. This complicates the network architecture especially at high data rate, since high speed burst mode Rx (10 Gbit/s) and burst-proof optical amplifiers are not commercially available (Townsend, P. D., et al., 2007). Besides, due to the large split ratio, ranging and synchronization among ONUs may be an issue.

Broadband wireless access is becoming more and more popular nowadays due to its flexibility and scalability. This can also be deployed in some areas where using optical fiber is expensive. In order to include the mobile feature provided by the wireless access, the integration of PON and wireless access is a potential solution for increasing the capacity and mobility, as well as decreasing the cost in the access networks. Recently, architectures of simultaneous modulation and transmission of a RF signal and a baseband signal have been demonstrated (Jia, Z., et al., 2006; Chang, G. K., Yu, J., Jia, Z., & Yu, J., 2006; Lin, C. T., Peng, W. R., Peng, P. C., Chen, J., Peng, C. F., Chiou, B. S., & Chi, S., 2006; Lin, C. T., Chen, J., Peng, P. C., Peng, C. F., Peng, W. R., Chiou, B. S., & Chi, S., 2007; Jia, Z., Yu, J., Boivin, D., Haris, M., & Chang, G. K., 2007; Yu, J., Jia, Z., Wang, T., & Chang, G. K., 2007; Chen, L., Shao, Y., Lei, X., Wen, H., & Wen, S., 2007; Chen, L., Wen, H., & Wen, S., 2006; Ma, J., Yu, J., Yu, C., Xin, X.,

Figure 9. BER of downstream and upstream OFDM-QAM signals at back-to-back and LR transmission. Inset: Constellation diagrams of (a) downstream and (b) upstream signals after 100 km SSMF transmission without dispersion compensation (Chow, C. W., Yeh, C. H., et al., 2008).

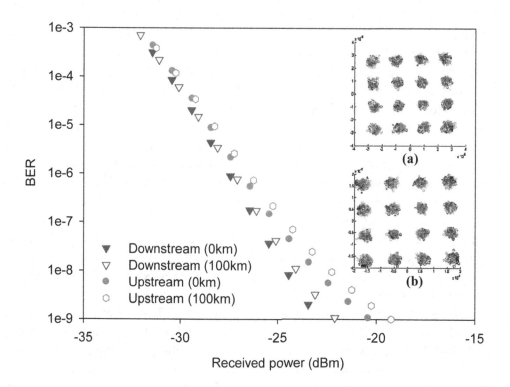

Zeng, J., & Chen, L., 2007). The transmission of RF signal over optical fiber has been actively investigated in realizing future broadband wireless networks. In these systems, the metro or access links between the central and base station are implemented by using optical fiber, and they are generally called radio-over-fiber (ROF) systems. Several optical RF signal generation approaches, such as using direct modulation, external optical modulation, optical double-frequency heterodyning, four-wave mixing, wavelength conversion and continuum light filtering have been summarized in (Ma, J., et al., 2007). It is worth to mention that conventional optical modulation formats, such as NRZ or DPSK may not be suitable for the ROF system if direct conversion of optical to RF signal at antenna sites is used. Because of this, the feasibility of deploying some of the modulation formats which are now commonly used in wireless communications, e.g. orthogonal frequency division multiplexed (OFDM) - quadrature amplitude modulation (QAM) and quadrature phase shift keying (QPSK), in optical fiber network have attracted much attention (Shieh, W., Yi, X., Ma, Y., & Tang, Y., 2007; Lowery, A., & Armstrong, J.; Chow, C. W., Yeh, C. H., Wang, C. H., Shih, F. Y., Pan, C. L., & Chi, S., 2008). However, waveform of the OFDM signal presents a high peak-to-average-ratio, hence, the Tx, amplifier and Rx used in the networks should have much higher linear response characteristics than conventional optical networks in order to avoid signal clipping. Other possible solutions are also under intensive studies nowadays.

CONCLUSION

PON is regarded as the most attractive and highly cost-effective optical access network architecture. The first generations of PONs are now standardized and commercially available. These TDM-based PONs have already been deployed to many millions of households in Japan, North America, and Korea to provide higher bandwidth services. However the bandwidth sharing nature of the TDM PONs means that they cannot support future high bandwidth demand. A strong upgrade of existing access network is required in order to cope with the exponentially increasing of bandwidth demand. Alternative architecture based on WDM PON has recently attracted attention because the sustained data rates of WDM PON do not drop with the number of simultaneous users. The main issue facing WDM PON is the cost of the wavelength specific optical components needed in the network. Thus colorless ONUs with centralized optical carrier distribution can be a cost-effective solution. In this chapter, different solutions of colorless ONUs have been discussed. Different RB components in the carrier distributed PON have been analyzed. RB mitigation schemes, including using dual-feeder fiber architecture and spectral reshaped modulation formats have been proposed and demonstrated. Finally, some novel PONs including signal remodulation PONs, long reach PONs and wireless/wired PONs have been pointed out.

ACKNOWLEDGMENT

The author would like to thank Prof. Paul Townsend, Dr. Andrew Ellis, Dr. Giuseppe Talli, Prof. Hon Tsang, Dr. Yang Liu, Prof. Chinlon Lin, Dr. Chien-Hung Yeh and Prof. Sien Chi for supports and discussions.

REFERENCES

Akanbi, O., Yu, J., & Chang, G. K. (2005, September). *A new bidirectional WDM-PON using DWDM channels generated by optical carrier suppression and separation technique.* Paper presented at European Conference and Exhibition on Optical Communication (ECOC), (We3.3.1), Glasgow, UK.

Arellano, C., Bock, C., Prat, J., & Langer, K. D. (2006, March). *RSOA-based optical network units for WDM-PON.* Paper presented at Optical Fiber Communication Conference (OFC), (OTuC1), Anaheim, California, USA.

Arellano, C., Polo, V., & Prat, J. (2007, July). *Effect of the multiplexer position in Rayleigh-limited WDM-PONs with amplified-reflective ONU.* Paper presented at International Conference on Transparent Optical Networks (ICTON), (Paper Mo.P.22), Rome, Italy.

Borghesani, A., Lealman, I. F., Poustie, A., Smith, D. W., & Wyatt, R. (2007, September). *High temperature, colourless operation of a reflective semiconductor optical amplifier for 2.5Gbit/s upstream transmission in a WDM-PON.* Paper presented at European Conference and Exhibition on Optical Communication (ECOC), (Paper 6.4.1), Berlin, Germany.

Chan, L. Y., Chan, C. K., Tong, D. T. K., Tong, F., & Chen, L. K. (2002). Upstream traffic transmitter using injection-locked Fabry-Perot laser diode as modulaotr for WDM access networks. *Electronics Letters, 38*(1), 43–45. doi:10.1049/el:20020015

Chang, G. K., Yu, J., Jia, Z., & Yu, J. (2006, March). *Novel optical-wireless access network architecture for simultaneously providing broadband wireless and wired services.* Paper presented at Optical Fiber Communication Conference (OFC), (OFM1), Anaheim, California. USA.

Chapuran, T. E., Wagner, S. S., Menendez, R. C., Tohme, H. E., & Wang, L. A. (1991). Broadband multichannel WDM transmission with super-luminescent diodes and LEDs. *Proc. of Global Telecommunications Conference (GLOBECOM)* (vol. 1, pp. 612-618).

Chen, L., Shao, Y., Lei, X., Wen, H., & Wen, S. (2007). A novel radio-over-fiber system with wavelength reuse for upstream data connection. *IEEE Photonics Technology Letters, 19*(6), 387–389. doi:10.1109/LPT.2007.891958

Chen, L., Wen, H., & Wen, S. (2006). A radio-over-fiber system with a novel scheme for millimeter-wave generation and wavelength reuse for up-link connection. *IEEE Photonics Technology Letters, 18*(19), 2056–2058. doi:10.1109/LPT.2006.883293

Choi, K. M., & Lee, C. H. (2005, September). *Colorless operation of WDM-PON based on wavelength locked Fabry-Perot laser diode.* Paper presented at European Conference and Exhibition on Optical Communication (ECOC), (We3.3.4), Glasgow, UK.

Chow, C. W. (2008). Wavelength remodulation using DPSK down-and-upstream with high extinction ratio for 10-Gb/s DWDM-passive optical networks. *IEEE Photonics Technology Letters, 20*(1), 12–14. doi:10.1109/LPT.2007.911009

Chow, C. W., Talli, G., Ellis, A. D., & Townsend, P. D. (2008). Rayleigh noise mitigation in DWDM LR-PONs using carrier suppressed subcarrier-amplitude modulated phase shift keying. *Optics Express, 16*(3), 1860–1865. doi:10.1364/OE.16.001860

Chow, C. W., Talli, G., & Townsend, P. D. (2007). Rayleigh noise reduction in 10-Gb/s DWDM-PONs by wavelength detuning and phase-modulation-induced spectral broadening. *IEEE Photonics Technology Letters, 19*(6), 423–425. doi:10.1109/LPT.2007.892899

Chow, C. W., Yeh, C. H., Wang, C. H., Shih, F. Y., Pan, C. L., & Chi, S. (2008). WDM extended reach passive optical networks using OFDM-QAM. *Optics Express, 16*(16), 12096–12101. doi:10.1364/OE.16.012096

Feuer, M. D., Wiesenfeld, J. M., Perino, J. S., Bums, C. A., Raybon, G., Shunk, S. C., & Dutta, N. K. (1996). Single-port laser-amplifier modulators for local access. *IEEE Photonics Technology Letters, 8*(9), 1175–1177. doi:10.1109/68.531827

Fujiwara, M., Kani, J., Suzuki, H., & Iwatsuki, K. (2006). Impact of backreflection on upstream transmission in WDM single-fiber loopback access networks. *Journal of Lightwave Technology, 24*(2), 740–746. doi:10.1109/JLT.2005.862429

Fujiwara, M., Suzuki, H., & Yoshimoto, N. (2006). Quantitative loss budget estimation in WDM single-fibre loopback access networks with ASE light sources. *Electronics Letters, 42*(22), 1301–1302. doi:10.1049/el:20062735

Gysel, P., & Staubli, R. K. (1990). Statistical properties of Rayleigh backscattering in single-mode fibers. *Journal of Lightwave Technology, 8*(4), 561–567. doi:10.1109/50.50762

Hann, S., Kim, T. Y., & Park, C. S. (2005, September). *Direct-modulated upstream signal transmission using a self-injection locked F-P LD for WDM-PON.* Paper presented at European Conference and Exhibition on Optical Communication (ECOC), (We3.3.3), Glasgow, UK.

Healey, P., Townsend, P., Ford, C., Johnston, L., Townley, P., & Lealman, I. (2001). Spectral slicing WDM-PON using wavelength-seeded reflective SOAs. *Electronics Letters, 37*(19), 1181–1182. doi:10.1049/el:20010786

Heismann, F., Gray, D. A., Lee, B. H., & Smith, R. W. (1994). Electrooptic polarization scramblers for optically amplified long-haul transmission systems. *IEEE Photonics Technology Letters, 6*(9), 1156–1158. doi:10.1109/68.324697

Hung, W., Chan, C. K., Chen, L. K., & Tong, F. (2003). An optical network unit for WDM access networks with downstream DPSK and upstream remodulated OOK data using injection-locked FP laser. *IEEE Photonics Technology Letters, 15*(10), 1476–1478. doi:10.1109/LPT.2003.818055

Ji, H. C., Yamashita, I., & Kitayama, K. I. (2008). Cost-effective colorless WDM-PON delivering up/down-stream data and broadcast services on a single wavelength using mutually injected Fabry-Perot laser diodes. *Optics Express, 16*(7), 4520–4528. doi:10.1364/OE.16.004520

Jia, Z., Yu, J., Boivin, D., Haris, M., & Chang, G. K. (2007). Bidirectional ROF links using optically up-converted DPSK for downstream and remodulated OOK for upstream. *IEEE Photonics Technology Letters, 19*(9), 653–655. doi:10.1109/LPT.2007.894961

Jia, Z., Yu, J., Yeo, Y. K., Wang, T., & Chang, G. K. (2006, September). *Design and implementation of a low cost, integrated platform for delivering super-broadband dual services simultaneously.* Paper presented at European Conference and Exhibition on Optical Communication (ECOC), (Tu1.6.6), Cannes, France.

Jiang, S., Bristiel, B., Jaouen, Y., Gallion, P., & Pincemin, E. (2007). Bit-error-rate evaluation of the distributed Raman amplified transmission systems in the presence of double Rayleigh backscattering noise. *IEEE Photonics Technology Letters, 19*(7), 468–470. doi:10.1109/LPT.2007.893027

Kim, S. J., Han, J. H., & Lee, J. S., Park, & C. S. (1999). Intensity noise suppression in spectrum-sliced incoherent light communication systems using a gain-saturated semiconductor optical amplifier. *IEEE Photonics Technology Letters, 11*(8), 967–1044. doi:10.1109/68.775315

Leclerc, O., Brindel, P., Rouvillain, D., Pincemin, E., Dany, B., & Desurvire, E. (1999). 40Gbit/s polarization-insensitive and wavelength-independent InP Mach-Zehnder modulator for all-optical regeneration. *Electronics Letters, 35*(9), 730–732. doi:10.1049/el:19990504

Lee, J. H. Lee, K., & Kim., C. H. (2007, September). *Continuous-wave supercontinuum-based bidirectional long reach WDM-POM incorporating FP-LD-based OLT and RSOA-based ONUs.* Paper presented at European Conference and Exhibition on Optical Communication (ECOC), (Paper 6.4.5), Berlin, Germany.

Lee, J. H., Kim, C. H., Han, Y. G., & Lee, S. B. (2006, September). *1.25 Gbit/s WDM PON upstream transmission using Fabry-Perot laser diodes injected by depolarised CW supercontinuum source.* Paper presented at European Conference and Exhibition on Optical Communication (ECOC), (Tu3.5.5), Cannes, France.

Lee, K. L., & Wong, E. (2006, September). *Directly-modulated self-seeding reflective SOAs in WDM-PONs: performance dependence on seeding power and modulation effects.* Paper presented at European Conference and Exhibition on Optical Communication (ECOC), (Tu4.5.2), Cannes, France.

Li, Z., Dong, Y., Wang, Y., & Lu, C. (2005). A novel PSK-Manchester modulation format in 10-Gb/s passive optical network system with high tolerance to beat interference noise. *IEEE Photonics Technology Letters, 17*(5), 1118–1120. doi:10.1109/LPT.2005.845663

Lin, C. T., Chen, J., Peng, P. C., Peng, C. F., Peng, W. R., Chiou, B. S., & Chi, S. (2007). Hybrid optical access network integrating fiber-to-the-home and radio-over-fiber systems. *IEEE Photonics Technology Letters, 19*(8), 610–612. doi:10.1109/LPT.2007.894326

Lin, C. T., Peng, W. R., Peng, P. C., Chen, J., Peng, C. F., Chiou, B. S., & Chi, S. (2006). Simultaneous generation of baseband and radio signals using only one single electrode Mach-Zehnder modulator with enhanced linearity. *IEEE Photonics Technology Letters, 18*(23), 2481–2483. doi:10.1109/LPT.2006.887233

Lowery, A., & Armstrong, J. Orthogonal-frequency-division multiplexing for dispersion compensation of long-haul optical systems. *Optics Express, 14*(6), 2079–2084.

Lu, G. W., Deng, N., Chan, C. K., & Chen, L. K. (2005, March). *Use of downstream IRZ signal for upstream data re-modulation in a WDM passive network.* Paper presented at Optical Fiber Communication Conference (OFC), (OFI8), Anaheim, California. USA.

Ma, J., Yu, J., Yu, C., Xin, X., Zeng, J., & Chen, L. (2007). Fiber Dispersion influence on transmission of the optical millimeter-waves generated using LN-MZM intensity modulation. *Journal of Lightwave Technology, 25*(11), 3244–3256. doi:10.1109/JLT.2007.907794

MacHale, E. K., Talli, G., Chow, C. W., & Townsend, P. D. (2007, September). *Reduction of signal-induced Rayleigh noise in a 10Gb/s WDM-PON using a gain-saturated SOA.* Paper presented at European Conference and Exhibition on Optical Communication (ECOC), (Paper 7.6.3), Berlin, Germany.

MacHale, E. K., Talli, G., & Townsend, P. D. (2006, March). *10Gb/s bidirectional transmission in a 116km reach hybrid DWDM-TDM PON.* Paper presented at Optical Fiber Communication Conference (OFC), (OFE1), Anaheim, California, USA.

MacHale, E. K., Talli, G., Townsend, P. D., Borghesani, A., Lealman, I., Moodie, D. G., & Smith, D. W. (2008, September). *Extended-reach PON employing 10Gb/s integrated reflective EAM-SOA.* Paper presented at European Conference and Exhibition on Optical Communication (ECOC), (Th.2.F.1), Brussels, Belgium.

Marki, C. F., Alic, N., Gross, M., Papen, G., Esener, S., & Radic, S. (2007, March) *Performance of NRZ and Duobinary modulation formats in Rayleigh and ASE-dominated dense optical links.* Paper presented at Optical Fiber Communication Conference (OFC), (OFD6), Anaheim, California. USA.

McCoy, A. D., Horak, P., Thomsen, B. C., Ibsen, M., Mokhtar, M. R., & Richardson, D. J. (2004, September). *Optimising signal quality in a spectrum-sliced WDM system using SOA-based noise reduction.* Paper presented at European Conference and Exhibition on Optical Communication (ECOC), (Tu4.6.4), Stockholm, Sweden.

Monroy, I. T., Tangdiongga, E., Jonker, R., & de Waardt, H. (2000). Interferometric crosstalk reduction by phase scrambling. *Journal of Lightwave Technology, 18*(5), 637–646. doi:10.1109/50.842077

Mun, S. G., Moon, J. H., Lee, H. K., Kim, J. Y., & Lee, C. H. (2008). A WDM-PON with a 40 Gb/s (32 × 1.25 Gb/s) capacity based on wavelength-locked Fabry-Perot laser diodes. *Optics Express, 16*(15), 11361–11368. doi:10.1364/OE.16.011361

Nesset, D., Davey, R. P., Shea, D., Kirkpatrick, P., Shang, S. Q., Lobel, M., & Christensen, B. (2005, September). *10 Gbit/s bidirectional transmission in 1024-way split, 110 km reach, PON system using commercial transceiver modules, Super FEC and EDC.* Paper presented at European Conference and Exhibition on Optical Communication (ECOC), (Tu1.3.1), Glasgow, UK.

Park, K. Y., Mun, S. G., Choi, K. M., & Lee, C. H. (2005). A theoretical model of a wavelength-locked Fabry–Perot laser diode to the externally injected narrow-band ASE. *IEEE Photonics Technology Letters, 17*(9), 1797–1799. doi:10.1109/LPT.2005.851886

Park, S. B., Jung, D. K., Shin, D. J., Shin, H. S., Yun, I. K., & Lee, J. S. (2007). Colorless operation of WDM-PON employing uncooled spectrum-sliced reflective semiconductor optical amplifiers. *IEEE Photonics Technology Letters, 19*(4), 248–250. doi:10.1109/LPT.2007.891197

Payne, D. B., & Davey, R. P. (2002). The future of fibre access systems? *BT Technology Journal, 20*(4), 104–114. doi:10.1023/A:1021323331781

Payoux, F. Chanclou, & P., Genay, N. (2007, March). *WDM-PON with colorless ONUs.* Paper presented at Optical Fiber Communication Conference (OFC), (OTuG5), Anaheim, California, USA.

Payoux, F., Chanclou, P., & Brenot, R. (2006, September). *WDM PON with a single SLED seeding colorless RSOA-based OLT and ONUs.* Paper presented at European Conference and Exhibition on Optical Communication (ECOC), (Tu4.5.1), Cannes, France.

Payoux, F., Chanclou, P., Moignard, M., & Brenot, R. (2005, September). *Gigabit optical access using WDM PON based on spectrum slicing and reflective SOA.* Paper presented at European Conference and Exhibition on Optical Communication (ECOC), (We3.3.5), Glasgow, UK.

Pepeljugoski, P. K., & Lau, K. Y. (1992). Interferometic noise reduction in fiberoptic links by superposition of high frequency modulation. *Journal of Lightwave Technology, 10*(7), 957–963. doi:10.1109/50.144919

Prat, J., Omella, M., & Polo, V. (2007, March). *Wavelength shifting for colorless ONUs in single-fiber WDM-PONs.* Paper presented at Optical Fiber Communication Conference (OFC), (OTuG6), Anaheim, California, USA.

Prat, J., Polo, V., Bock, C., Arellano, C., & Olmos, J. J. V. (2005). Full-duplex single fiber transmission using FSK downstream and IM remote upstream Modulations for fiber-to-the-home. *IEEE Photonics Technology Letters, 17*(3), 702–704. doi:10.1109/LPT.2004.840930

Ruffin, A. B., Downie, J. D., & Hurley, J. (2008). *Purely passive long reach 10 GE-PON architecture based on duobinary signals and ultra-low loss optical fiber.* Paper presented at Optical Fiber Communication Conference (OFC), (OThL4), San Diego, California, USA.

Shea, D. P., Ellis, A. D., Payne, D. B., Davey, R. P., & Mitchell, J. E. (2003, September). *10 Gbit/s PON with 100 km reach and x1024 split.* Paper presented at European Conference and Exhibition on Optical Communication (ECOC), (We.P.147), Rimini, Italy.

Shea, D. P., & Mitchell, J. E. (2006, March). *Experimental upstream demonstration of a long reach wavelength-converting PON with DWDM backhaul.* Paper presented at Optical Fiber Communication Conference (OFC), (OWL4), Anaheim, California.

Shea, D. P., & Mitchell, J. E. (2007). A 10-Gb/s 1024-way-split 100-km long-reach optical-access network. *Journal of Lightwave Technology, 25*(3), 685–693. doi:10.1109/JLT.2006.889667

Shieh, W., Yi, X., Ma, Y., & Tang, Y. (2007). Theoretical and experimental study on PMD-supported transmission using polarization diversity in coherent optical OFDM systems. *Optics Express, 15*(16), 9936–9947. doi:10.1364/OE.15.009936

Shin, D. J., Jung, D. K., Shin, H. S., Kwon, J. W., Hwang, S., Oh, Y., & Shim, C. (2005). Hybrid WDM/TDM-PON with wavelength-selection-free transmitters. *Journal of Lightwave Technology*, *23*(1), 187–195. doi:10.1109/JLT.2004.840031

Shin, D. J., Keh, Y. C., Kwon, J. W., Lee, E. H., Lee, J. K., & Park, M. K. (2006). Low-cost WDM-PON with colorless bidirectional transceivers. *Journal of Lightwave Technology, 24*(1), 158–165. doi:10.1109/JLT.2005.861122

Shin, H. S., Jung, D. K., Kim, H. S., Shin, D. J., Park, S. B., Hwang, S. T., et al. (2005, September). *Spectrally pre-composed ASE injection for a wavelength-seeded reflective SOA in a WDM-PON*. Paper presented at European Conference and Exhibition on Optical Communication (ECOC), (We3.3.7), Glasgow, UK.

Shin, H. S., Jung, D. K., Shin, D. J., Park, S. B., Lee, J. S., Yun, I. K., et al. (2006, March). *16 x 1.25 Gbit/s WDM-PON based on ASE-injected R-SOAs in 60°C temperature range*. Paper presented at Optical Fiber Communication Conference (OFC), (OTuC5), Anaheim, California, USA.

Staubli, R. K., & Gysel, P. (1991). Crosstalk penalties due to coherent Rayleigh noise in bidirectional optical communication systems. *Journal of Lightwave Technology, 9*(3), 375–380. doi:10.1109/50.70015

Talli, G., Chow, C. W., MacHale, E. K., & Townsend, P. D. (2006, September). *High split ratio 116km reach hybrid DWDM-TDM 10Gb/s PON employing R-ONUs*. Paper presented at European Conference and Exhibition on Optical Communication (ECOC), (Mo4.5.2), Cannes, France.

Talli, G., Chow, C. W., MacHale, E. K., & Townsend, P. D. (2007). Rayleigh noise mitigation in long-reach hybrid DWDM-TDM PONs. *Journal of Optical Networking, 6*(6), 765–776. doi:10.1364/JON.6.000765

Talli, G., Chow, C. W., & Townsend, P. D. (2008). (accepted for publication). Modeling of modulation formats for interferometric noise mitigation. *Journal of Lightwave Technology*.

Talli, G., Cotter, D., & Townsend, P. D. (2006). Rayleigh backscattering impairments in access networks with centralised light source. *Electronics Letters, 42*(15), 877–878. doi:10.1049/el:20061546

Talli, G., & Townsend, P. D. (2006). Hybrid DWDM-TDM long reach PON for next generation optical access. *Journal of Lightwave Technology, 24*(7), 2827–2834. doi:10.1109/JLT.2006.875952

Tian, Y., Su, Y., Yi, L., Leng, L., Tian, X., He, H., & Xu, X. (2006, September). *Optical VPN in PON based on DPSK erasing/rewriting and DPSK/IM formatting using a single Mach-Zehnder modulator*. Paper presented at European Conference and Exhibition on Optical Communication (ECOC), (Tu4.5.6), Cannes, France.

Townsend, P. D., Talli, G., Chow, C. W., MacHale, E. M., Antony, C., Davey, R., et al. (2007, October). *Long reach passive optical networks*. Paper presented at IEEE LEOS Annual Meeting, Lake Buena Vista, Florida, USA.

Urban, P. J., Koonen, A. M. J., Khoe, G. D., & de Waardt, H. (2007, September). *Coherent crosstalk-suppression in WDM Access networks employing reflective semiconductor optical amplifiers*. Paper presented at European Conference and Exhibition on Optical Communication (ECOC), (Paper 6.4.2), Berlin, Germany.

Wong, E., Zhao, X., Chang-Hasnain, C. J., Hofmann, W., & Amann, M. C. (2006, March). *Uncooled, optical injection-locked 1.55 mm VCSELs for upstream transmitters in WDM-PONs*. Paper presented at Optical Fiber Communication Conference (OFC), Anaheim, California, USA.

Wong, E., Zhao, X., Chang-Hasnain, C. J., Hofmann, W., & Amann, M. C. (2007). Rayleigh backscattering and extinction ratio study of optically injection-locked 1.55 lm VCSELs . *Electronics Letters, 43*(3), 182–183. doi:10.1049/el:20073446

Yoshida, T., Kimura, S., Kimura, H., Kumozaki, K., & Imai, T. (2006). A new single-fiber 10-Gb/s optical loopback method using phase modulation for WDM optical access networks. *Journal of Lightwave Technology, 24*(2), 786–796. doi:10.1109/JLT.2005.862441

Yu, J., Jia, Z., Wang, T., & Chang, G. K. (2007). A novel radio-over-fiber configuration using optical phase modulator to generate an optical mm-wave and centralized lightwave for uplink connection. *IEEE Photonics Technology Letters, 19*(3), 140–142. doi:10.1109/LPT.2006.890087

Zhao, J., Chen, L. K., & Chan, C. K. (2007, March). *A novel re-modulation scheme to achieve colorless high-speed WDM-PON with enhanced tolerance to chromatic dispersion and re-modulation misalignment,"* Paper presented at Optical Fiber Communication Conference (OFC), (OWD2), Anaheim, California. USA.

Section 3
Technical Challenges and Determinants for Further Growth

Chapter 6
Bandwidth Allocation Methods in Passive Optical Access Networks (PONs)

Noemí Merayo
University of Valladolid, Spain

Patricia Fernández
University of Valladolid, Spain

Ramón J. Durán
University of Valladolid, Spain

Rubén M. Lorenzo
University of Valladolid, Spain

Ignacio de Miguel
University of Valladolid, Spain

Evaristo J. Abril
University of Valladolid, Spain

ABSTRACT

Passive Optical Networks (PONs) are very suitable architectures to face today's access challenges. This technology shows a very cost saving architecture, it provides a huge amount of bandwidth and efficiently supports Quality of Service (QoS). In PON networks, as all subscribers share the same uplink channel, a medium access control protocol is required to provide a contention method to access the channel. As the performance of Time Division Multiplexing Access (TDMA) protocol is not good enough because traffic nature is heterogeneous, Dynamic Bandwidth Allocation (DBA) algorithms are proposed to overcome the problem. These algorithms are very efficient as they adapt the bandwidth assignment depending on the updated requirements and traffic conditions. Moreover, they should offer QoS by means of both class of service and subscriber differentiation. Long-Reach PONs, which combine the access and the metro network into only one by using 100 km of fibre, is an emergent technology able to reach a large number of far subscribers and to decrease the associated costs.

DOI: 10.4018/978-1-60566-707-2.ch006

EVOLUTION OF THE ACCESS NETWORK: THE FIRST MILE

The access network, also called the first mile or last mile, connects the service provider central offices to residential or business customers. The demanded services are quite different depending on the type of customer. Residential users demand applications related to leisure activities, such as broadband Internet, television or interactive games, whereas companies demand multimedia services for the bidirectional transmission of all kind of information.

In the recent years, the network traffic has been increasing at very high rates, which has caused an important evolution in the transport network. However, the access network has not suffered any important evolution or change. Besides, the new emerging services and the growth of Internet traffic have accentuated the lack of access network capacity. The deployed technology, DSL (Digital Subscriber Line) and coaxial cable, are not able to cover the bandwidth necessary to support these new high demanding services. Up to now, the access network has been associated to the type of delivered information, the pair copper for telephony and the coaxial cable for television. However, from now on the new access network should be unique and should deliver voice data and video under the same platform, which is called "Triple play service".

Moreover, as Internet traffic has increased highly its presence, operators have deployed different technologies to support such demand. In this way, telephone operators tended to deploy the Digital Subscriber Line (DSL) technology, which uses the same twisted pair as telephony lines and therefore it requires a DSL modem at the customer premises and a Digital Subscriber Line Access Multiplexer (DSLAM) in the central office. The data rate offered by DSL technologies is not enough to support integrated voice, data and video services. Furthermore, the physical

area that can be covered by DSL is limited to the distance, which means that it cannot reach every contracted subscriber.

On the other hand, the cable television enterprises integrate data services over their coaxial cable networks. Usually, these architectures combine both fibre and coaxial cable, resulting in a Hybrid Fibre Coaxial (HFC) network, where the fibre reaches the head-end to a curbside optical node, and the coaxial cable covers the rest of the path to the final subscriber. The main problem of this architecture is that each shared node has a limited effective data throughput, which is divided among many homes, each subscriber obtaining a very slow speed during peak hours.

The new emerging services and the insufficient deployed access technologies, help to increase the existing "bottle neck" in the access. Thus, another access technology that could be simple, scalable and capable to transport voice, data and video over the same network is strongly needed. In this way, optical fibre is expected to be the best option to deal with the existing first mile challenges.

The deployment of fibre in the access network is known as **FTTX**. However the FTTX term can group different categories depending on the portion of fibre included in the access network. Therefore, the most typical FTTX infrastructures are:

- **FTTH**: Fibre To The Home.
- **FTTB**: Fibre To The Business.
- **FTTmdu**: Fibre To The multi-tenant building.
- **Deep Fibre**: The fibre ends in one point near the subscriber, and another technology such as copper pair is used to reach the end subscriber. The most important infrastructures are Fibre To the Node (FTTN) and Fibre To the Curb (FTTC).

As fibre is viewed as the most suitable transmission medium in the access network, many topology alternatives were studied, such as the point-to-point, the Curb Switched and the point-to-multipoint architectures.

In the point-to-point topology it is extended one fibre per user and therefore it needs N or $2N$ fibres and $2N$ optical transceivers, being N the number or users. On the other hand, the Curb-Switched technology uses only one trunk fibre which eliminates fibre in the Central Office (CO), but it needs $2N+2$ optical transceivers and electrical power in the field. Finally, the point-to-multipoint is the most cost saving architecture as it uses one trunk fibre from the CO to a splitter and it only needs $N+1$ optical transceivers. Besides, it requires no electrical power in the field as it uses an optical splitter to divide the signal into so many branches as the number of users connected to the network. This network infrastructure is called Passive Optical Access Network (**PON**), which is a very attractive solution to treat the problem in the first mile, as it can provide both high bandwidth and class of service differentiation. This access technology is mainly based on a bidirectional communication between the Optical Line Termination (OLT) located inside the Central Office and several Optical Network Units (ONUs) located inside or near the end subscribers (Kramer, Mukherjee & Maislos, 2003; Lung, 1999; Pesavento & Kelsey, 1999).

The chapter is organized as follows. Section 2 describes the existing PON standards and the main deployed topologies. In Section 3 is presented the main challenges in PON networks regarding contention methods used in PON architectures and the supported quality of service. Then, Section 4 explains the future trends based on the deployment of large split extended PONS, also called Long-Reach PONs. Finally, in Section 4, the most relevant conclusions obtained in this chapter are shown.

PON DESCRIPTION

PON Standards

There are some alternatives which are intended to standardize PON networks, such as APON, BPON, GPON and EPON. Among them, the most important are GPON and EPON. The first one, GPON, defined by the FSAN (Full Service Access Network) consortium, could be viewed as a good future solution, as it can support multiple classes of service with extreme efficiency. However, Ethernet PONs (EPONs), based on the Ethernet protocol, are considered a good choice for a today's optimized access network, as Ethernet is a well-known cheap technology and interoperable with a variety of legacy equipment. It makes that nowadays many studies are focused on such technology.

APON Standard

The **APON** networks are based on the ATM protocol and they are defined in the ITU-T Recommendation G.983.1 (International Telecommunication Union, 2005). The most important characteristic of this protocol is the use of fixed length cells of 53 bytes. In the downstream direction cells are transmitted in a multicast way towards all ONUs. Therefore, each ONU should extract those cells which belong to it. Each downstream APON frame consists of 54 data cells plus 2 PLOAM (Physical Layer Operation, Administration and Maintenance) cells (Figure 1).

In the upstream direction, the frame consists of 53 data packets, each of them of 56 bytes. As each cell has 53 bytes of data, the 3 remaining bytes are used for the synchronization with the OLT. Moreover, each frame includes one management slot called Multiburst Slot (MBS).

As in the upstream direction all users share the same access channel, APON typically uses the TDMA (Time Division Multiplexing Access) protocol to avoid possible collisions. In this way,

Figure 1. Basic format of the downstream and the upstream frame in an APON network

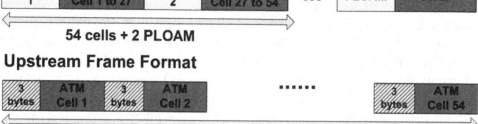

ONUs demand bandwidth by means of the MBS cells, whereas the OLT informs ONUs about the new allocated bandwidth using the PLOAM cells. On the other hand, since downstream data are broadcasted to all ONUs, APON implements the security mechanism to encrypt data called DES (Data Encryption Standard).

Regarding the transmission rates, the standard permits the next bit rates: symmetric 155 Mbps and 622 Mbps, or asymmetric 622 Mbps in the downstream and 155 Mbps in the upstream.

BPON Standard

The APON term was not very suitable, as it was thought that only ATM services could be supported by the network. Therefore, the APON term was changed for the new Broadband PON (BPON) nomenclature. These networks are able to support many broadband services and Ethernet and video services. Moreover, a new encryption mechanism was implemented, more sophisticated than the previous DES algorithm used in the APON standard.

GPON Standard

GPON (Gigabit PON) is defined in the G.984.1 Recommendation (International Telecommunica-

tion Union, 2003). It supports several bit rates for both the upstream and downstream channels. In particular, it can operate at several combinations of asymmetric or symmetric bit rates, from 155.52 Mbps to 2.5 Gbps.

In the GPON specification, the frame format is defined for both channels. In the downstream direction it is used a fixed frame of 125 µs synchronous with a clock of 8 KHz. This downstream frame periodicity is used to maintain the global synchronization of the whole system. In the upstream direction the length of the frame is also fixed to 125 µs.

In the upstream, each frame contains a number of transmissions from one or more ONUs, but one ONU can be allocated a maximum time slot of 125 µs, that is the maximum length of one frame (Figure 2). Each burst inside the upstream frame contains at minimum the Physical Layer Overhead (PLOu). Besides the payload, it may also contain the Physical Layer Operation and Administration and Management (PLOAMu), the Power Leveling Sequence upstream (PLSu) and the Dynamic Bandwidth Report upstream (DBRu) sections. The DBRu field is used when ONUs report their buffer status to the OLT in order to demand bandwidth for next transmissions. In particular, the GPON specification defines three ways in which one ONU can inform the OLT

Figure 2. Upstream and Downstream Frame format in the GPON specification

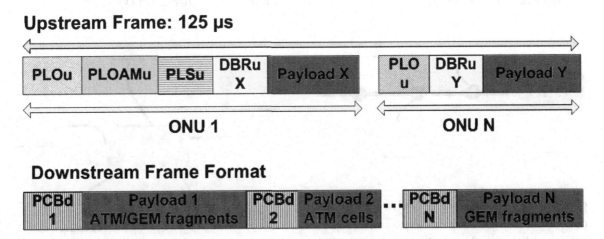

about its status: sending piggy-backed reports in the upstream DBRu field, using status indication bits in the PLOu field, or including an optional ONU report in the DBA payload.

On the other hand, the downstream frame consists of the Physical Control Block downstream (PCBd) field, the ATM partition and the GEM partition (Figure 2). The downstream frame provides the common time reference for the PON and also the signalling control for the upstream. Moreover, the OLT sends pointers in the PCBd field, each of them indicating the time at which each ONU starts and ends its upstream transmission. This performance allows that only one ONU accesses the shared channel at the same time, avoiding traffic collisions between packets from different ONUs.

One important issue of the GPON specification is that packets contained inside each frame can be encapsulated using ATM cells or using the GPON Encapsulation Method (GEM). The ATM method is an evolution of the APON/BPON standard. However, with the GEM method all type of traffic can be supported by the GPON network (native TDM, SONET, SDH, ATM or Ethernet traffic) using a variant of SONET/SDH Generic Framing Procedure (GFP). The most relevant traffic encapsulation is the Ethernet, which are mapped in

the GEM frame without including some overhead such as the Preamble and SDF bytes. Moreover, GPON permits the fragmentation of Ethernet packets across multiple GEM frames, contrary to the EPON standard which does not allow it. The packet fragmentation process inherent in the GEM encapsulation can be used to serve time sensitive traffic. Thus, the sensitive data encapsulated into GEM frames are always sent at the beginning of each payload. Due to the fixed 125 μs frame periodicity, it will be ensured that sensitive data are provided with very low latency.

EPON Standard

The main standardization force behind **EPON** (Ethernet PON) is the IEEE 802.3ah Task Force (IEEE 802.3ah Ethernet in the First Mile Task Force, 2004). It was designed to support a symmetrical bit rate of 1 Gbps in the upstream and downstream channels.

As it is based on the Ethernet protocol, the EPON transmission unit is equal to a standard Ethernet frame with some modifications in the preamble field. In the preamble of each Ethernet frame the Logical Link ID (LLID) label is inserted, which locally identifies each ONU inside the EPON network. In this way, ONUs are polled

Figure 3. Example of the bus topology for a Passive Optical Network (PON)

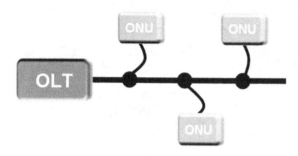

Figure 4. Example of the star topology for a Passive Optical Network (PON)

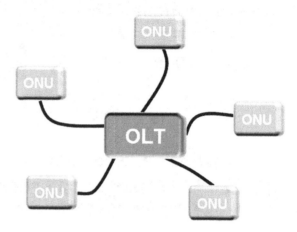

periodically to check if some ONUs are recently connected, and therefore some parameters have to be established such as the local address (LLID) inside the network.

On the other hand, the EPON standard does not allow frame fragmentation, and hence the time slot length allocated to each ONU should match the packet lengths to fully use the total capacity of the allocated time slots.

In the downstream direction, packets are broadcasted from the OLT to every ONU. Thus, as every ONU received the same data, its MAC (Medium Control Access) layer has to differentiate between packets which belong to it or not. In order to carry out this task, the MAC layer of the ONU extracts the LLID label of every Ethernet frame and checks if the ONU address corresponds to its ONU. In the upstream direction, each ONU sends its data typically following a TDMA scheme. At the end of every cycle each of them reports its buffer status in order to demand bandwidth for the next cycle. The OLT distributes the available bandwidth for every ONU and sends to them a separate grant message to inform them about the slot length for the next cycle and the time each of them is able to transmit.

Passive Optical Network Topologies

In order to deploy a PON network there are many topologies which can be applied, such as:

- Bus topology. It takes advantage of the optical fibre as it provides a point-to-multipoint connectivity between the OLT and the ONUs (Figure 3). Since this architecture does not need a splitter, it allows a less centralized distribution and it permits more distance coverage between the OLT and the ONUs. However, any fault in the bus causes that users become disconnected from the network, which is the main handicap of this topology.

- Star topology. This topology supports a point-to-point connectivity between the OLT and the ONUs, as it deploys an optical fibre per user (Figure 4). This configuration provides very high bandwidth to end users keeping very low cost for the operation, administration and maintenance (OAM) functions. However, this solution requires a large optical fibre deployment, which implies a very expensive implementation.

- Ring topology. This architecture offers the advantage of a point-to-multipoint connectivity between the OLT and the ONUs (Figure 5). Moreover, in this type of

Figure 5. Example of the ring topology for a Passive Optical Network (PON)

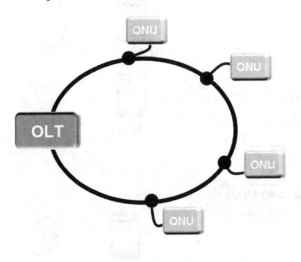

architectures it is very easy to implement protection mechanisms. However, this topology shows important difficulties related to the OAM functions.

- Tree topology. This is a Point-to-Multipoint architecture which offers the advantage that the infrastructure is shared by all users. This implies an important reduction in the implementation and maintenance costs in the access network (Figure 6). This architecture is the most typical and widespread and the majority of studies related to PON networks are based on this topology.

Figure 6. Example of the tree topology for a Passive Optical Network (PON)

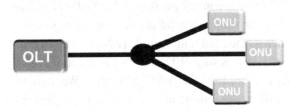

MAIN CHALLENGES IN PASSIVE OPTICAL NETWORKS (PONS)

Medium Access Control Protocols in PON Networks

Passive Optical Networks are typically Point-to-MultiPoint (P2MP) high capacity access networks based on a tree topology between the Optical Line Terminal (OLT) and the Optical Network Units (ONUs) (Figure 7). In the downstream direction (from the OLT to the ONUs) the OLT controls the traffic which arrives from outside the network. In this direction packets do not suffer any problem as the OLT decides which packets are sent to the different ONUs in a broadcast way. Therefore, all ONUs receive the same traffic and each of them has to differentiate if packets belong to it, or on the contrary discard them.

The main problem in PON networks occurs in the upstream direction (from ONUs to the OLT), because all users share the same wavelength, and a Medium Access Control protocol (MAC) is necessary to avoid collisions between packets from different ONUs. There are some medium access control protocols which can be applied to PON networks, such as WDMA (Wavelength Division Multiple Access), O-CDMA (Optical Code Division Multiple Access) and TDMA (Time Division Multiple Access). In relation with O-CDMA, current technology is not prepared enough to apply it in PON networks. On the other hand, WDMA is a very expensive alternative for today's access networks as it requires several wavelengths operating in the upstream, but it is a promising alternative for future PONs. Finally, TDMA is the today's most widespread control scheme in these networks as it allows a single upstream wavelength and it is very easy to implement.

Figure 7. PON network operation in both directions, upstream and downstream

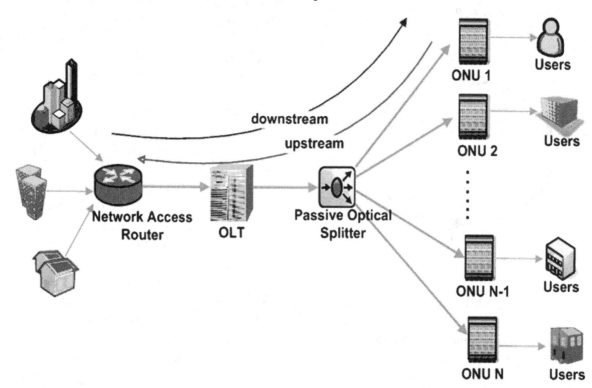

O-CDMA (Optical Code Division Multiple Access) Protocol

The **O-CDMA** control protocol in PON networks (Figure 8) assigns an orthogonal code to each subscriber in the access network. Then, each subscriber uses this code to modulate its data and transmit them in the upstream channel. In order to correctly receive the transmitted data, the OLT has to correlate the received signal with the set of orthogonal codes. When the codes are correct high correlation peaks can be observed, but on the contrary these peaks are very low (Zhang, Kun & Bo, 2007; Lee, Sorin & Yoon Kim, 2006). The implementation of O-CDMA in a PON network permits all ONUs to transmit in the upstream channel using the same wavelength. However, this technique requires a very strong synchronization, which implies a handicap in its implementation. One possible alternative to this problem is to apply asynchronous CDMA, but it leads to a penaliza-

tion in the correlation level. Another possibility is the use of several wavelengths to generate the orthogonal codes, but it is expensive as it requires multiple light sources at each ONU.

WDMA (Wavelength Division Multiple Access) Protocol

WDMA-PON architectures are viewed as a solution for the scalability problem of a traditional PON. In these traditional PONs, the offered power limits the maximum number of ONUs connected to one OLT and as only one wavelength is shared by all users, the available bandwidth to each subscriber decreases with the number of them.

There are many proposed WDM-PON architectures which depend on network components such as the used remote node (RN) in the external field. In this way, WDM-PON architectures tend to use splitters or AWG (Arrayed Waveguide Gratings) as remote nodes. Among those WDM-

Figure 8. Implementation of the O-CDMA control access protocol in a PON network

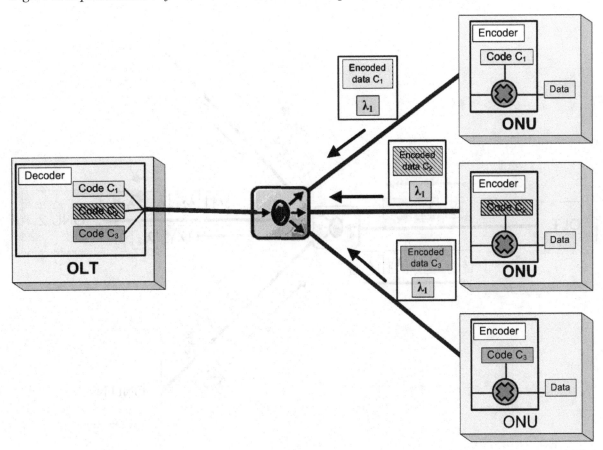

PON technologies based on splitters, one of the simplest architectures is shown in Figure 9. In this configuration the sources of the OLT and the ONUs are able to transmit at several wavelengths. This solution permits a progressive update of the architecture, as new wavelengths can be added to those subscribers with higher bandwidth demand, whereas the remaining subscribers continue with the previous configuration (McGarry, Reisslein & Maier, 2006).

Another more sophisticated architecture based on splitters is shown in Figure 10. This configuration permits to increase the number of users and the offered bandwidth. As the splitter limits the number of subscribers typically to 64, this solution proposes the deployment of new TDM-PON infrastructures. However, if all TDM-

PONs operate with the same wavelengths, the available bandwidth would significantly decrease. In order to avoid this problem, this architecture incorporates an AWG in the OLT, which allows an independent management of each deployed TDM-PON infrastructure.

The previous configuration based on only one AWG improves efficiently the available bandwidth in a PON network. However, there are important technological limits that do not allow the span of the network. In order to solve this problem, an architecture based on several levels of AWG (typically two) has been proposed. This novel architecture takes advantages of the cyclic properties of the AWG in order to simultaneously increase the available bandwidth and the number of users in the PON. Then, the AWG of the first level

Figure 9. Example of a WDM-PON architecture with a splitter as remote node (RN)

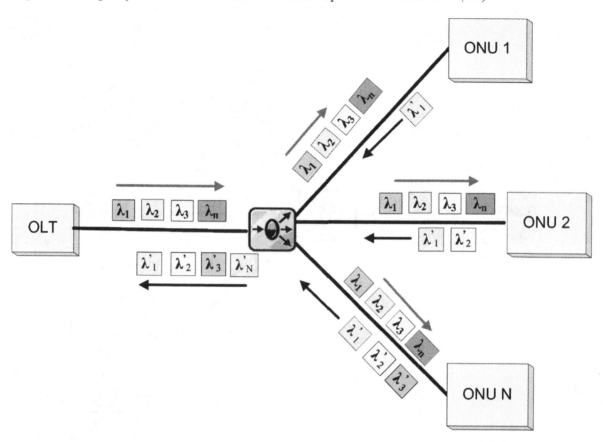

(KxM) is located in the OLT and it distributes the generated laser signals towards each M branch of the external network infrastructure. On the other hand, the second level of the AWG (LxN) routes each of the N wavelengths to the corresponding ONU (Mayer, Martinelli, Pattavina & Salvadori 2000; Bock, Prat & Walker, 2005).

TDMA (Time Division Multiple Access) Protocol

In a **TDMA** scheme, time is divided into periodic cycles, which at the same time are divided into so many time slots as the number of ONUs which share the channel. Therefore, each slot is dedicated to one ONU and one cycle is organized in such a way that one slot transports packets from one ONU periodically. This TDMA behaviour

has been studied in some works (Brunnel, 1986; Ko & Davis, 1984), which model this scheme by means of a switch. When one specific ONU starts its transmission, it can be viewed as if the ONU is visited by a server, which stays inside the ONU taking out packets during the time slot in that cycle. However, TDMA may be inefficient in the use of the bandwidth when the traffic nature is not homogeneous (Byun, Nho & Lim, 2003; Luo, & Ansari, 2005), as it always allocates the same bandwidth independently of the demand of each ONU. In fact, it is well known that real traffic is not homogeneous and not continuous, and therefore TDMA is not the most suitable scheme to apply. To deal with this problem, algorithms which distribute the available bandwidth in a dynamic way have been proposed, called Dynamic Bandwidth Allocation algorithms (**DBA**). They

Figure 10. Example of a WDM-PON architecture with a splitter as remote node (RN) and an incorporated AWG

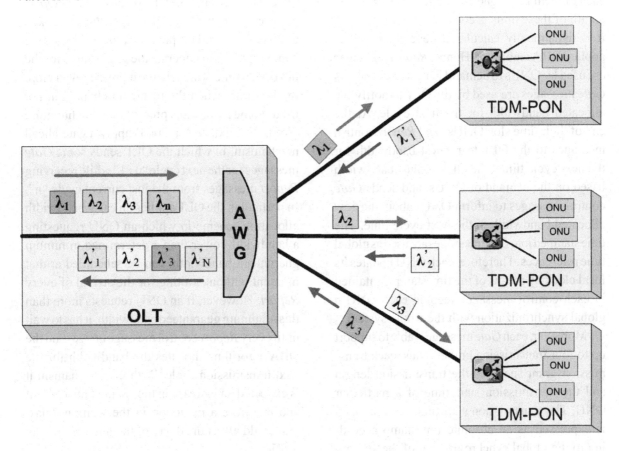

adapt the network capacity to traffic conditions by changing the distribution of the bandwidth assigned to each ONU depending on the current requirements.

Dynamic Bandwidth Allocation Algorithms in PON Networks

In the downstream direction a PON may be viewed as a point-to-multipoint network, and in the upstream direction as a multipoint-to-point network. This performance makes that the upstream and downstream channels are not communicated with each other. In order to treat this situation, the IEEE 802.3ah Task Force developed the Multipoint Control Protocol (MPCP), which controls the communication between the downstream and the upstream channels. Therefore, the application of a

dynamic bandwidth assignment requires the use of the MPCP protocol to dynamically distribute the upstream bandwidth to each ONU, controlling the data transmission from the ONUs to the OLT.

This protocol is implemented in the Medium Access Control (MAC) layer inside the OLT and it carries out some other tasks such as the auto-discovery, registration and ranging (distance from each ONU to the OLT) operations of new added ONUs to the PON network. The MPCP protocol uses five control messages: *Register, Register_ack, Register_req* messages are used to make the auto-discovery, registration and ranging tasks, whereas, *Report* and *Gate* control messages arbitrate the communication between the OLT and the ONUs in order to allocate bandwidth to each of them. Therefore, the upstream bandwidth is divided into bandwidth units via TDMA and each of them is assigned to a different ONU. The length of these units is considered not fixed, but it is dynamically calculated depending on the applied DBA algorithm. Hence, *Report* messages are used by ONUs to demand bandwidth whereas *Gate* messages are used by the OLT to notify the allocated bandwidth for the next cycle. At the end of their time slot, ONUs send *Report* control messages to the OLT to request bandwidth for the next cycle time. The OLT assigns bandwidth based on the demand of ONUs, and sends *Gate* control messages to inform ONUs about the new allocated bandwidth in the next cycle time. The time stamped into *Gate* messages are used as global time references. Therefore, each ONU updates its local clock by means of the timestamp contained in each control message, keeping each ONU a global synchronization with the whole system.

Moreover, each *Gate* message is able to support up to four transmission grants. Thus, each transmission grant specifies the transmission length and the transmission start time of a particular ONU. The transmission start time of each ONU is expressed as an absolute timestamp according to the global synchronization of the system. Hence, each ONU sends packets during its time slot according to its intra-ONU scheduler, which control the packet transmission from various local queues belonging to different applications.

Online and Offline Scheduling Methods in DBA Algorithms

Dynamic Bandwidth Allocation algorithms can be classified into Offline (centralized) or Online (polling) scheduling methods. In centralized algorithms (Figure 11) the OLT distributes the bandwidth for the next cycle once it receives the updated requirements of every ONU in that cycle. This behaviour causes a waste of time during this waiting period (Assi, Ye, Dixit & Ali, 2003; Chang, Kourtessis & Senior, 2006; Choi & Huh, 2002; Sherif, Hadjiantonis, Ellinas, Assi & Ali, 2004), although it allocates bandwidth for the next cycle

knowing the total demand of every connected ONU. Besides, once the OLT receives every *Report* message, the considered DBA algorithm is invoked and it needs some computation time to allocate the bandwidth and to generate the grants table for the next cycle time. This scheme increases even more the idle time when the upstream channel in not used. Some algorithms presented in the literature (Assi, Ye, Dixit & Ali, 2003) apply a gate-ahead mechanism, in which the OLT sends some *Gate* messages for the next cycle "n+1" while receiving *Report* messages from the just current cycle "n". In particular, the OLT applies an early bandwidth allocation scheme in which an ONU requesting a bandwidth lower than a determined minimum guaranteed bandwidth, can be scheduled at that moment without waiting for the arrival of every *Report*. However, if an ONU requests more than this minimum guaranteed bandwidth, it has to wait until every *Report* control message arrives and the DBA algorithm allocates the bandwidth for the next transmission cycle. With this mechanism it is expected an increase in the channel throughput and therefore a reduction in the waiting delay, that could affect the delay of the highest priority traffic.

On the other hand, in polling or online schemes, the OLT allocates bandwidth to each ONU for the next cycle before the last packet of the previous one arrives, resulting in efficient upstream channel utilization (Byun, Nho & Lim, 2003; Kramer, Mukherjee & Pesavento, 2002; Kramer, Mukherjee, Ye, Dixit & Hirth, 2002). However, these algorithms do not consider the global state of the network. Among polling or online algorithms, one of the most efficient is the Interleaved Polling with Adaptive Cycle Time (IPACT), an adaptive cycle time approach in which ONUs are polled individually in a round robin fashion, as it is shown in Figure 12 (Kramer, Mukherjee & Pesavento, 2002; Kramer, Mukherjee, Ye, Dixit, & Hirth, 2002).

Figure 11. PON network performance implemented a DBA algorithm based on a centralized scheme

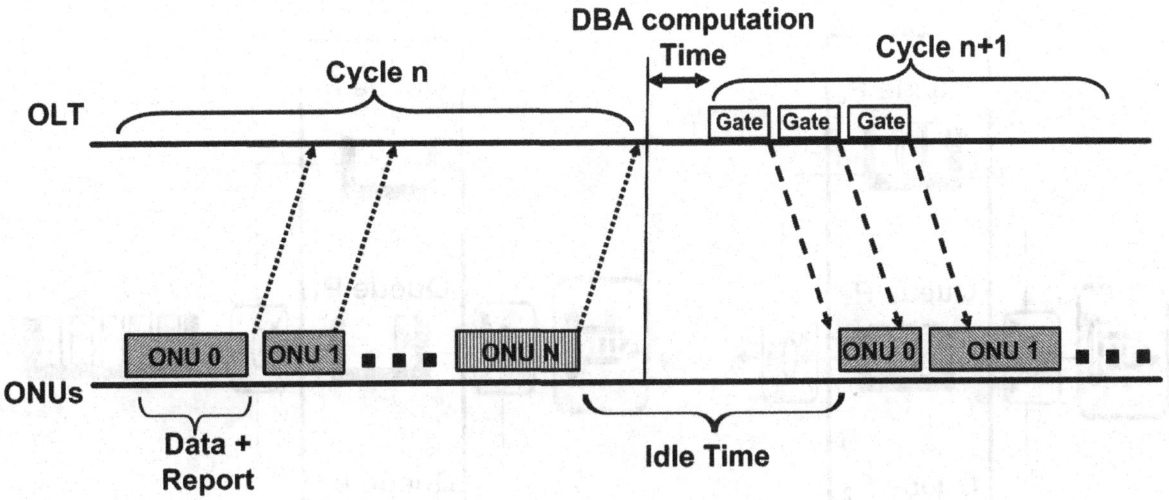

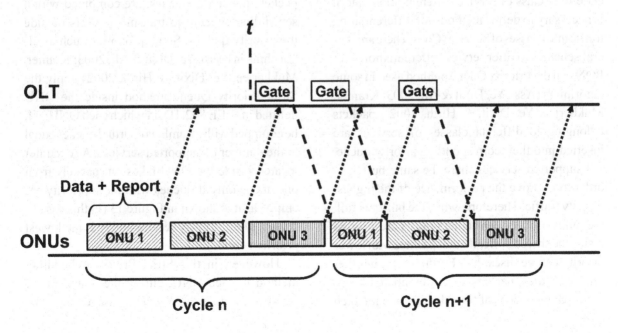

Quality of Service in PON Networks (QoS)

One of the most important challenges in PON networks is that they should offer Quality of Service (QoS). Available resources have to be used efficiently and the network has to provide preferential treatment to certain type of services and customs. Hence, networks providers and operators have to offer a minimum guarantee to customers in relation with the offered services. In particular, a PON network has to support both

Figure 12. PON network performance implemented the IPACT algorithm based on an interleaved polling policy

Figure 13. ONU scheduler equipped with three priority queues, P_0, P_1 and P_2 which applies the strict priority queue method

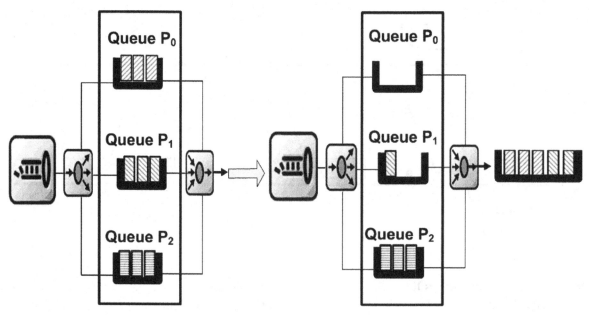

class of service differentiation and subscriber differentiation.

Class of Service Differentiation (CoS)

Dynamic Bandwidth Allocation algorithms should cope with class of service differentiation, and it is necessary to define methods for differentiating traffic into Classes of Service (**CoS**). There are several schemes to offer service differentiation over PONs. In the Priority Queue method used in some algorithms (Assi, Ye, Dixit & Ali, 2003; Kramer, Mukherjee, Ye, Dixit, & Hirth, 2002), packets belonging to different classes of services are inserted into their corresponding priority queue. All supported services share the same buffer, in order to improve the performance of the highest priority traffic. Therefore, when the buffer is full, incoming packets with higher priority are inserted inside the shared buffer if other packets with lower priority can be discarded. Furthermore, in other related studies, packets are differentiated keeping separate queues of fixed length, one for each

class of service (Nowak, Perry & Murphy, 2004). Hence, when queues are full, incoming packets will be dropped independently of their priority, as they are not permitted to replace packets with lower priority. These previous methods, the Priority Queue and separate queues used to categorize packets inside the queues, are combined with a scheduling scheme to transmit packets outside the priority queues. Some proposals such as algorithms in (Assi, Ye, Dixit & Ali, 2003; Kramer, Mukherjee, Ye, Dixit, & Hirth, 2002) apply the Strict Priority Queue method inside the ONU defined in IEEE 802.1D, in which each ONU will be equipped with a number of virtual queues equal to the number of supported services. A scheduler located inside the ONU takes out packets from one queue only if queues with more priority are empty, as it is shown in Figure 13 (in this case P_0 is the highest priority service and P_2 is the lowest priority service).

However, in this Strict Priority scheduling method in each ONU, during the interval time between sending the *Report* message and starting the transmission in the corresponding time slot, new higher priority packets arrive to the ONU. Thus, these priority packets arriving in the waiting time will be sent out ahead of the lower priority packets reported in the previous cycle time. This performance leads to an increase in the mean packet delay and packet loss of the lower priority traffic.

In order to reduce this problem other works implement the Fair Nonstrict Priority Scheduling scheme (Sherif, Hadjiantonis, Ellinas, Assi & Ali, 2004), which works as Strict Priority Queue, but in this case only packets reported in the last updated message can be sent out in the actual cycle. Hence, if higher priority packets arrive to one ONU just after the beginning of its current time slot, they have to wait until the reported lower priority packets are transmitted. Therefore, if the time slot can be able to send out more traffic, it will be used to transmit packets of high priority. This method provides fairness for every supported class of service.

Moreover, in algorithms presented in (Choi & Huh, 2002; Luo & Ansari, 2005), it is the OLT which allocates in a centralized way the exact bandwidth to each service, depending on both the priority and the demand of them, as it is represented in Figure 14 (in this case P_0 is the highest priority service an P_2 is the lowest priority service). In particular, in the dynamic bandwidth allocation algorithm with multiple services presented in (Luo & Ansari, 2005), each class of service is set a maximum bandwidth permitted at each cycle. Then, the OLT first allocates bandwidth to the highest services, as they require bandwidth guarantees, and the remaining bandwidth is given to the lowest priority traffic. The bandwidth allocated to the highest priority services is upper bounded by the smaller value of its request and its maximum bandwidth permitted, which controls that they do not forcefully consume the upstream channel. Besides, the service scheduling scheme presented in (Choi & Huh, 2002) assumed a fixed bandwidth assignment for the highest priority

services regardless of whether or not they have packets to send. Then, the remaining bandwidth is distributed between the medium priority traffic depending on the updated demand of such service in every ONU. Finally, if some bandwidth remains, it will be allocated to the lowest priority services in the same way as the medium priority traffic.

Customer Differentiation: Service Level Agreement (SLA)

Access networks have also to face customer differentiation. End users contract a service level agreement (**SLA**) with a provider, which contains technical specifications called service level specifications (SLSs). It forces the network to treat each SLA subscriber in a different way and dynamic algorithms have to take into account the requirements for every SLA subscriber. Some related studies are focused on a fair bandwidth distribution between various service providers which offer the same services through the same shared upstream channel to different users (Banerjee, Kramer & Mukherjee, 2006).

On the other hand, other works are related to an unique service provider which offers multi-service levels according to subscribers' requirements (Assi, Ye, Dixit, & Ali, 2003; Chang, Kourtessis & Senior, 2006; Ma, Zhu, & Cheng, 2003). As an example, the Bandwidth Guaranteed Polling (BGP) method proposed in (Ma, Zhu, & Cheng, 2003) divides ONUs into two disjoint sets: bandwidth guaranteed ONUs and best effort ONUs. Bandwidth guaranteed ONUs always receive the guaranteed bandwidth determined by their Service Level Agreement (SLA), whereas the others receive the remaining bandwidth. The associated SLA specifies the quantity of bandwidth that has to be guaranteed. Therefore, the upstream channel is divided into equal bandwidth units, although these units are chosen in such a way that the number of obtained units have to be larger than the supported ONUs inside the access network. Moreover, the OLT keeps two entry tables, one for

Figure 14. ONU scheduler equipped with three priority queues, P_0, P_1 and P_2 where the OLT allocates the exact bandwidth for every class of service

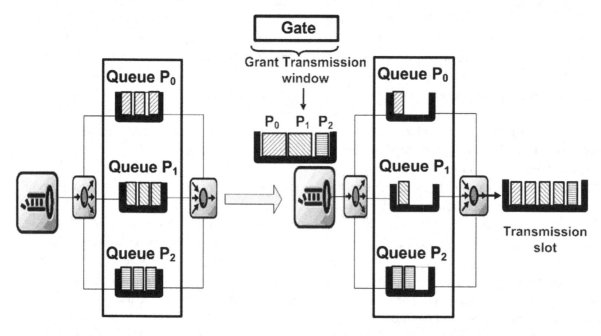

bandwidth guaranteed ONUs (Entry Table) and one for the best effort ONUs. The former table has as many entries as bandwidth units obtained in the total upstream channel, whereas the table for the best effort ONUs is not fixed in size. Therefore, those bandwidth units which are not used by the guaranteed ONUs can be dynamically allocated to the best effort ONUs. Each bandwidth guaranteed ONU can be allocated one or more bandwidth units according to their SLA or their bandwidth demand. Therefore, guaranteed ONUs with more than one entry will be polled by the OLT more than once in a round of polling. The OLT polls every ONU in the order of the Entry Table by a pointer indicating the current entry. If one entry is not assigned to one guaranteed ONU, it will be offered to a non-guaranteed ONU in the order listed in the best effort table.

Another common way to provide client differentiation is to use a weighted factor assigned to each ONU associated with one specific SLA. Then, the bandwidth is allocated depending on these fixed weights. In the method presented in

(Assi, Ye, Dixit & Ali, 2003), each ONU is assigned a minimum guaranteed bandwidth based on its associated weight, so that the upstream channel is divided among the ONUs in proportion to their SLAs. In particular, the weighted factors are assigned depending on the priority of the SLA of each ONU, and the sum of the weights has to be one. Therefore, the sum of every minimum guaranteed bandwidth is equal to the total available bandwidth contained in the considered cycle time. In this way, at the end of every transmission cycle, each ONU requests bandwidth for the next cycle. If the requested bandwidth is lower than the minimum guaranteed bandwidth, the ONU will be allocated the demanded bandwidth, and the remaining will be pooled together with the excess bandwidth of all other ONUs which do not request their permitted minimum bandwidth. Then, in the second stage of the algorithm the total excess bandwidth is distributed among those ONUs whose bandwidth demand is higher than their guaranteed bandwidth. Finally, this remaining bandwidth will be allocated depending on the bandwidth request of each high loaded ONU.

Besides, in the Dynamic Minimum Bandwidth (DMB) algorithm (Chang, Kourtessis & Senior, 2006), the OLT distributes the available bandwidth assigning different weights to each client depending on its SLA. Therefore, ONUs associated with a higher weight will be allocated more bandwidth. In this algorithm, the bandwidth allocation at every cycle is made in two stages. In the former step, a guaranteed minimum bandwidth is assigned to each ONU depending on the priority of its SLA, and it is composed of two types of bandwidths. The first bandwidth component, called basic bandwidth is a fixed bandwidth allocated to every ONU independently of the priority of its service level. The second component, the extra bandwidth is calculated taking into account both the priority of the service level and the bandwidth request of every ONU. Once the first step is completed, if some ONUs do not used their entire guaranteed minimum bandwidth, in the second step the algorithm allocates this remaining bandwidth among those ONUs which demand more bandwidth than their guaranteed minimum bandwidth.

As quality of service can be offered by means of class of service or client differentiation, a suitable combination of both of them is necessary in order to support full quality of service in the upstream channel. However, none of the previous proposed methods which apply client differentiation controls that the supported services comply with the standard restrictions. Thus, they do not consider the final behaviour of the offered services by every subscriber. However, this is a very important aspect that should be covered by a DBA algorithm as service providers have to guarantee that the highest priority traffic complies with constrains in terms of mean packet delay or packet loss rate. In this way, the algorithm proposed in (Merayo, Durán, Fernández, Lorenzo, de Miguel, Aguado & Abril, 2008) provides both service and client differentiation but, unlike other previous works, it dynamically controls that each subscriber satisfies every class of service require-

ment. Thus, the algorithm ensures that the most sensitive services keep the mean packet delay below the maximum upper bound permitted for every priority customer. The algorithm is based on a set of weights to distribute the bandwidth as it is done in other works (Assi, Ye, Dixit, & Ali, 2003; Chang, Kourtessis & Senior, 2006) but in contrast with them, it dynamically changes the value of the weights in order to achieve the best performance under every load situation.

LONG-REACH PASSIVE OPTICAL NETWORKS

There is an increasing interest in the development of larger split extended optically amplified PONs, also called **Long-Reach PONs**. In particular, many components have been developed for deploying optical access networks, i.e. optical amplifiers which extend the optical access network coverage from 20 km to 100 km. Moreover, advances in WDM permit to offer more bandwidth that only can efficiently be used if more users are integrated in the access network, which may be possible by means of Long-Reach architectures. The wavelength-division multiplexing (WDM) technology permits more wavelengths to be multiplexed on a fibre, each fibre transmitting at speeds of 40-100 Gpbs. Therefore, this WDM technology used in a traditional PON network (WDM-PON) can allocate far more bandwidth than end subscribers' demand due to the limited number or users in the coverage area. In order to serve more users in an optical access network, there is a need to increase the network span to cover more subscribers, and therefore it exits a strong tendency to use a Long-Reach PON to increase its initial reach up to 100 km (Song, Barnerjee, Kim & Mukherjee, 2007).

These network architectures provide high cost efficiency by simplifying the network as the access and the metro networks can be combined into only one by using 100 km of fibre instead of 20 km

Figure 15. Example of a Long-Reach PON (LR-PON) in a ring topology

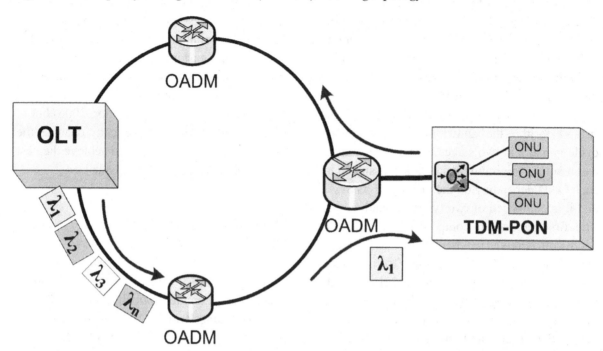

of a common PON. Hence, the cost associated to electronic interfaces between the access and the outer metro backhaul network is eliminated. Then, the OLT of the traditional PON can be replaced at the local exchange by some elementary hardware, such as optical amplifiers. As a result, a network operator may need few major central offices, reducing the associated cost (Song, Barnerjee, Kim & Mukherjee, 2007).

One example of deployment of a Long-Reach PON based on a ring topology is shown in Figure 15. This configuration is perfect for network resilience and bidirectional transmission. At each node of the ring an Optical Add-Drop Multiplexer (OADM) is used to insert and drop one wavelength to the end subscribers and also to mitigate the signal power loss along the long-reach transmission.

As the OADMs add some power supply, the interfaces become not totally passive, and therefore the term "Long-Reach PON" is not fully suitable. In this way, the SuperPON term is widely used in the literature to determine such PON systems which include the use of limited active components. Moreover, in this novel architecture, at each node of the ring the OADM extracts one wavelength assigned to one independent TDM PON, although all of them are combined in the same fibre in order to span the distance up to 100 km. Therefore, each TDM PON can be treated as an independent Long-Reach PON in terms of access level.

Other Long-Reach PON architectures are based on the combination of DWDM-TDM (Talli & Townsend, 2006) which use DWDM (Dense Wavelength Division Multiplexing) to permit several TDM PONs. Each TDM PON works in a different wavelength, but all of them share the same amplifier and backhaul fibre. This hybrid architecture shows a significant improvement in the deployed access networks, as it combines a long reach transmission due to the optical amplifiers used, and it allows an increment in the number of users due to DWDM. In these DMDM-TDM architectures, a number of powered PONs, each

Figure 16. Example of the TSD (Two-State DMB) algorithm in a Long-Reach PON

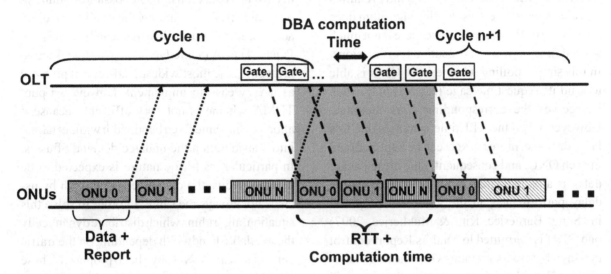

one with a specific wavelength assigned for the upstream, are combined using DWDM multiplexers, and therefore each of them can be viewed as an independent Long-Reach PON.

Several prototypes of Long-Reach PON and SuperPON have been developed in recent years, such as ACTS-PLANET (Van Deventer, Van Dam, Peters & Vermaerke, 1997), SuperPON by British Telecom (Davey, Payne, et al, 2006), and Hybrid PON (Talli & Townsend, 2006) by the University College in Cork.

Long-Reach PONs, as well as PONs are based on a tree topology between the OLT and the ONUs. In the upstream direction as the transmission is Multipoint-to-Point, it needs some medium access control protocol. These protocols have to be very efficient due to the high increment in the propagation delay from the OLT to ONUs. In order to overcome this high propagation delay it has been proposed several dynamic bandwidth allocation algorithms for Long-Reach PON, some of them based on a centralized policy, and others based on a polling scheme.

Regarding centralized schemes, as the OLT allocates the bandwidth for the next cycle once it receives the updated demand of every ONU in that cycle, ONUs have to wait a time equal to

the round trip time (RTT) to transmit in the next cycle. Since in Long-Reach PONs the distance between the OLT and the ONUs is increased up to 100 km, the RTT time in which ONUs cannot transmit is increased to 1 ms. In order to take advantage of this wasted time, it is necessary to implement some scheme that allows ONUs to transmit during it. The proposed algorithm TSD (Two-State DMB) (Chang, Merayo, Kourtessis & Senior, 2007), considers this round trip time as virtual fixed cycles where ONUs are able to transmit. Since the real bandwidth demand is not available at the beginning of every virtual cycle, the OLT distributes the bandwidth in these cycles applying a prediction of the required bandwidth of every ONU by means of traffic estimation, as it can be seen in Figure 16. Therefore, centralized methods in Long-Reach PONs become more complex and require traffic prediction which may be complicated due to the bursty user traffic.

In other algorithms based on a polling policy such as the proposed in (Jiménez, Merayo, Durán, Fernández & Lorenzo, 2008) the OLT allocates bandwidth for each ONU before the transmission from the previous one has finished, which leads to an efficient bandwidth utilization. This scheme applied in a Long-Reach PON is much simpler than centralized ones, and consequently requires less computing time. Besides, the algorithm does not require traffic prediction and there is not much wasted time between consecutive transmissions. In this simple polling scheme, each ONU is able to send its request message (*Report*) only when it receives the corresponding *Gate* message. However, when the RTT time increases due to a large distance, more packets can be kept buffered in each ONU, and consequently the mean packet delay is also increased. In order to improve the delay performance, in the algorithm proposed in (Song, Barnerjee, Kim & Mukherjee, 2007) one ONU is permitted to send its Report control message before its previous Gate message is received. Thus, this scheme creates a new "thread" of signalling between the ONU and the OLT and it lets parallel "polling processes" run at the same time. The number of threads is not limited, and it depends on the system environment.

Besides, the computation complexity of the multi-thread polling idea is not increased if compared with the simple thread scheme. This similarity happens because in both cases the *Report* messages arrive at the OLT at a comparable rate due to the trade-off between increased RTT and increased threads in the Long-Reach PON.

CONCLUSION

The new emerging services and the insufficient deployed access technologies have accentuated the lack of access capacity. Optical fibre is expected to satisfactory deal with the existing first mile challenges. In this way, the Passive Optical Network (PON) is the most promising technology to be implemented in the access segment network. PON network shows typically a point-to-multipoint tree topology between the OLT and the final subscribers (ONUs), resulting in a very easy and cost saving infrastructure. However, in the upstream direction as all users share the same channel some medium control access protocol is

needed to avoid collisions of possible simultaneous transmissions. Among the studied medium access control protocols which can be applied in PONs, TDMA (Time Division Multiple Access) protocol is the most widespread control protocol as it very easy to implement. However, a pure TDMA scheme is not very efficient because it allocates the same fixed bandwidth without taking into consideration the updated demand of users. In particular, as traffic nature is expected to be heterogeneous, the demanded bandwidth by users is not always the same. In order to solve this situation, algorithms which distribute dynamically the available bandwidth depending on the traffic demand of each ONU have been proposed. These algorithms are called Dynamic Bandwidth Allocation (DBA) algorithms, and they are able to adapt the network capacity to traffic conditions.

Every DBA algorithm applies an Offline (centralized) policy or an Online (polling) policy in order to allocate bandwidth to ONUs. In the former, the OLT allocates bandwidth having the knowledge of the network state, but this policy shows a wasted time as the OLT needs the reception of every updated demand. However, in the Online policy one ONU is able to transmit just after the end of the previous one, which results in efficient channel utilization, but the OLT does not know the total network demand.

Independently of the allocation method followed, one of the most important challenges in PONs is the support of Quality of Service (QoS). Network operators have to provide a minimum guarantee to customers in relation to the priority of the offered services (CoS) and the priority of the customer (SLA). Therefore, the access network has to support efficiently both class of service differentiation and subscriber differentiation. In this way, DBA algorithms have to offer subscriber differentiation but at the same time to guarantee that the highest priority traffic complies with the standard constrains independently of the priority of the subscriber.

Finally, there is an increasing interest in the

development of larger split extended optically amplified PONs called Long-Reach PONs. These network architectures provide high cost efficiency by simplifying the network as the access and the metro networks are combined into only one by using 100 km of fibre. In this way, protocols applied to Long-Reach PONs should be very efficient due to the high increment of the propagation delay from the OLT to ONUs. Therefore, several dynamic bandwidth allocation algorithms have been proposed in the literature for Long-Reach PONs focused on overcoming this high increase in the propagation delay (up to 1 ms for networks of 100 km) have been proposed.

REFERENCES

Assi, C., Ye, Y., Dixit, S., & Ali, M. A. (2003). Dynamic Bandwidth Allocation for Quality-of-Service over Ethernet PONs. *IEEE Journal on Selected Areas in Communications, 21*(9), 1467–1477. doi:10.1109/JSAC.2003.818837

Banerjee, A., Kramer, G., & Mukherjee, B. (2006). Fair Sharing Using Dual Service-Level Agreements to Achieve Open Access in an Ethernet Passive Optical Network (EPON). *IEEE Journal on Selected Areas in Communications, 24*(8), 32–43.

Bock, C., Prat, J., & Walker, S. D. (2005). Hybrid WDM/TDM PON Using the AWG FSR and Featuring Centralized Light Generation and Dynamic Bandwidth Allocation. *Journal of Lightwave Technology, 23*(12), 3981–3988. doi:10.1109/JLT.2005.853138

Brunnel, H. (1986). Message Delay in TDMA Channels with contiguous output. *IEEE Transactions on Communications, 34*(7), 681–684. doi:10.1109/TCOM.1986.1096608

Byun, H.-J., Nho, J.-M., & Lim, J.-T. (2003). Dynamic bandwidth allocation algorithm in ethernet passive optical networks. *Electronics Letters, 39*(13), 1001–1002. doi:10.1049/el:20030635

Chang, C.-H., Kourtessis, P., & Senior, J. M. (2006). GPON service level agreement based dynamic bandwidth assignment protocol. *Electronics Letters, 42*(20), 1173–1174. doi:10.1049/el:20062326

Chang, C.-H., Merayo, N., Kourtessis, P., & Senior, J. M. (2007). Dynamic Bandwidth assignment for Multi-service access in long-reach GPON. *Proceedings of the 33rd European Conference and Exhibition on Optical Communication (ECOC 2007)*. Berlin, Germany.

Choi, S.-I., & Huh, J. (2002). Dynamic bandwidth allocation algorithm for multimedia services over ethernet PONs. *ETRI Journal, 24*(6), 465–468.

Davey, R. P., & Payne, D. B. (2006). DWDM Reach Extension of a GPON to 135 km. *IEEE/OSA . Journal of Lightwave Technology, 24*(1), 29–31. doi:10.1109/JLT.2005.861140

IEEE. 802.3ah Ethernet in the First Mile Task Force, (2004). IEEE 802.3ah Standard. Retrieved November 20, 2007, from http://www.ieee802.org/3/efm/public.

International Telecommunication Union (ITU-T). (2003). Gigabit-capable Passive Optical Networks (GPON): General characteristics, in Series G: Transmission systems and media, Digital sections and digital line systems-optical line systems for local and access networks, Telecommunication Standardization Sector of ITU. Retrieved February 10, 2008, from http://www.itu.int/rec/T-REC-G.984.1-200303-S /en.

International Telecommunication Union (ITU-T). (2005). Recommendation G.983.1, Broadband optical access systems based on Passive Optical Networks (PON), in Series G: Transmission systems and networks, Digital sections and digital line systems-optical line systems for local and access networks, Telecommunication Standardization Sector of ITU. Retrieved January 15, 2008, from http://www.itu.int/rec/T-REC-G.983.1-200501-I/en.

Jiménez, T., Merayo, N., Durán, R. J., Fernández, P., & Lorenzo, R. M. (2008). Adaptive allocation algorithm to support Multi-Service Level Agreements in a Long-Reach EPON. *Proceedings of the 13rd European Conference and Optical Communications* (pp. 101-107). Krems (Austria).

Ko, K.-T., & Davis, B. R. (1984). Delay Analysis for a TDMA Channel with contiguous output and Poisson Message Arrival. *IEEE Transactions on Communications, 32*(6), 707–709. doi:10.1109/TCOM.1984.1096126

Kramer, G., Mukherjee, B., & Maislos, A. (2003). Ethernet Passive Optical Networks. In Sudhir Dixit (Ed.), *Multiprotocol over DWDM: Building the Next Generation Optical Internet* (pp. 229-275). John Wiley & Sons.

Kramer, G., Mukherjee, B., & Pesavento, G. (2002). Interleaved Polling with Adaptive Cycle Time (IPACT): A Dynamic Bandwidth Distribution Scheme in an Optical Access Network. *Photonic Network Communications, 4*(1), 89–107. doi:10.1023/A:1012959023043

Kramer, G., Mukherjee, B., Ye, Y., Dixit, S., & Hirth, R. (2002). Supporting differentiated classes of service in Ethernet passive optical networks. *Journal of Optical Networking, 1*(8), 280–298.

Lee, C.-H., Sorin, V., & Yoon Kim, B. (2006). Fiber to the Home Using a PON Infraestructure. *Journal of Lightwave Technology, 24*(12), 4568–4583. doi:10.1109/JLT.2006.885779

Lung, B. (1999). PON architecture Futureproofs FTTH. *Lightwave, 16*(10), 104–107.

Luo, Y., & Ansari, N. (2005). Bandwidth allocation for multiservice access on EPONs. *IEEE Communications Magazine, 43*(12), 16–21. doi:10.1109/MCOM.2005.1561907

Ma, M., Zhu, Y., & Cheng, T.-H. (2003). A bandwidth guaranteed polling MAC protocol for ethernet passive optical networks. *Proceedings of the Twenty Second Annual Joint Conference of the IEEE Computer and Communications Societies (INFOCOM 2003), 1*, 22–31, San Francisco.

Mayer, G., Martinelli, M., Pattavina, A., & Salvadori, E. (2000). Design and cost performance of the multistage WDM PON access networks. *Journal of Lightwave Technology, 18*(2), 125–143. doi:10.1109/50.822785

McGarry, M. P., Reisslein, M., & Maier, M. (2006). WDM Ethernet passive optical networks. *IEEE Communications Magazine, 44*(2), 15–22. doi:10.1109/MCOM.2006.1593545

Merayo, N., Durán, R. J., Fernández, P., Lorenzo, R. M., de Miguel, I., Aguado, J. C., & Abril, E. J. (2008). EPON algorithm to provide service and client differentiation based on a dynamic weight adaptation. *Proceedings of the 13rd European Conference and Optical Communications* (pp. 129-136), Krems (Austria).

Nowak, D., Perry, P., & Murphy, J. (2004). A Novel Service Level Agreement Based Algorithm for Differentiated Services Enabled Ethernet PONs. In IEICE Press, *3rd International Conference on Optical Internet: Vol. 1* (pp. 598-599), Japan.

Pesavento, M., & Kelsey, A. (1999). PONs for the Broadband Local Loop. *Lightwave, 16*(10), 68–74.

Sherif, S. R., Hadjiantonis, A., Ellinas, G., Assi, C., & Ali, M. (2004). A novel decentralized Ethernet-Based PON Access Architecture for Provisioning Differentiated QoS . *Journal of Lightwave Technology, 22*(11), 2483–2497. doi:10.1109/JLT.2004.836757

Song, H., Banerjee, A., Kim, B.-W., & Mukherjee, B. (2007). Multi-Thread Polling: A Dynamic Bandwidth Distribution Scheme in Long-Reach PON. *In Proceedings of the IEEE Globecom '07,* Washington, DC.

Talli, G., & Townsend, P. D. (2006). Hybrid DWDM-TDM Long-Reach PON for Next-Generation Optical Access. *IEEE/OSA . Journal of Lightwave Technology, 24*(7), 2827–2834. doi:10.1109/JLT.2006.875952

Van Deventer, M. O., Van Dam, P., Peters, P., & Vermaerke, F. (1997). Evolution Phases to an Ultra-Broadband Access Network: Results from ACTS-PLANET. *IEEE Communications Magazine, 35*(12), 72–77. doi:10.1109/35.642835

Zhang, C., Kun, Q., & Bo, X. (2007). Passive optical networks based on optical CDMA: Design and system analysis. [English Edition]. *Chinese Science Bulletin, 52*(1), 118–126. doi:10.1007/s11434-007-0020-8

Chapter 7
Dynamic Bandwidth Allocation for Ethernet Passive Optical Networks

Jun Zheng
Southeast University, China

Hussein T. Mouftah
University of Ottawa, Canada

ABSTRACT

Bandwidth allocation is one of the critical issues in the design of Ethernet passive optical networks (EPONs). In an EPON system, multiple optical network units (ONUs) share a common upstream transmission channel for data transmission. To efficiently utilize the limited bandwidth of the upstream channel, a system must dynamically allocate the upstream bandwidth among multiple ONUs based on the instantaneous bandwidth demands and quality of service requirements of end users. This chapter gives an introduction of the fundamental concepts on bandwidth allocation in an EPON system, discusses the major challenges in designing a polling protocol for bandwidth allocation, and presents an overview of the state-of-the-art dynamic bandwidth allocation (DBA) algorithms proposed for EPONs.

INTRODUCTION

With the explosive growth of the Internet and ever-increasing users' demand for various broadband applications, such as Internet telephony, high-definition television (HDTV), interactive games, and video on demand, subscriber access networks, which cover the "last mile" area, and serve numerous residential and small business users, have become a bandwidth bottleneck in providing broadband services to subscriber users (Zheng &

Mouftah, 2004]. In recent years, subscriber access networks have been extensively upgraded with the deployment of innovative xDSL and CaTV technologies. However, these technologies have their own limitations and are insufficient to meet the ever-increasing bandwidth demand of subscriber users. To alleviate this bottleneck, fiber to the home/curb/building (FTTH/FTTC/FTTB) technologies have been long envisioned as a preferred solution, and passive optical networks (PONs) have been widely considered as a promising technology for implementing various FTTx solutions.

DOI: 10.4018/978-1-60566-707-2.ch007

As one of the promising solutions, EPON has attracted a great interest from both industry and academia in recent years. EPON combines low-cost Ethernet equipment and low-cost passive optical components, and thus has a number of advantages over traditional access networks, such as larger bandwidth capacity, longer operating distance, lower equipment and maintenance cost, and easier update to higher bit rates (Kramer & Pesavento, 2002). An EPON system is a point-to-multipoint fiber optical network with no active elements in the transmission path from source to destination. It can use different multipoint topologies, such as bus, ring, and tree (Kramer & Pesavento, 2002), and different network architectures (Foh, 2004; Kramer & Pesavento, 2002; Shami, 2005; Sherif, 2004). The standard PON architecture is based on a tree topology and consists of an optical line terminal (OLT), a 1:N passive star coupler (or splitter/combiner), and multiple optical network units (ONUs), as shown in Figure 1. The OLT resides in a central office (CO) that connects the access network to a metropolitan area network (MAN) or a wide area network (WAN), and is connected to the passive star coupler through a singe optical fiber. The passive coupler is located a long distance away from the CO but close to the subscriber premises. Each ONU is located either at curbs or at subscriber premises, and is connected to the passive coupler through a dedicated short optical fiber. The distance between the OLT and each ONU typically ranges from 10 to 20 km. In an EPON system, all transmissions are performed between the OLT and the ONUs. In the downstream direction, an EPON is a point-to-multipoint network, in which the OLT broadcasts data to each ONU through the 1:N splitter, where N is typically between 4 and 64. Each ONU extracts the data destined for it based on its media access control (MAC) address. In the upstream direction, a PON is a multipoint-to-point network, in which multiple ONUs transmit data to the OLT through the 1:N passive combiner. The line data rate from an ONU to the OLT and the user access rate from a user to an ONU do not necessarily have to be equal and the line data rate is usually much higher than the user access rate. Since all ONUs share the same upstream transmission medium with limited bandwidth, an EPON system must employ a MAC mechanism to arbitrate the access to the shared medium in order to avoid data collisions in the upstream direction and at the same time to efficiently share the upstream transmission bandwidth among all ONUs. For this reason, bandwidth allocation is critical for ensuring the quality of network services in an EPON system and a large amount of research work has been conducted on this critical issue in recent years. The purpose of this chapter is to give an introduction of the fundamentals on bandwidth allocation in an EPON system and present an overview of the state-of-the-arts dynamic bandwidth allocation (DBA) algorithms proposed thus far for EPON systems.

The remainder of the chapter is organized as follows. In Section II, we introduce some fundamental concepts on bandwidth allocation in an EPON system. In Section III, we discuss the major challenges in designing a polling protocol for bandwidth allocation. In Section IV, we present an overview of the state-of-the arts DBA algorithms proposed for EPONs. In Section V, we conclude this chapter.

FUNDAMENTALS ON BANDWIDTH ALLOCATION IN EPONS

In this section, we introduce some fundamental concepts on bandwidth allocation in an EPON system, including channel separation, multiple access, and bandwidth negotiation and allocation.

Channel Separation

To increase the transmission efficiency of an EPON system, the upstream and downstream transmission channels should be separated appropriately. A

Figure 1. EPON architecture

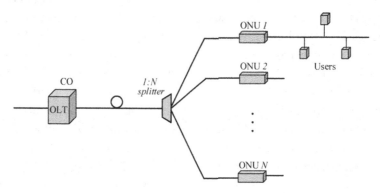

simple solution is to use space division multiplexing, where two separate optical fibers and passive couplers are used, one for upstream transmission and the other for downstream transmission. To reduce network cost, a more attractive solution is to use a single coupler and a single fiber for both directions with one wavelength for upstream transmission and another for downstream transmission. Currently, the most popular solution to channel separation is to use a 1550 nm wavelength for downstream transmission and another 1310 nm wavelength for upstream transmission (Kramer & Pesavento, 2002).

Multiple Access

In the upstream direction, multiple ONUs transmit data packets to the OLT through a common passive combiner and share the same optical fiber from the combiner to the OLT. Due to the directional property of a passive combiner, data packets from an ONU can only reach the OLT but not the other ONUs. For this reason, conventional contention-based multiple access, e.g., the carrier sense multiple access with collision detection (CSMA/CD) protocol, is difficult to implement because the ONUs are unable to easily detect a collision that may occur at the OLT. Although the OLT is able to detect a collision and inform the ONUs by sending a collision message, the transmission efficiency would be largely reduced because of

considerable propagation delay between the OLT and the ONUs. To address this problem, an optical looping-back technique was proposed in (Desai, 2001) to achieve high channel efficiency with CSMA/CD. With this looping-back technique, a portion of the upstream signal power transmitted by each ONU is looped back to the other ONUs at the star coupler by using a 3×N coupler and connecting two ports of the coupler together through an isolator, as shown in Figure 2. If two or more ONUs transmit data simultaneously, collisions will be detected at each ONU and all data transmissions will be stopped immediately. The optical CSMA/CD protocol is applied to all upstream transmissions (Chae, 2002). The OLT will receive the data packets transmitted by each ONU and will discard those packets with collisions. However, to implement the optical CSMA/CD protocol, each ONU has to use an additional receiver operating at the upstream wavelength and a carrier sensing circuit, which would largely increase the network cost. On the other hand, contention-based multiple access is unable to provide guaranteed bandwidth to each ONU and thus is difficult to support any form of quality of service (QoS). For these reasons, contention-based multiple access is currently not a preferred solution to the upstream multiple access.

Another possible solution is to use wavelength division multiplexing (WDM) technology and allow each ONU to operate at a different wavelength,

Figure 2: PON using looping-back star coupler

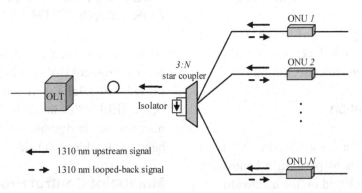

thus avoiding interference with the transmissions of the other ONUs. This solution is simple to implement but requires either a tunable receiver or a receiver array at the OLT to receive the data transmitted in multiple channels. In particular, it also requires each ONU to use a fixed transmitter operating at a different wavelength, which would result in an inventory problem. Although the inventory problem can be solved by using tunable transmitters, such devices are costly at the current stage.

Compared with contention-based multiple access and WDM, time division multiplexing (TDM) on a single wavelength is more attractive for upstream transmission. In this solution, each ONU is allocated a timeslot or transmission window for data transmission by the OLT. Each timeslot is capable of carrying several Ethernet packets. Packets received from one or more users are buffered in an ONU until the timeslot for that ONU arrives. Upon the arrival of its timeslot, the ONU will send out its buffered packets at the full transmission rate of the upstream channel. Accordingly, TDM avoids data collisions from different ONUs. Moreover, it requires only a single wavelength for all ONU transmissions and a single transceiver at the OLT, which is highly cost-effective.

Bandwidth Management

In TDM, bandwidth management is important for efficiently utilizing the bandwidth of the shared upstream wavelength. It involves two main tasks: bandwidth negotiation and bandwidth allocation.

Bandwidth Negotiation

Bandwidth negotiation is to exchange information between the OLT and each ONU in order for each ONU to report its bandwidth demand to the OLT and for the OLT to send its bandwidth allocation decision to each ONU. For this purpose, IEEE 802.3ah defines a multipoint control protocol (MPCP) to support bandwidth negotiation between the OLT and ONUs in EPON, including two 64-bytes MAC control messages: REPORT and GATE. The REPORT message is generated by each ONU to report its queue status to the OLT. The OLT allocates bandwidth for each ONU based on the queue status information contained in the received REPORT message, and uses the GATE message to deliver its bandwidth allocation decision to each ONU.

One way to deliver the bandwidth request or REPORT message is to dedicate a very short timeslot in the upstream channel. This requires twice laser on/off for one upstream transmission from each ONU. Another way is to piggyback the bandwidth request or REPORT message at the end of a data timeslot, which reduces laser on/off times into one per transmission for each ONU, and thus reduces the physical-layer power overhead and the inter-frame guard (Luo, 2007).

Bandwidth Allocation

To allocate bandwidth (or a timeslot) for each ONU, the OLT needs to perform a bandwidth allocation algorithm based on the bandwidth requests from each ONU as well as some allocation policy and/or service level agreement (SLA). In this context, there are many bandwidth allocation algorithms proposed in the literature, which can be classified into two broad categories: static bandwidth allocation (SBA) and dynamic bandwidth allocation (DBA). With SBA, each ONU is allocated a timeslot with a fixed length, which does not require bandwidth negotiation and is thus simple to implement. However, due to the bursty nature of the network traffic, it may result in a situation in which some timeslots are overflowed even under very light load, causing packets being delayed for several timeslots, while other timeslots are not fully used even under very heavy traffic, leading to the upstream bandwidth being under-utilized. For this reason, static allocation is not preferred. To increase bandwidth utilization, the OLT must dynamically allocate a variable timeslot to each ONU based on the instantaneous bandwidth demand of the ONUs. To implement DBA, polling has been widely used (Kramer, 2002). With polling, the OLT can dynamically allocate bandwidth for each ONU and flexibly arbitrate the transmissions of multiple ONUs, which can significantly increase bandwidth utilization and improve network performance.

POLLING PROTOCOL DESIGN FOR BANDWIDTH ALLOCATION

In this section, we first introduce the multipoint control protocol (MPCP) being standardized by the IEEE 802.3ah Ethernet in the First Mile Task Force (IEEE P802.3ah, 2004) and then discuss the major issues in designing a polling protocol for bandwidth allocation using MPCP.

Multipoint Control Protocol (MPCP)

MPCP is a signaling protocol for facilitating dynamic bandwidth allocation and arbitrating the transmissions of multiple ONUs. It resides at the MAC control layer and has two operation modes: normal mode and auto-discovery mode. In the normal mode, MPCP relies on two Ethernet control messages, GATE and REPORT, to allocate bandwidth to each ONU. The GATE message is used by the OLT to allocate a transmission window to an ONU. The REPORT message is used by an ONU to report its local conditions to the OLT. In the auto-discovery mode, the protocol relies on three control messages, REGISTER, REGISTER_REQUEST, and REGISTER_ACK, which are used to discover and register a newly connected ONU, and to collect relevant information about that ONU, such as the round-trip delay and MAC address.

In its normal operation, MPCP in the OLT gets a request from the higher MAC client layer to transmit a GATE message to a particular ONU. Upon getting such a request, MPCP will timestamp the GATE message with its local time and then send the message to the ONU. The GATE message typically contains a granted start time, a granted transmission window, and a 4-byte timestamp, which is used to calculate the round-trip time between the OLT and the ONU. Once the ONU receives the GATE message, it programs its local register with the values contained in the GATE message. Meanwhile, it also updates its local clock to that of the timestamp extracted from

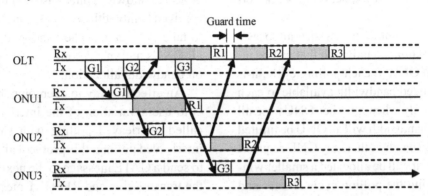

the received GATE message in order to maintain synchronization with the OLT. At the granted start time, the ONU will start to transmit data for up to the window size. The transmission may include multiple data packets, depending on the window size and the queue length in the ONU. No packet fragmentation is allowed during the transmission. If the next packet cannot be transmitted in the current window, it will be deferred to the next window.

A REPORT message is sent by an ONU in the allocated transmission window together with a data packet. It can be transmitted automatically or on demand either at the start or at the end of a window. A REPORT is generated at the MAC client layer and is time-stamped at the MAC layer. It typically contains the bandwidth demand of an ONU based on the instantaneous queue length of that ONU. The ONU should also account for additional overhead in its request, including a 64-bit frame preamble and a 96-bit inter-frame gap associated with each Ethernet packet. Once a REPORT message is received by the OLT, it is passed to the MAC client layer, which is responsible for bandwidth allocation and recalculation of the round-trip delay to the source ONU. Figure 3 illustrates a flow of GATE (G) messages and REPORT (R) messages for upstream transmission of three ONUs.

It should be pointed out that MPCP is not concerned with any particular bandwidth allocation scheme and transmission scheduling algorithm, and allows them to be vendor-specific. To design an efficient polling protocol based on MPCP, several problems must be considered, including maximum bandwidth constraint, channel utilization, and packet scheduling.

Maximum Bandwidth Constraint

A polling protocol typically operates on a cycle-based basis. In each polling cycle, each ONU is polled once and is allocated a transmission window based on its bandwidth demand. If the OLT allows each ONU to send all its buffered packets in one transmission, ONUs with high traffic load may monopolize the entire bandwidth of the upstream channel. This is unfair to those ONUs with low traffic load. To address this problem, the OLT should limit the maximum transmission bandwidth of each ONU. The maximum window size can be either fixed based on some criterion, such as a SLA, or variable based on instantaneous network conditions. Under high traffic load, the maximum window size determines the maximum polling cycle. In general, making the maximum polling cycle too long will result in larger delay for all packets under high traffic load, including those high-priority packets. On the other hand, making the maximum cycle too short will result in more bandwidth being wasted by inter-frame gaps (or guard times). Accordingly, the maximum window size has a great impact on network performance.

While the maximum window size imposes a limit on the maximum bandwidth that can be allocated to each ONU in each polling cycle, it is also the guaranteed bandwidth available to each ONU. In fact, only when all other ONUs use all their available bandwidth will an ONU be limited to its guaranteed bandwidth. If any ONU requests less bandwidth, it will be allocated a smaller window size, making the polling cycle shorter and thus increasing the actual bandwidth available to all other ONUs.

Channel Utilization

Since the upstream channel is shared by multiple ONUs, it is important to efficiently utilize the bandwidth of the common channel. A polling protocol can poll multiple ONUs for transmission based on different policies. A simple policy, called poll-and-stop polling, is to send a GATE message to an ONU and then stop for the data and REPORT message to come back from that ONU before the OLT sends a GATE message to the next ONU, as shown in Figure 4(a). Obviously, this protocol wastes a lot of bandwidth on the upstream channel, which would largely reduce channel utilization and increase packet delay.

A more efficient way is to use the interleaved polling (Kramer, 2002), which allows the OLT to send a GATE message to the next ONU before the data and REPORT message(s) from the previous polled ONU(s) arrive, as shown in Figure 4(b). This is feasible because the upstream channel and downstream channel are separated, and the OLT maintains relevant information about each ONU in a polling table, including the bandwidth demand of each ONU and the round-trip time to each ONU. The results obtained in (Chae, 2002) indicate that the interleaved polling protocol can significantly improve the network performance in terms of channel utilization and average packet delay. However, this protocol allows the OLT to

allocate bandwidth only based on those already-received bandwidth demands. The OLT is unable to take into account the bandwidth demands of all ONUs and make a more intelligent decision on bandwidth allocation.

An effective way to overcome this drawback is to use a variation of the interleaved polling, called interleaved polling with stop. Like the interleaved polling, this protocol allows the OLT to send a GATE message to the next ONU before the transmission and REPORT message(s) from the previous polled ONU(s) arrive. Unlike the interleaved polling, the OLT does not start the next polling cycle before the transmissions and REPORT messages from all ONUs are received. This allows the OLT to perform bandwidth allocation based on the bandwidth demands of all ONUs at the end of each polling cycle and thus make a more intelligent decision. However, such intelligence is obtained at the cost of upstream channel utilization because the upstream channel is not utilized from the instant the transmission of the last polled ONU in the previous cycle is completed to the instant the transmission of the first polled ONU in the next cycle starts. Figure 4(c) illustrates an example of the control message flows with the interleaved-polling-with-stop protocol.

Transmission Scheduling

To ensure efficient transmission, a polling protocol must schedule the transmissions of multiple ONUs in a manner that avoids data collisions from different ONUs. This is not difficult to implement because such scheduling is based on the granted window size and the round-trip time to each ONU. Since the OLT knows the granted window size and the round-trip time to the last polled ONU, it can calculate the transmission start time and window size for the next ONU. Note that, to allow the receiver in the OLT to prepare for receiving the transmissions, a minimum gap or guard time is usually required between the transmissions of

Figure 4: Polling policies: (a) poll-and-stop polling; (b) interleaved polling; (c) interleaved polling with stop

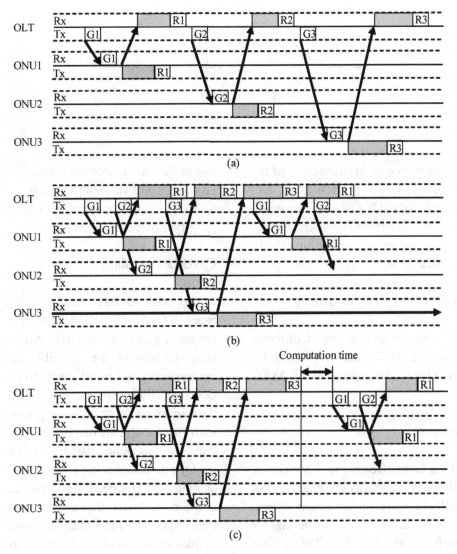

different ONUs.

On the other hand, the OLT must also be responsible for scheduling the transmission order of different ONUs, which may have a great impact on network performance. This is not difficult to implement because the order of the transmissions is usually determined one cycle ahead by performing a scheduling algorithm. The most widely-used scheduling algorithm is round-robin (RR), which has been adopted by many polling protocols. RR schedules the transmissions of different ONUs in the order of their indexes in the polling table and is simple to implement. However, it does not take into account the instantaneous traffic conditions at each ONU and thus may not be able to provide the best performance in terms of packet delay and data loss. To improve network performance, it is desirable to use an adaptive scheduling algorithm that can dynamically schedule the order of different ONU transmissions based on the instantaneous traffic conditions at each ONU. For example, an adaptive scheduling algorithm can schedule the

Figure 5: Priority queuing and intra-ONU scheduling SM

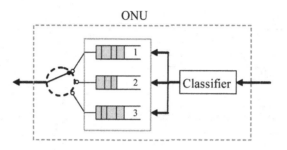

ONU transmissions in a descending order of the instantaneous queue length of each ONU, i.e., the longest queue first (LQF), or in an ascending order of the arrival time of the first packet queuing in each ONU, i.e., the earliest packet first (EPF) (Zheng & Mouftah, 2005).

Quality of Service Provisioning

To meet the service requirements of different network users, an EPON system must consider differentiated QoS provisioning in its MAC design.

Priority Queuing

Priority queuing is an effective way to support differentiated QoS. With priority queuing, network traffic is classified into a set of classes with diverse QoS requirements and for each traffic class a priority queue is maintained at each ONU. Figure 5 illustrates an example of priority queuing, in which an ONU maintains three priority queues that share the same memory buffer of fixed size. Data packets from end users are first classified by checking the type-of-service (ToS) field of the IP packets encapsulated in the Ethernet packets and then buffered in corresponding priority queues. If a higher priority packet finds the buffer full at the time of its arrival, it can preempt a lower priority packet. If a lower priority packet arrives and finds the buffer full, it will be dropped. As a result, lower priority traffic may experience very high packet

loss and even resource starvation. To address this problem, an ONU should perform some kind of traffic policing to control the amount of higher priority traffic from each end user.

ONU Scheduling

In supporting differentiated QoS, there are two types of ONU scheduling strategies: inter-ONU scheduling and intra-ONU scheduling. Inter-ONU scheduling is responsible for arbitrating the transmissions of different ONUs, and intra-ONU scheduling is responsible for arbitrating the transmissions of different priority queues in each ONU. There are two strategies to implement these two scheduling paradigms. One is to allow the OLT to perform both inter-ONU scheduling and intra-ONU scheduling. In this case, the OLT is the only device that arbitrates the upstream transmissions. Each ONU can request the OLT to allocate bandwidth for each traffic class. For this purpose, an ONU must report the status of its individual priority queues to the OLT through REPORT messages. MPCP specifies that each ONU can report the status of up to eight priority queues. The OLT can then generate multiple grants, each for a specific traffic class, to be sent to the ONU using a single GATE message. The format of the 64-byte MPCP GATE message can be found in (IEEE P802.3ah, 2004).

The other strategy is to allow the OLT to perform inter-ONU scheduling whereas to allow each ONU to perform intra-ONU scheduling. In this case, each ONU requests the OLT to allocate bandwidth for it based on its buffer occupancy status. The OLT only allocates the requested bandwidth to each ONU. Each ONU will divide the allocated bandwidth among different classes of services based on their QoS requirements and schedule the transmissions of different priority queues within the allocated bandwidth. For intra-ONU scheduling, there are two types of scheduling algorithms: strict priority scheduling and non-strict priority scheduling. In strict priority scheduling, a lower-priority queue is scheduled only if all queues with higher priority are empty. Obviously, this will potentially result in infinite packet delay and high packet loss for low-priority traffic. To address the problem, a non-strict priority scheduling algorithm was proposed in (Ma, 2003). In non-strict priority scheduling, only those packets that were reported are transmitted first as long as they can be transmitted within the allocated timeslot. The transmission order of different priority queues is based on their priorities. If the packets that were reported are all scheduled and the current timeslot can still accommodate more packets, those newly-arriving packets that were not reported are also transmitted based on their priorities. As a result, all traffic classes can have access to the upstream channel within the allocated timeslot as reported to the OLT while their priorities are maintained, which ensures fairness in scheduling.

DYNAMIC BANDWIDTH ALLOCATION ALGORITHMS FOR EPONS

Due to the limited upstream bandwidth, an EPON system may not always be able to provide sufficient bandwidth to meet the bandwidth demand of all end users. To better serve the end users, it is desirable to dynamically allocate bandwidth to each ONU based on the instantaneous traffic conditions of the ONUs in order to improve the network performance in terms of packet delay, packet loss, and throughput. For this purpose, a variety of DBA algorithms have been proposed in the literature. In this section, we will present an overview of the major DBA algorithms for EPONs.

Interleaved Polling with Adaptive Cycle Time (IPACT)

IPACT (Kramer, 2002) is the first DBA algorithm proposed for EPON. It employs a resource negotiation process to facilitate queue report and bandwidth allocation. The OLT polls ONUs and grants timeslots to each ONU in a round-robin fashion. The timeslot granted to an ONU is determined by the queue status reported from that ONU. Therefore, the OLT is able to know the dynamic traffic load of each ONU and allocate the upstream bandwidth in accordance with the bandwidth demand of each ONU. Moreover, it also employs the SLAs of end users to upper bound the allocated bandwidth to each ONU.

In IPACT, several bandwidth allocation schemes are investigated, including limited allocation, constant credit, linear credit, and elastic allocation. In limited allocation, the OLT simply grants an ONU the number of bytes the ONU requested, but not exceeding a maximum window size. This is the most conservative scheme because it assumes that no more packets arrived after the ONU sent its request. In practice, however, because of the round-trip time between the OLT and each ONU, there might be more packets arriving between the instant an ONU sends a REPORT message and the instant the ONU receives a GATE message. In this case, those newly-arriving packets may not be able to be transmitted in the current cycle, resulting in increased average packet delay. To address this problem, the constant credit scheme and the linear credit scheme were proposed.

In constant-credit allocation, a credit is added to the requested window size and is considered in the granted window size. The size of the credit is constant no matter how large the requested window size is. Once an ONU receives a GATE message, it can send packets for up to the requested window size plus the constant credit. The choice of the credit size may have an impact on the network performance. A too small size will not be able to improve packet delay a lot. A too large size will reduce the bandwidth utilization of the upstream channel. The choice should be based on the traffic characteristics or some empirical data.

In linear-credit allocation, a similar credit is added to the requested window size. However, the size of the linear credit is proportional to the requested window size. The basis behind this scheme is that network traffic usually has a certain degree of predictability. This means that if a long burst of data is observed, this burst is very likely to continue for longer time.

In elastic allocation, there is no limit imposed on the maximum window size. The only limit is the maximum cycle time. The maximum window size W_{max} is granted in such a way that the accumulated size of last N grants (including the one being granted) does not exceed $N \times W_{max}$, where N is the number of ONUs. In this way, if only one ONU has data to send, it may get a granted window size up to $N \times W_{max}$.

Among all the above bandwidth allocation schemes, limited allocation exhibits the best performance (Kramer & Pesavento, 2002). However, IPACT does not consider the multi-service needs of subscribers. To meet such the multi-service needs, a variety of DBA algorithms have been proposed for supporting differentiated services.

Estimation-Based DBA

Byun *et al.* (2003) proposed an estimation-based DBA algorithm, which can reduce the queue length of each ONU and thus the average packet delay by estimating the packets arrived at an ONU during the waiting time and incorporating the estimation in the grant to the ONU. In the algorithm, a control gain is used to adjust the estimation based on

the difference between the departed and arrived packets in the previous transmission cycle. The simulation results show that the proposed DBA algorithm can reduce the average packet delay as compared to IPACT.

Bandwidth Guaranteed Polling (BGP)

Bandwidth guaranteed polling (BGP) (Ma, 2003) is a DBA algorithm proposed for providing bandwidth guarantees in EPONs. In BGP, all ONUs are divided into two groups: bandwidth guaranteed and bandwidth non-guaranteed. The OLT performs bandwidth allocation by using a couple of polling tables. The first polling table divides a fixed-length polling cycle into a number of bandwidth units and each ONU is allocated a certain number of such bandwidth units. The number of bandwidth units allocated to an ONU is determined by the bandwidth demand of that ONU based on its SLA with a service provider. A bandwidth guaranteed ONU with more than one entry in the poling table has its entries spread through the table. This can reduce the average queuing delay because the ONU is polled more frequently. However, it leads to more grants in a cycle and thus requires more guard times between grants, which reduces channel utilization. On the other hand, it can potentially lead to lower channel utilization because an Ethernet frame cannot be fragmented in transmission. If a frame is too large to fit in the remainder of the current bandwidth unit, it will have to wait for the next bandwidth unit and a portion of the current bandwidth unit is thus wasted. To address this problem, BGP allows an ONU to communicate to the OLT its actual use of a bandwidth unit. If the unused portion of the bandwidth unit is large enough, this portion will be granted to a bandwidth non-guaranteed ONU. Otherwise, the next bandwidth guaranteed ONU is polled. However, this mechanism is largely limited by the propagation delays between the OLT and ONUs.

The unused portions of the bandwidth units for the bandwidth guaranteed ONUs are distributed to the bandwidth non-guaranteed ONUs in the order of their positions in the second polling table. The construction of the second polling table is different from that of the first table. Each entry is dynamically created as a bandwidth non-guaranteed ONU requests a grant. The analytical and simulation results show that the bandwidth-guaranteed ONUs with multiple bandwidth units in BGP have better performance than the ONUs with no differentiated bandwidth requirements in IPACT. BGP can provide differentiated services to different users with various bandwidth requirements, and can achieve better performance than IPACT.

Fair Sharing with Dual SLAs (FSD-SLA)

Banerjee *et al.* (2006) proposed a fair sharing with dual SLAs (FSD-SLA) algorithm, which employs dual SLAs in IPACT to manage the fairness for both subscribers and service providers. The primary SLA specifies those services whose minimum requirements must be guaranteed with a high priority. The secondary SLA describes the service requirements with a lower priority. This algorithm first allocates timeslots to those services with the primary SLA to guarantee their upstream transmissions. After the services with the primary SLA are guaranteed, the next round is to accommodate the secondary SLA services. If the bandwidth is not sufficient to accommodate the secondary SLA services, the max-min policy is adopted to allocate the bandwidth with fairness. If there is excessive bandwidth, FSD-SLA will allocate the bandwidth to the primary SLA entities first and then to the secondary SLA entities, both by using max-min fair allocation. The simulation results show that FSD-SLA can deliver much better fairness in terms of bandwidth allocation and packet delay than traditional fair-queuing algorithms such as Deficit Round Robin (Shreedhar, 1996).

Class-of-Service-Oriented Packet Scheduling (COPS)

Naser and Mouftah (2006) proposed a class-of-service-oriented packet scheduling (COPS) algorithm to support differentiated services. COPS uses two groups of leaky bucket credit pools on the OLT side to regulate the traffic of each ONU and each class-of-service (CoS). One group contains k credit pools, corresponding to k CoSs in the EPON system. Each pool is used to control the average rate of certain CoS traffic from all ONUs to the OLT. The other group contains m credit pools, corresponding to m ONUs in the system. Each pool is used to control the usage of the upstream channel by an ONU. In allocating and granting bandwidth or timeslots, the OLT begins with the highest CoS and ends with the lowest CoS, which is performed in two rounds. In the first round, each ONU with the traffic of the current CoS is granted up to the number of credits available for that ONU. If a request is granted, the granted bytes are subtracted from the corresponding credit pool. At the end of the first round, the unused credits are pooled together and are distributed to those ONUs whose bandwidth demands were not fully satisfied. As long as there are credits available in the pools, a new request will be accommodated. Simulation results show that COPS has lower average and maximum delay for all CoSs except the highest-priority one as compared to IPACT with limited allocation.

Hybrid Granting Protocol (HGP)

Shami *et al.* (2005) proposed a hybrid granting protocol (HGP) to support differentiated QoS provisioning by guaranteeing bandwidth and minimizing jitter. In HGP, traffic is classified into three categories: Assured Forwarding (AF), Best Effort (BE), and Expedited Forwarding (EF). For the EF traffic, HGP employs a queue prediction mechanism to size the grant to an ONU to accommodate all the traffic in the queue of that ONU at the point of granting. This is because the EF traffic has a constant bit rate and thus can be easily predicted. For the AF and BE traffic, it sizes the grant to an ONU only based on the REPORT message from that ONU. A transmission cycle consists of two sub-cycles: EF sub-cycle and AF/BF sub-cycle, and begins with the EF sub-cycle followed by the AF/EF sub-cycle. The EF sub-cycle carries the EF traffic while the AF/BF sub-cycle carries AF and BF traffic for each ONU. The length of the EF sub-cycle is predetermined while that of the AF/BF sub-cycle depends on the traffic load of each ONU. Accordingly, there are two grants for each ONU in every transmission cycle. The status of the AF and BF queues in an ONU is not reported until the end of the EF grant for that ONU, which allows the ONU to report up-to-date queue status to the OLT. In this way, HGP guarantees the bandwidth to the EF traffic and thus minimizes the jitter experienced by the EF traffic, while keeping QoS support for the AF and BF traffic with flexible bandwidth allocation. The simulation results show that HGP has smaller queuing delay under higher traffic load as compared to a regular DBA algorithm (Assi, 2003). Under lower traffic load, the regular DBA algorithm has smaller queuing delay because of the increased number of guard time per cycle. HGP can significantly improve the network performance in terms of packet delay, delay jitter, and buffer utilization for the EF service without degrading QoS support for the AF and BE services.

Dynamic Bandwidth Allocation with Multiple Services (DBAM)

Luo and Ansari (2005) proposed a dynamic bandwidth allocation with multiple services (DBAM) algorithm to accommodate different types of traffic in EPONs. Instead of providing multiple services among ONUs and among end users separately, DBAM incorporates both of them into the REPORT/GATE mechanism with class-based bandwidth allocation. It applies pri-

ority queuing to the EF, AF, and BE frames, and employs priority-based scheduling to schedule the buffered frames. Moreover, DBAM uses limited bandwidth allocation to arbitrate bandwidth allocation among ONUs, thus prohibiting aggressive bandwidth scrambling. In addition, it employs class-based traffic prediction to take into account the traffic that arrives during the waiting period, which ranges from sending the queue status report to sending the traffic buffered in each ONU. Such prediction is based on the actual traffic received in the previous waiting period. The OLT serves all ONUs in a fixed round-robin fashion in order to facilitate traffic prediction. The simulation results show that the fixed service order can increase the accuracy of traffic prediction, which can improve the network performance in terms of frame delay, queue length, and frame loss as compared to fixed allocation (Kramer, 2001), class-based allocation (Choi & Huh, 2002), and limited allocation with excessive distribution (Assi, 2003).

Limited Sharing with Traffic Prediction (LSTP)

Luo and Ansari (2005) proposed a limited sharing with traffic prediction (LSTP) algorithm, which employs an adaptive filter to predict the traffic that arrives during the waiting period and thus more accurately grant bandwidth to each ONU. For each class of traffic, LSTP estimates the data that arrive during the waiting period based on the data of this class that actually arrived in previous transmission cycles by using a linear predictor. The bandwidth demand of an ONU is thus the reported queue length plus the estimation. The OLT arbitrates the upstream bandwidth using this estimation and reserves a portion of the upstream bandwidth for transmitting the estimated data in the earliest transmission cycle, thus reducing packet delay and loss. In addition, LSTP facilitates service differentiation by using different SLA parameters to restrict different classes of traffic. The simulation results show that it improves the network performance in terms of packet delay and packet loss as compared to fixed allocation (Kramer, 2001), limited allocation (Kramer, 2002), and limited allocation with excess distribution (Assi, 2003).

Two-Layer Bandwidth Allocation (TLBA)

Xie *et al*. (2004) proposed a two-layer bandwidth allocation (TLBA) algorithm for supporting differentiated services in EPONs. TLBA is a hierarchical allocation algorithm that allocates bandwidth in two layers. In the first layer, the transmission cycle is partitioned or the upstream bandwidth is allocated among differentiated service classes, which is called class-layer allocation. In the second layer, the partition or bandwidth allocated to each class is distributed to all ONUs within the same class based on a max-min fairness policy, which is called ONU-layer allocation. The OLT allocates bandwidth based on the instantaneous demand of each ONU and does not limit the size of each demand. Accordingly, an ONU is allowed to report the lengths of all its queues to the OLT and the OLT allocates the upstream bandwidth to meet all the demands as much as possible. To avoid any class from monopolizing the available bandwidth in a cycle, a per-class threshold is introduced. The bandwidth threshold guarantees a minimum bandwidth for a class under high traffic load. Any excessive bandwidth from the classes that need less than their thresholds is distributed among the classes that need more than their thresholds. The excessive bandwidth distribution is performed based on the weights that are assigned to each class. The simulation results show that even under very high traffic load, TLBA can ensure a minimum bandwidth for each service class based on its bandwidth threshold, and can reduce the average queuing delay of high-priority and medium-priority traffic as compared to the two stage-queue scheme (Kramer, 2002).

Summary

In all the above DBA algorithms, IPACT is the first DBA algorithm proposed for EPON and can incorporate several different bandwidth allocation schemes. As the pioneering DBA algorithm, it has been used as a benchmark for performance evaluation by most DBA algorithms proposed later. The estimation-based DBA algorithm improves the average packet delay performance of IPACT by estimating the packets arrived at an ONU during the waiting time and incorporating the estimation in the grant to the ONU. However, both algorithms do not support differentiated services.

The other algorithms provide differentiated services and QoS support by employing various techniques, such as dual SLAs, leaky bucket credit pools, priority queuing, and traffic prediction. FSD-SLA employs dual SLAs in IPACT to improve the fairness in terms of bandwidth allocation and packet delay, and can achieve a delay performance comparable with that of IPACT. BGP supports differentiated services by providing bandwidth guarantees for high-demand ONUs while serving low-demand ONUs with best-effort service, and can achieve better performance than IPACT. COPS supports differentiated services by employing two groups of leaky bucket credit pools. It can achieve smaller average and maximum delay for all CoSs except the highest one, which experiences slightly larger average delay, as compared with IPACT with limited allocation. HGP employs queue prediction for the EF traffic, but not for the AF and BE traffic. As a result, it can significantly improve the QoS performance for the EF service without degrading the performance of the AF and BE services. DBAM and LSTP are based on traffic prediction. Both of them can improve the average packet delay and packet loss probability as compared with IPACT. However, it is unclear how much the traffic prediction contributes to the performance improvement. TLBA supports differentiated services by employing a two-layer hierarchical allocation and has shown to have better performance than some existing algorithm. It should be pointed out that most of these DBA algorithms use IPACT as the benchmark in evaluating their performance. A comprehensive performance comparison between these DBA algorithms is still yet found in the literature.

CONCLUSION

Bandwidth allocation is a critical issue in the design of an EPON system. Because multiple ONUs share a common upstream channel, an EPON system must efficiently utilize the limited upstream bandwidth in order to meet the bandwidth demands and QoS requirements of end users. For this purpose, the OLT should dynamically allocate the upstream bandwidth among all ONUs based on their instantaneous bandwidth demands and a variety of DBA algorithms have been proposed in the literature. In this chapter, we introduced the fundamental concepts on bandwidth allocation in an EPON system and presented an overview of the major DBA algorithms proposed for EPON systems. With recent advances in enabling technologies, optical devices that are previously costly are becoming more affordable, which makes it economically feasible to use multiple upstream channels in an EPON system and has presented many new challenges in the design of DBA algorithms for multi-channel EPON systems.

REFERENCES

Assi, C., Ye, Y., Dixit, S., & Ali, M. (2003). Dynamic bandwidth allocation for quality-of-service over Ethernet PONs. *IEEE Journal on Selected Areas in Communications, 21*(9), 1467–1477. doi:10.1109/JSAC.2003.818837

Banerjee, A., Kramer, G., & Mukherjee, B. (2006). Fair sharing using dual service-level agreements to achieve open access in a passive optical network. *IEEE Journal on Selected Areas in Communications, 24*(8), 32–44.

Byun, H.-J., Nho, J.-M., & Lim, J.-T. (2003). Dynamic bandwidth allocation algorithm in Ethernet passive optical networks. *IEEE Electronics Letter, 39*(13), 1001–1002. doi:10.1049/el:20030635

Chae, C.-J., Wong, E., & Tucker, R. S. (2002). Optical CSMA/CD media access scheme for Ethernet over passive optical network. *IEEE Photonics Technology Letters, 14*(5), 711–713. doi:10.1109/68.998734

Choi, S., & Huh, J. (2002). Dynamic bandwidth allocation algorithm for multimedia services over Ethernet PONs. *ETRI Journal, 24*(6), 465–468.

Desai, B. N., Frigo, N. J., Smiljanic, A., Reichmann, K. C., Iannone, P. P., & Roman, R. S. (2001). An optical implementation of a packet-based (Ethernet) MAC in a WDM passive optical network overlay. *Proc. of 2001 Optical Fiber Communication Conference (OFC'01), Vol. 3,* WN5-1-WN5-3.

Foh, C., Andrew, L., Wong, E., & Zukerman, M. (2004). FULL-RCMA: a high utilization EPON. *IEEE Journal on Selected Areas in Communications, 22*(8), 1514–1524. doi:10.1109/JSAC.2004.830459

IEEE. P802.3ah Ethernet in the First Mile Task Force (2004). IEEE IEEE Std 802.3ah. http://www.ieee802.org/3/efm/index.html.

Kramer, G., Mukherjee, B., Dixit, S., Ye, Y., & Hirth, R. (2002). Supporting differentiated classes of service in Ethernet passive optical networks. *OSA Journal of Optical Networking, 1*(8/9), 280–298.

Kramer, G., Mukherjee, B., & Perawnto, G. (2001). Ethernet PON (ePON): design and analysis of an optical access network. *Photonic Network Communications, 3*(3), 307–319. doi:10.1023/A:1011463617631

Kramer, G., Mukherjee, B., & Pesavento, G. (2002). IPACT: a dynamic protocol for an Ethernet PON (EPON). *IEEE Communications Magazine, 40*(2), 74–80. doi:10.1109/35.983911

Kramer, G., & Pesavento, G. (2002). Ethernet passive optical network (EPON): building a next-generation optical access network. *IEEE Communications Magazine, 40*(2), 66–73. doi:10.1109/35.983910

Luo, Y., & Ansari, N. (2005). Bandwidth allocation for multiservice access on EPONs. *IEEE Communications Magazine, 43*(2), S16–S21. doi:10.1109/MCOM.2005.1391498

Luo, Y., & Ansari, N. (2005). Limited sharing with traffic prediction for dynamic bandwidth allocation and QoS provisioning over EPONs. *OSA Journal of Optical Networking, 4*(9), 561–572. doi:10.1364/JON.4.000561

Luo, Y., Yin, S., Ansari, N., & Wang, T. (2007). Resource management for broadband access over time-division multiplexed passive optical networks. *IEEE Network, 21*(5), 20–27. doi:10.1109/MNET.2007.4305168

Ma, M., Zhu, Y., & Cheng, T. (2003). A bandwidth guaranteed polling MAC protocol for Ethernet passive optical networks. *Proc. of 2003 IEEE Conference on Computer Communications (INFOCOM'03), Vol. 1*, 22-31.

Naser, H., & Mouftah, H. T. (2006). A joint-ONU interval-based dynamic scheduling algorithm for Ethernet passive optical networks. *IEEE/ACM Transactions on Networking, 14*(4), 889-899.

Shami, A., Bai, X., Assi, C., & Ghani, N. (2005). Jitter performance in Ethernet passive optical networks. *IEEE/OSA . Journal of Lightwave Technology, 23*(4), 1745–1753. doi:10.1109/JLT.2005.844510

Shami, A., Bai, X., Ghani, N., Assi, C., & Mouftah, H. T. (2005). QoS control schemes for two-stage Ethernet passive optical access networks. *IEEE Journal on Selected Areas in Communications, 23*(8), 1467–1478. doi:10.1109/JSAC.2005.852185

Sherif, S., Hadjiantonis, A., Ellinas, G., Assi, C., & Ali, M. (2004). A novel decentralized Ethernet-based PON access architecture for provisioning differentiated QoS. *IEEE/OSA . Journal of Lightwave Technology, 22*(11), 2483–2479. doi:10.1109/JLT.2004.836757

Shreedhar, M., & Varghese, G. (1996). Efficient fair queuing using deficit round robin. IEEE/ACM Transactions on Networking, 4(3), 375-385.

Xie, J., Jiang, S., & Jiang, Y. (2004). A dynamic bandwidth allocation scheme for differentiated services in EPONs. *IEEE Communications Magazine, 42*(8), S32–S39. doi:10.1109/MCOM.2004.1321385

Zheng, J., & Mouftah, H. T. (2005). An adaptive MAC polling protocol for Ethernet passive optical networks (EPONs). *Proc. of 2005 IEEE International Conference on Communications (ICC'05), Vol. 3*, 1874-1878.

Zheng, J., & Mouftah, H. T. (2004). *Optical WDM Networks: Concepts and Design Principles*. Hoboken, New Jersey: Wiley-IEEE Press.

Zheng, J., & Mouftah, H. T. (2006). Efficient bandwidth allocation algorithm for Ethernet passive optical networks. *IEE Proceedings. Communications, 153*(3), 464–468. doi:10.1049/ip-com:20050358

Chapter 8
Multicast Routing in Optical Access Networks

Miklós Molnár
IRISA-INSA, France

Fen Zhou
IRISA-INSA, France

Bernard Cousin
IRISA-Université de Rennes I, France

ABSTRACT

Widely available broadband services in the Internet require high capacity access networks. Only optical networking is able to efficiently provide the huge bandwidth required by multimedia applications. Distributed applications such as Video-Conferencing, HDTV, VOD and Distance Learning are increasingly common and produce a large amount of data traffic, typically between several terminals. Multicast is a bandwidth-efficient technique for one-to-many or many-to-many communications, and will be indispensable for serving multimedia applications in future optical access networks. These applications require robust and reliable connections as well as the satisfaction of QoS criteria. In this chapter, several access network architectures and related multicast routing methods are analyzed. Overall network performance and dependability are the focus of our analysis.

INTRODUCTION

Recent advances in communication technology have resulted in multicast applications playing an important part in everyday Internet traffic. Data transmission generated by multicast multimedia services as Video-On-Demand, High Definition TV diffusion, Video-Conferences, Distance Learning and Online-Games requires large bandwidth, while QoS (Quality of Service) parameters such as end-to-end delay and jitter must be tolerated. From the white paper of the European Information & Communications Technology Industry Association (EICTA) on Next Generation Networks and Next Generation Access, high speed network access is characterized as (a) the availability of symmetrical access (b) instant communication (no latency) and (c) simultaneous applications (EICTA, 2008). All-optical networks show promise as an infrastructure that can guarantee dependability, flexibility, high bandwidth and QoS

DOI: 10.4018/978-1-60566-707-2.ch008

for users of multicast applications. All-optical networks have optical access network component directly connected to the mesh optical backbone. The huge capacity of fibers and light based routing in optical switches provide end-users with large bandwidth connections to the network. The most promising technology corresponds to wavelength division multiplexing (WDM). The transmission of data can be organized in either a connection based or a burst switched manner (Qiao & Yoo, 1999). Aggregation techniques and time division multiplexing can be applied to enhance overall network performance. In currently implemented solutions the optical switch configuration is performed via an independent control plane or a fixed-tuned wavelength channel for control messages. This control plane enables precise and thus efficient management of the optical network.

From the point of view of network operators and access providers, access network technology should offer a flexible solution at low cost. Low cost can be achieved with the use of passive equipment and a simple topology (for example a star). The huge capacity of an optical infrastructure currently allows wastage of network resources. However, in the long run a better utilization of network resources may be an important operator objective. The network should thus offer the possibility to manage resources and to balance network load. The dependability of the network is also a fundamental property for operators and users. Currently, optical access network technology is widely based on PONs (Passive Optical Networks), but Ethernet point-to-point and active Ethernet solutions are also present in the market. PONs contain passive elements. They are simple, easy to install and do not require an electrical power supply. A typical FTTx access network implemented with PONs is star based and contains splitters. The most significant drawback of star topologies is their vulnerability. Absolute dependability is a critical and fundamental requirement for modern communication networks. Dependable network services cannot be provided without redundancies

in the network topology. Thus, dependable access networks must contain, at least in their core part, redundant edges and nodes, thus producing cycle or mesh topologies.

Multicast routing is not specifically analyzed for current access networks. However, the coexistence of many multicast sessions raises some important problems. For instance, in a star topology the intelligent allocation of wavelengths among multicast sessions can optimize the use of network resources (Sheu & Huang, 1997; Sivalingam, Bogineni, & Dowd, 1992). In a mesh topology the light-tree structure can be introduced. Dependable multicasting is made possible using light-trees because they can be replaced entirely or partially when some network elements fail. In our analysis, we suppose that future optical access networks will be heterogeneous and meshed. This implies that the network topology has some active and configurable switches, and provides sufficient redundancy to offer dependable services with a high level of flexibility for efficient resource management. Moreover, precise configuration of the lightpaths and light-trees enhances the security of the network because data is not broadcast as in a star topology. Multicast routing in heterogeneous and mesh optical access networks (which contain active switches and passive elements) can play an important role in the all-optical networks of the future. Optical switch architectures and optical fiber characteristics introduce some specific constraints which must be taken into account by the routing algorithm. Moreover, the throughput of the network depends strongly on the efficiency of the routing algorithms. For these reasons, we propose a survey of multicast routing algorithms under the typical physical constraints of wavelength switched optical access networks.

This chapter presents the underlying problems as follows: following a description of typical access network architectures and routing (scheduling) methods, we describe the main constraints on optical switches and fibers. Since dependable network architectures correspond to mesh networks, the constraints have a large impact on performance, restrain the routing algorithm, and influence the multicast structures. Multicast routes usually correspond to partial spanning trees, but due to optical constraints the light-tree structure must be adapted. Generally, light-trees and light-forests are proposed to support multicast communications. These two types of structures allow the various constraints in all-optical networks to be satisfied. Even when splitters are available in the network, splitting diminishes strongly (at least proportionally) the light power, thus several specific energy aware light-trees may be required to minimize splitting. Wavelength conversion capabilities can dramatically enhance the performance of multicast routing. In consequence, several multicast routing algorithms are presented which satisfy the various requirements of heterogeneous optical access networks.

OPTICAL ACCESS NETWORK ARCHITECTURES

To be competitive access network operators have to install cost-effective networks with a large enough bandwidth capacity to serve end-user requirements. For instance FTTH architectures and the deployment of physical fiber infrastructure in British Telecom experiments are discussed in (Mayhew, Page, Walker, & Fisher, 2002). This analysis illustrates very well the difficulties that an operator encounters when determining future network investments. On one hand the high bandwidth requirement of future services limits the choice of technology. Only optical fiber network offers the capability of cost-effective wide-scale provision of the full range of future broadband services. On the other hand, the cost and the profitability of the access network limit operator investment. The authors demonstrate that applying an FTTH network offers smaller potential revenue from a residential service than from a service in the business area. This leads to a need for cost

optimization of both the transmission system and the fiber infrastructure.

Star Topologies

The most frequently proposed topology is the passive star, where a PSC (Passive Star Coupler) links the access nodes. This configuration was developed to realize Ethernet-based PON (EPON) technology in access networks (Kramer, Mukherjee, & Pessavento, 2001). All communication in an EPON is performed between an optical line terminal (OLT) and optical network units (ONUs). The OLT connects the optical access network to the backbone. In the downstream direction (from OLT to ONUs), a PON corresponds to a point-to-multipoint network, and in the upstream direction it is a multipoint-to-point network (*Figure 1*).

Star based PONs are simple, easy to install in existing infrastructures and easy to maintain. In these access networks, each ONU has a dedicated short optical fiber and shares a long distribution trunk fiber to the OLT with the other ONUs. For downstream traffic the EPON implements a broadcast and select scheme using the splitting capacity of the central coupler. For upstream communication an appropriate Dynamic Bandwidth Allocation (DBA) algorithm is used to assign time slots to end-users.

Generally, multicast communication in the star can be realized easily, but as illustrated in the next section, the medium access control protocol can be very specific. The main disadvantage of star topologies is their vulnerability. If the coupler fails the entire access segment is hampered. The fiber connecting the core network to the PSC (Passive Star Coupler) via the optical line terminal (OLT) also represents a highly vulnerable joint of the access network. Moreover, it is difficult to resolve security and confidentiality issues. Last but not least, due to the fact that messages are broadcast in PONs, network capacity is wasted.

Figure 1. Tree and ring based topologies

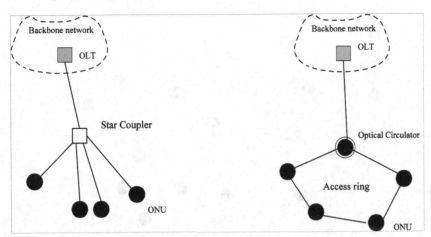

Ring Topologies

To improve private network capacity, a ring-based PON architecture is proposed to implement the LAN (Hossain, Dorsinville, Ali, Shami, & Assi, 2006). In this architecture, a long standard trunk fiber connects the OLT to the ONUs, which are interconnected within a short distribution fiber ring (*Figure 1*). The feeder fiber and the ring are connected using a 3-port optical circulator.

The links into the ring are unidirectional: both upstream and downstream communication use the same rotational direction. The ONUs apply the *Tap-and-Continue* (*TaC*) function to separate a part of the downstream flow for eventual local utilization. The downstream flow is then removed at the end of the ring to avoid its useless retransmission. The upstream transmission is based on a TDMA scheme and ONU-ONU communication is merged with the upstream traffic within the same pre-assigned time slot. To remove useless ONU-ONU and upstream communications from the upstream flow, a special removing, regenerating and retransmitting function is present in each ONU. This solution supports a fully distributed control plane among the ONUs as well as upstream communication to the OLT but does not ensure fault tolerance. This ring based architecture is improved in (Hossain, Erkan,

Hadjiantonis, Dorsinville, Ellinas, & Ali, 2008), where a two-fiber self-healing PON is proposed. This improved architecture provides simple and cost-effective fully distributed resilience capabilities against most kinds of networking failures. This solution also supports a truly shared LAN capability among end users. The control plane contains distributed fault detection and recovery mechanisms as well as a decentralized dynamic bandwidth allocation scheme. The proposed decentralized automatic protection switching technique is capable of protecting against both node (ONU) and fiber failures (distribution and trunk) through active participation of ONUs. Another simple, self-restoring and ring-based PON with two fiber-rings and TDMA option for bandwidth sharing can be found in (Yeh, Lee, & Chi, 2008). Optical line terminals and optical network units are used to protect against the occurrence of fiber failure in the optical access network; a protection technique is proposed for fast restoration of the access network in the case of failures.

New General Architectures

A project supported by DARPA has proposed a very flexible metropolitan and access network architecture (Kuznetsov, Froberg, Henion, Rao, Korn, Rauschenbach, Modiano, & Chan, 2000).

Figure 2. The hyper-channel concept in SMART

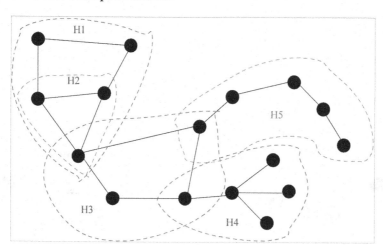

The proposition is based on the coexistence of WDM and IP routing, leveraging the advantages of both solutions. The different access networks are connected to the optical backbone using a (generally SONET based) double ring. The access network may correspond to an arbitrary topology (star, ring, bus or meshed). Electronic IP routing is combined with optical flow switching in the WDM domain using heterogeneous access node architecture in the OLTs. These OLTs allow IP routing and also to bypass it by optically switching the high capacity connections of some high-end users, while using electronic management for all other communication. Beyond its flexibility, an incontestable advantage of this network proposition resides in the dependability aspect of the access network design. The authors foresee protection switching and service restoration functions which are unavoidable elements of a dependable access networking.

A significantly different and new Internet architecture, called SMART (Scalable Multi-Access Reconfigurable Transport), for end-to-end optical networking is proposed in (Zheng & Gumaste, 2006). The suggested network organization can be applied in WAN, MAN, LAN and also access networks using the same basic idea. The proposition is based on light-trails which are extensions of lightpaths. A light-trail corresponds to an arbitrary optical bus connecting several nodes. The architecture requires a reconfigurable (active) node architecture. The abstract model of the network topology corresponds to a hyper graph (or "hyper-network"), where nodes are connected with hyper-edges (or hyper-channels). Using k wavelengths on a bus corresponds to k hyper-channels between the given set of nodes. A hyper-path connecting a pair of source and destination nodes is a sequence of hyper-channels. To connect hyper-channels, SMART also proposes O/E/O junctions containing electronic router or switch components. With the help of reconfigurable nodes, the hyper-channels can be configured statically or dynamically. On the base of a given reference physical topology, this configuration can produce arbitrary hyper-network topologies which can be used from WANs to access networks anywhere (cf. *Figure 2*).

Hyper-channels are considered as shared medium, single-hop optical subnets. For a hyper-channel to be efficient, bandwidth allocation among traffic components using the channel must be provided. Fixed scheduling coordination (such as TDMA) or dynamic scheduling mechanisms can be used to ensure channel efficiency. The proposed mixed (optical-electrical) node architecture allows expensive optical components such as wavelength converters and splitters to be replaced by the cheapest electronic routers. The authors state that the solution is universal and scalable. So, a hyper-channel can be seen as a linearly arranged PON which is more scalable than a star-coupler based PON (Zheng & Gumaste, 2006).

In conclusion, cheap solutions are based on broadcast and select networks. However user-friendly and efficient these optical network architectures are, they should offer dependability and the possibility of dynamic reconfiguration. To provide dependable services, robust, general and efficient solutions should be used. This is increasingly true for new and forthcoming multicast-based multimedia services.

MULTICASTING IN SIMPLE ACCESS NETWORKS

To demonstrate the capacities and limits of star topologies we first present a brief introduction to the most important communication scheduling and multicast routing methods used in them. Because dependable multicasting needs route redundancies, the mesh topology appears to be the best solution. Due to its importance the problem of multicast routing in mesh networks is presented in the last section.

Channel Sharing in Passive Star Networks

Access nodes in optical access networks can have a number of tunable or tuned transmitters and receivers. Frequently there is only one transmitter and receiver in the access nodes and different configurations may exist depending on the tuning situation. For example, an FT-TR configuration indicates a fixed-tuned transmitter and a tunable receiver in the node. The optical channels of the fibers are distributed according to static or dynamic (tuned) wavelength allocation. Moreover,

time or code division based multiplexing can be used to improve channel sharing. In the following, we assume that the network operates in a slotted mode. Generally, messages should be *queued* at the source nodes for scheduling. To manage message transmissions, two main approaches have been proposed: single-hop and multi-hop.

Single Hop Networks

In the single-hop approach, each communication uses only one lightpath (or light-tree) from the source node to the destination(s) (Mukherjee, 1992a). The passive star coupler based architectures suppose tunable receivers and/or transmitters in the nodes. In some cases a reserved bidirectional control channel between a central scheduler and each node is required.

To transmit the queued messages, the network resources (transmitter of the source, receivers of the destinations and the wavelengths) should be allocated and configured in conformance with the communication requests. Numerous channel access methods are proposed. These access methods are often classified as random access based, reservation based and pre-allocation based methods.

Random scheduling implements a simple but efficient scheduling scheme. In random methods, when a channel becomes available, the scheduler randomly selects a source node that is waiting to send data. A given node may correspond to the destination of several messages at the same time. If the destination has only one tunable receiver, then a collision will occur and some transmissions will fail. Generally, in the case of failure the messages are retransmitted. In the case of multicast messages, all destinations should receive the messages. An analysis of two random scheduling of multicast requests can be found in (Modiano, 1999). In the proposed model, at each time slot, the W channels of the star network can simultaneously be used to transmit multicast messages, each channel intended for k randomly chosen nodes.

The optimum case is when the receivers of the destinations are tuned to the chosen wavelength, in other cases they are tuned to other wavelengths. In the first proposed strategy, a selected message is continuously retransmitted until it is received by all of its intended destinations. A second strategy consists of the introduction of a random delay before the retransmission of a message that was not received by all recipients. Performance evaluations show that this second strategy is more advantageous for overall network throughput (Modiano, 1999). Performance can be improved when several messages arrive at a node by tuning the receiver to receive the multicast message having the least number of destinations. Network utilization can also be significantly enhanced with multiple receivers in the nodes.

Reservation based scheduling dedicates channels exclusively for data transmission. For example, in (Wu, Ke, & Huang, 2007) potential senders use an ALOHA based random MAC scheme to send reservation requests to the central node. As reservation requests may collide and be lost, the reservation process needs an explicit confirmation. The scheduler (using its knowledge of the tuning time and delays) organizes asynchronous data transmissions between senders and destinations. A multicast scheduling algorithm called LBQA (Look Back Queue Access) is proposed. This algorithm favors multicast messages which can be sent immediately to all destinations. When there are no more all-receiver messages to transmit and while there are available data channels, the algorithm schedules also partitioned multicast messages (for an available subset of the destinations). The authors state that this scheduling algorithm can also be applied in PONs. The proposed architecture and the scheduling have some drawbacks. The scheduled time slot must allow sufficient time to tune the concerned transmitter and the receivers before data communication can start. This delay limits network performance. A large number of nodes in the domain can lead to heavy collisions

on reservation control channels. To diminish the number of collisions an architecture with two star coupler subnets bridged by two tunable pass band filters is proposed. The separation of the nodes into two sub-networks reduces the control load on each of them and improves the wavelength reuse possibilities in both sub-networks.

In *pre-allocation based channel access methods* the data channels (*i.e.* wavelengths) for transmission and reception are assigned to the nodes in advance. Thus, a control channel is not needed for resource allocation. Different communication flows using the same channel may share it via TDM-like multiplexing. The objective of the channel access method is to assign time slots of the different channels to the communication flows. If the bandwidth demand is uniformly distributed between flows, the simplest solution, a simple round-robin algorithm (each communication has a slot in a frame) results in very good performance (cf. the scheduling of *unicast* requests in (Bogineni, Sivalingam, & Dowd, 1993)). When the various communication flows need different bandwidths, the problem is finding an optimal scheduling which satisfies communication flow QoS requirements by minimizing the overall network mean packet delay. This optimization corresponds to an *NP*-hard load balancing problem. A typical scheduling algorithm for the pre-allocation method is presented in (Borella & Mukherjee, 1996). An efficient approximated algorithm is proposed for an arbitrary traffic pattern on any number of channels assuming an arbitrarily large transmitter tuning time.

The allocation problem for *multicast* traffic in WDM/TDM based star networks is presented in (Bianco, Galante, Leonardi, Neri, & Nucci, 2003). In the proposed broadcast and select network, transmitters operate on fixed wavelengths, while receivers can be tuned to any available wavelength. If there are more source nodes than available wavelengths, several communication flows share a wavelength. Wavelengths are slotted and synchronized; each slot on a wavelength can transmit one packet. By dynamically allocating the available slots (wavelengths are assigned to the sources), full connectivity can be achieved among nodes. The tuning times are assumed to be non-negligible with respect to the fixed size slot time. The problem is formulated as follows. The traffic pattern is given by a slot allocation request matrix R. An element $r_{s,D}$ corresponds to the number of packets which should be transmitted from the source s to the destination set D. The scheduling algorithm aims to find a time/wavelength assignment that satisfies the requests while minimizing the requested frame length. This latter corresponds to the total time necessary for the requested data transmission. It is immediately apparent that the tuning time of receivers has an impact on the optimal solution. The overall network throughput can be improved by minimizing the number of times each receiver must be tuned within a frame. Since the scheduling problem is *NP*-hard, the author proposes a heuristic algorithm based on the Tabu Search. Of course, the algorithm solves the off-line scheduling problem but cannot react quickly enough to assure the allocation on a packet-by-packet basis; only a slow variation of bandwidth can be tolerated in this solution.

Thus multicasting in star networks with passive couplers corresponds to a particular scheduling problem. The main difficulty with multicast is that the receivers of the destinations should be available (together or separately) to transmit multicast messages successfully. Large multicast trees can overload the network: reservations and/or retransmissions can block other requests. Moreover, dynamic tuning for every time-slot and the resultant latencies decrease overall network performance. Let us also notice that the messages have to be queued for scheduling purposes at the nodes of the PON. At the end users, this is not problematic: messages can be buffered electronically. At the OLT side, storage requires O/E/O conversion and as a result the communication between end points becomes opaque.

Multi-Hop Networks

Based on a physical star topology, virtual multi-hop topologies for optical access networks are proposed in (Mukherjee, 1992b). In these networks, the transmitters and the receivers of the access nodes are tuned in a fixed manner. Since the transmitter of the source node of a given communication flow can be tuned to a different wavelength from that used by the receiver of the destination(s), a route may contain different hops (lightpaths). In a multi-hop path, the wavelength of a (first) segment should be converted according to the tuning of the receiver(s). This wavelength conversion can be performed using O/E/O conversions of ONUs. The retransmission of the incoming light after conversions uses the transmitter of the ONU which is tuned for a different wavelength. So, the route from an arbitrary source to a destination may correspond to a multi-hop route and the virtual topology is a meshed graph. The diameter of this graph is limited. To perform multicast, multi-hop trees can be built by assembling the concerned hops in the directed virtual topology.

Figure 3 shows a physical star topology of seven nodes. Using two wavelengths in each direction and in each fiber, the regular virtual topology illustrated in *Figure 3(b)* can be configured using only fixed tuned receivers and transmitters. The used wavelengths are indicated with numbers between 1 and 14. A two-hop path from the node *a* to the node *g* is indicated with dotted lines.

Improvement of Access Network Performance

The optimal scheduling for heterogeneous unicast and multicast communications is a *NP*-hard problem. Moreover, the tuning time of transmitters/receivers and the synchronization requirement for multicast communication (*i.e.* all destinations should be available and tuned at the moment of data transmission) create scheduling difficulties. Some important propositions have been formu-

Figure 3. Physical (a) and virtual (b) topology of a multi-hop network

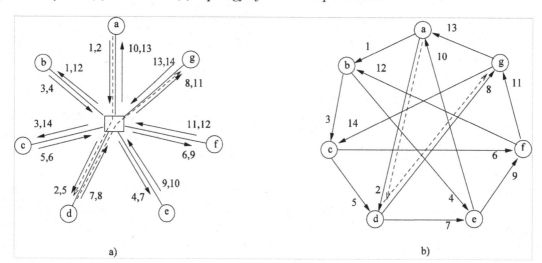

a)   b)

lated to resolve these problems.

To enhance the throughput of the network, the technique of *wormhole scheduling* can be applied. With this scheduling approach, several packets (and not only one) can be scheduled in order to minimize the overall tuning time. This technique can be applied both for unicast and multicast communications. Another idea to improve network performance is *pipelining of the tuning latency* by permitting data transmission for some nodes and transmitter/receiver tuning for others (Borella & Mukherjee, 1996; Tridandapani, Meditch, & Somani, 1994). The synchronization of all receivers belonging to a given multicast group can also have a significant affect on latency. Partitioning multicast communication (Jue & Mukherjee, 1997) aims to reduce this latency. In this solution, the multicast message is sent even if all the destinations are not ready. To cover the entire group, the same message is transmitted several times until all destinations have received the message.

Traffic is frequently burst transmitted which occasionally leads to some idle wavelengths, while other wavelengths are overloaded. This results in an inefficient use of network capacity. In (Papadimitriou & Pomportsis, 1999), the authors propose the application of tunable wavelength

converters placed at the network hub and a new MAC protocol which is capable of scheduling the incoming packets to the available wavelengths. With the help of wavelength conversions, the packet load is balanced between the wavelengths and consequently network performance is improved.

Multicast Routing in Ring Topologies

To perform multicast communications in ring topologies, splitters are not needed as long as *Tap-and-Continue* capability (*TaC*) (Ali & Deogun, 2000) exists in all ONUs. Generally, with the *TaC* capability of ONUs, one lightpath per multicast group is sufficient to cover any destination of ONUs. Using the TDMA scheme, several multicast groups can share the same lightpath to exploit its capacity. Let us notice that routes in a ring topology can be easily protected if the ring can be used in both directions.

Figure 4 illustrates a ring access network topology. Let us suppose that nodes *d, f* and *g* belong to a given multicast group. Taking advantage of the *TaC* capabilities of the traversed member nodes, the lightpath indicated with arrows is sufficient to supply all members in the ring.

Figure 4. Multicasting in a Ring Topology

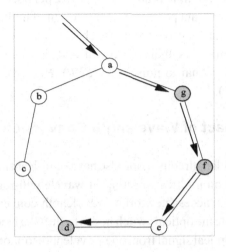

Multicast Using Active Elements

Active components in the last segment of optical networks can improve overall network performance. PON developers focus on integrating high performance active components into OLTs and ONUs that are located at both ends of the access network. For example in (Han, Kim, & Chung, 2001) the authors state that the scalability of multi-purpose fiber-optic access network can be improved significantly by using active components at the remote nodes. Unlike passive access networks, an architecture which includes active end nodes can support a large number of optical network units.

The authors in (Kim, Choi, Im, Kang, & Kevin Rhee, 2007) propose a switching system using fast time-slotted passive switching with O/E/O conversion and shared buffers. This system provides more flexible routing and significantly reduces the blocking probability by using electrical buffers. This optical access network facilitates multicast routing because the electrical buffer equipped switches can split the messages arbitrarily and any of these nodes may correspond to a branching node of multicast trees. Moreover optical amplification is not needed in this kind of node. The proposed switching system is believed to be a

techno-economically feasible and implementable solution for both optical packet and burst switching with current optical technologies.

Multicast routing in redundant (ring and mesh) networks provides a dependable solution for multicasting even in access networks. In these kinds of topology, efficient multicast routing is a challenging task which must also take into account the physical constraints of the optical network. The most important and specific constraints in optical routing and the most common algorithms are presented in the following sections.

CONSTRAINTS OF MULTICAST ROUTING IN ALL OPTICAL NETWORKS

Impact of Multi-Optical Channels

Wavelength-rooted networks operate based on the concept of lightpath and light-tree (He, Chan & Tsang, 2002). A lightpath is an all-optical communication channel between two end nodes, established by allocating the same wavelength throughout the route of the transmitted data. The light-tree is an extension of the lightpath which consists of multiple lightpaths on the same wavelength from the source to several destinations. The use of multiple wavelength channels on mesh topology precludes the use of several conventional multicasting techniques in IP networks (Hamad, Wu, Kamal, & Somani, 2006). Firstly, in the absence of a wavelength conversion device, it is required that the same wavelength be employed over the entire route (*i.e.* on the lightpath and light-tree). This is known as the wavelength continuity constraint (Mukherjee, 2000). Notice here that wavelength continuity must be satisfied both in depth due to signal propagation on the lightpath, and in breadth due to multicasting and signal branching in the light-tree. Channels on different fibers therefore cannot be treated independently, as is the case for multicasting in conventional IP networks. Secondly, two or more lightpaths and light-trees traversing the same fiber link cannot be assigned the same wavelength; otherwise they will interfere with one another. This requirement is known as the distinct wavelength constraint (Mukherjee, 2000). However, two lightpaths can share the same wavelength if they use disjoint sets of fiber links. This property is known as wavelength reuse (Murthy & Gurusamy, 2002).

Impact of Light Splitting

The capability to split light is a key enabling technology for multicast communication in wavelength-routed networks (Hamad et *al.*, 2006). Light splitting can be realized with a passive optical device called an optical power splitter (Mukherjee, 1997) which is able to replicate the incoming light signal in the optical domain and thus transmit it to several outgoing ports simultaneously without any O/E/O conversion. Splitters maintain optical signal transparency and also eliminate the need for the buffers usually required for data duplication in the electronic domain. However, the power splitter degrades signal power and causes crosstalk. Furthermore, due to the complicated architecture and expensive components, optical switches with power splitters are always more costly to build than those without. Hence, typically only a subset of optical switches support light splitting and such a network is characterized as a *sparse splitting* network (Malli, Zhang, & Qiao, 1998). Usually, an optical node with a light splitting capability is called a *Multicast Capable* (*MC* node), otherwise it is called a *Multicast Incapable* node (*MI* node) (Malli et *al.*, 1998). In addition, the *TaC* capacity is assumed to be available at all *MI* nodes. This refers to tapping a small amount of the power for signal detection from the incoming light signal, and forwarding the light signal to only one outgoing port.

Moreover, the splitting fanout also influences multicast routing in wavelength-routed optical access networks. The splitting fanout is the maximum number of light branches supported per node. It is an important parameter in the design of multicast trees and it also impacts the choice of the number of amplifiers, their placement, and also the value of the signal-to-noise ratio (*SNR*) (Hamad et *al.*, 2006).

Impact of Wavelength Conversion

Wavelength conversion also has a significant influence on multicast routing in wavelength-routed optical access networks. Wavelength converters enable the optical switch nodes to shift the incoming optical signal from one wavelength to another. Wavelength conversion functionality provides flexibility in network operation and simplifies multicast routing, since wavelength continuity is no longer a strict requirement if converters are used. All-optical wavelength converters (Elmirghani & Mouftah, 2000), however, are still very expensive and immature. As is the case with the power splitter, the architectures of optical switches equipped with all-optical wavelength converters are very complicated to design and therefore costly. This hinders the full deployment of wavelength converters. Hence, an optical access network where only some nodes are equipped with full wavelength conversion capability is more practical, and is referred to as a *sparse wavelength conversion* network. As a result, the limited availability of wavelength conversion restricts the construction of multicast trees.

Impact of Optical Amplification

When a light signal passes through a k-out power splitter, it is equally divided into k light beams and forwarded to different outgoing ports. The power of the light signal at each output port is only $1/k$ of the incoming light signal and thus is significantly degraded. In addition, power loss is also caused by power attenuation during light propagation. For a multicast light signal to be detected by all session members its transmission power must be carefully designed to guarantee a satisfactory *SNR* at the end users. Otherwise the maximum distance from the source to the destinations, namely the diameter of the multicast trees, will be affected and bounded due to light power loss. To minimize the impact of power loss when constructing multicast trees, active optical amplification devices such as the erbium-doped fiber amplifier (*EDFA*) (Desurvire, 1991) are required. However, optical amplifiers are expensive to fabricate and introduce many problems which complicate network management such as *Gain Dispersion, Gain Saturation* and *Noise* (Yan, Deogun, & Ali, 2003). Moreover, placing amplifiers on fiber links will increase the possible number of potential multicast receivers. However, the total number of amplifiers in the network can be reduced by an optimal placement strategy. To solve the optimal amplifier placement problem, at least two parameters, namely signal power and source-destination distance, should be given.

Due to the optical constraints discussed above, multicast routing algorithms in all-optical mesh access networks are different from those in traditional data networks; thus a great deal of research has been done to solve this challenging problem.

MULTICAST ROUTING IN OPTICAL MESH CORE AND ACCESS NETWORKS

Wavelength-division multiplexing (WDM) is an effective technique to exploit the large bandwidth of optical fibers and to meet the explosive growth of bandwidth demand in networks (He, Chan, & Tsang, 2002). Furthermore, the light signal in all-optical networks is optically switched without any O/E/O transition, thus optical routing results in very low latency. WDM networks therefore have the capability to support bandwidth-driven and time sensitive multicast multimedia services with a high level of QoS. The light-tree concept was first proposed in (Sahasrabuddhe & Mukherjee, 1999). However, due to the physical constraints

discussed in the previous section it is very hard to build such an all-optical light-tree. Then, in (Zhang, Wei, & Qiao, 2000), the light-forest is employed to solve the multicast problem in sparse splitting WDM networks. However, a recent work shows that more advantageous routing structures can be obtained using *light-hierarchies* (Molnár, 2008).

It is proved that the computation of the optimal multicast tree under optical constraints is *NP*-hard. Therefore, many heuristics have been proposed for the formation of light-trees to satisfy specific requirements. Typically, the network resource utilization and the power budget are taken into account.

Costs & Delay Sensitive Multicast Routing

Many existing routing algorithms focus on network costs and delay. For simplicity, the same cost is assumed for different wavelengths on different links, and hence hop count is used to calculate the wavelength channels and the delay, etc. Generally, these algorithms are evaluated in terms of link stress (the number of wavelengths required), wavelength channel cost (the number of wavelength channels used), average delay (the average hop counts from the source to the destinations) and the diameter of the multicast light-trees (the maximum number of hop counts from the source to the destinations). Existing multicast algorithms can be classified into two categories according to the technique used to construct the multicast tree. The first technique could be called the post processing or adaptation method. Firstly, it constructs a multicast tree for the multicast members without considering any constraints. It is always a shortest path tree or a tree approximated to the Steiner tree. Then, some adaptations will be made to this tree in order to satisfy the optical constraints. The second technique could be called the direct method, which takes into account optical constraints when building the multicast tree.

This kind of routing algorithm directly produces a light-tree or a light-forest.

Adaptation Methods

The post processing method always divides the construction of the multicast tree into three stages: (*i*) construct a multicast tree without considering any constraint, (*ii*) check the splitting capability of the nodes on the tree and (*iii*) reconnect the multicast forest. Three typical post processing methods namely Re-route-to-Source, Re-route-to-Any (Zhang, Wei, & Qiao, 2000) and Avoidance-of-MIB-Nodes (Zhou, Molnár & Cousin, 2008a) will be discussed in the following with some illustrative examples.

Re-route-to-Source & Re-route-to-Any
Initially, a multicast tree is constructed using any existing algorithm (*e.g.* a shortest path tree formed by Dijkstra's algorithm). Then, its nodes are checked one by one in the breadth-first or the depth-first order. If node *v* is an *MI* node and it has at least two children in the tree, then only one of them is kept (no heuristic is specified to choose which branch to keep in the algorithm (Zhang et al., 2000)) while all the other downstream branches are cut from *v*. The affected children of *v* re-join the forest either via the nearest *Virtual Source* node (*VS*, a Virtual Source node is capable of both splitting and wavelength conversion) along the reverse shortest path to the source (Re-route-to-Source), or via any other path leading to a *MC* node or a leaf *MI* node already in the cut tree (Re-route-to-Any).

Avoidance-of-MIB-Nodes
The adaptation algorithm proposed in (Zhou et al., 2008a) has three important advantages:

i.   It results in a shortest path tree with fewer Multicast Incapable Branching (MIB) nodes (decreasing up to 38% in some networks). This gain is obtained with the help of an enhanced version of Dijkstra's algorithm where *MC* nodes have a higher priority to compute shortest path than the other candidate nodes at the same *level* (*i.e. Candidate* nodes are at the same level when they are at the same shortest distance to the source), and with the help of a special *Node Adoption* procedure. In the adoption procedure, when all *Candidate* nodes at the same *level* are permanently labeled, a child is adopted from an *MI Candidate* node with several children to another leaf *MI Candidate* node without children at the same level if possible (cf. our example).

ii.   It aims to reduce link stress. In the second phase, when *MIB* nodes are processed, some branches of the tree are kept even if their root is an MIB node. If an *MIB* node is a critical articulation node of a branch (this can be very important when some nodes in the network fail), then this branch is kept, otherwise the deepest downstream branch will be kept. So, critical branches are left untouched.

iii. In the reconstruction phase, distance priority mechanisms are employed to reduce delay and diameter. An example in the well known NSF network is now considered (*Figure 5*). Let

$m=\{source: 10 \,|\, members: 1\text{-}14\}$ be a multicast session, where nodes 1, 8 and 10 are *MC* nodes. The traditional Dijkstra algorithm may produce a shortest path tree like that in *Figure 6*(a). There are 2 *MIB* nodes (node 6 and 12) in this shortest path tree. They are only able to feed one branch and the other branches must be cut. According to Re-route-to-Source, the affected nodes 3 and 13 should be connected to the source using the shortest path on another wavelength, thus two light-trees respectively using wavelengths w_1 (dash and dot line) and w_2 (dot line) can be obtained in *Figure 6(a)*. Meanwhile, with the Re-route-to-Any algorithm, the light-tree shown in *Figure 6(b)* may

Figure 5. NSF network

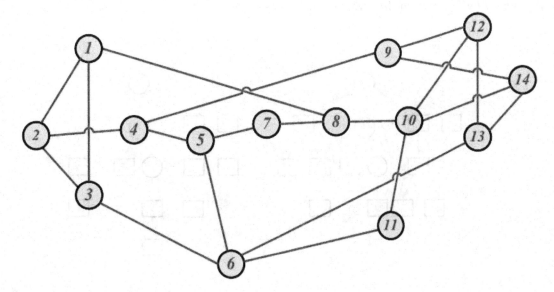

be constructed (for instance, node 2 is the closest connect node to node 3 and node 14 is the closest connect node to node 13). The Avoidance-of-*MIB*-Nodes algorithm can produce an even better result. In the shortest path tree shown in *Figure 6(a)*, we can see that nodes 1, 6, 7, 9 and 13 have the same shortest distance to source node 10. So, they can be viewed as *candidate* nodes. And, if node 1 (*MC* node) is raised to a higher priority and is chosen to be permanently labeled first, followed by 7, 9, 13 and 6, then the new shortest path tree of *Figure 7(a)* is produced which has only one *MIB* node. It is obvious that nodes 11, 12 and 14 have the same shortest distance to source node 10. Hence, they can be viewed as candidate nodes. When all of them have been permanently labeled, we can see that node 12 is an *MIB* node and node 14 is a leaf *MI* node. Note that nodes 13 or 9 can reach source node 10 by the shortest path through both nodes 12 and 14. One of them can be adopted by node 14, and a new shortest path tree without an *MIB* node is obtained in *Figure 7(b)*. Its link stress is 1 and cost is 13, while it is 2 and 16 respectively for Re-route-to-Source (in *Figure 6(a)*). Its average delay and the diameter of tree (2.0 and 3 respectively) are also better than

those of Re-route-to-Any (2.1 and 4 respectively, in *Figure 6(b)*).

Among these three adaptation algorithms, the Re-route-to-Source algorithm is able to produce the optimal average delay and the minimal diameter for the multicast tree. However, its total cost and the link stress are the worst. The Avoidance-of-*MIB*-Nodes algorithm outperforms the Re-route-to-Any algorithm in terms of link stress, average delay and multicast tree diameter.

Direct Methods

In the direct method, the light splitting and wavelength conversion capabilities of nodes are considered while spanning the multicast tree. The resulting trees already satisfy the optical constraints, thus no adaptation processing is required. However, while respecting the optical constraints, it is possible that a single light tree may not always be able to span all the destinations. As a result, several light-trees may be required to accommodate a multicast session. Here, we present three direct light-forest constructions, where the light-trees are constructed one by one: the Member-Only algorithm (Zhang et *al.*, 2000)),

Figure 6. (a) Shortest path tree, and the multicast tree constructed by Re-route-to-Source. (b) The multicast tree built by Re-route-to-Any. (Zhou et al., 2008a).)

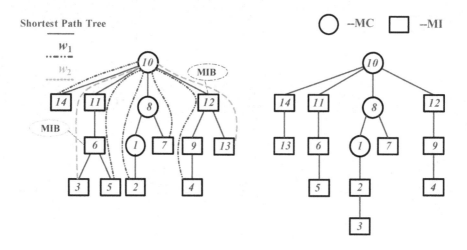

the Distance-Priority-Based algorithm (Zhou, Molnár & Cousin, 2008b) and the Virtual-Source-Capacity-Based algorithm (Sreenath, Satheesh, Mohan & Siva Ram Murthy, 2001). In order to facilitate the description of the algorithms, a number of notations should be introduced first. Let *MC_SET* denote the set of *MC* nodes and leaf *MI* nodes that are currently on the multicast tree under construction. The nodes in *MC_SET* may be used to connect unvisited (not yet spanned) destinations to the tree, because their splitting capability has not been exhausted. Let *MI_SET* be the set of non-leaf *MI* nodes on the current

multicast subtree. They are not capable of connecting any other node to the current subtree due to their splitting limitation. *VS_SET* consists of the virtual source nodes on the current multicast subtree. These nodes have both light splitting and wavelength conversion capacities. Finally, *UD* consists of the unvisited destination nodes of the multicast session.

Member-Only

The Member-Only algorithm is an adaptation of the famous *Minimum Path Heuristic* (Takahashi & Matsuyama, 1980) that respects the splitting

Figure 7. (a) Priority assignment. (b) Node adoption. (Zhou et al., 2008a)

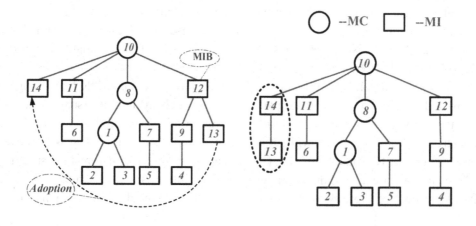

capability constraint on the optical nodes. It begins to build the multicast tree from the source and includes the destination nodes one by one. At each step, the nearest destination to the tree under construction is selected and joined to the tree through the shortest path as long as this path does not pass through any node in *MI_SET*. Since the nodes in *MI_SET* have no capacity to connect other nodes to the current tree, the algorithm only needs to try to find the shortest path $SP(c,d)$, where $c \in MC\_SET$, $d \in UD$, which does not involve any node in *MI_SET*. If such a shortest path is found, it is added to the subtree and the node sets are updated along the path; all *MC* nodes are added to *MC_SET* and the formerly leaf *MI* node is removed from it; all non-leaf *MI* nodes are added to *MI_SET*, and the newly added destination is removed from *UD*. When no shortest path satisfying the constraints can be found the current multicast tree is finished, and another multicast tree is started using the same procedure until no destination remains in *UD*.

Distance-Priority-Based Algorithm

This algorithm improves the Member-Only algorithm by attempting to diminish the average delay and diameter of the multicast trees while keeping almost the same link stress. It introduces two distance priority mechanisms in the construction of multicast light-trees. At each step of the Member-Only algorithm there can be several constraints-satisfied shortest paths found, say $SP_i(c_i,d_i)$ and $SP_j(c_j,d_j)$, where $c_i, c_j \in M\_SET$, $d_i, d_j \in UD$ and $dist(SP_i) = dist(SP_j)$. The choice of the path to be joined to the multicast tree will in fact greatly affect the final tree. Unlike Member-Only, where the nearest destination is selected randomly when several nearest destinations are found, this algorithm preferentially connects the candidate destination, say $d_{nearest}$, to T earlier, which is the nearest to the source (*destination priority*). Furthermore, at each step, there may exist several connector nodes in *MC_SET* at an equal distance to the selected candidate destination $d_{nearest}$. It is

advantageous to connect the destination $d_{nearest}$ via the connector node closest to the source in the tree (*connect node priority*). The algorithm results in a great reduction of the average delay and the diameter of light-trees, for instance up to 50% and 51% respectively in the USA Longhaul network (Zhou et al., 2008b).

Here, we use a simple example to show the difference between the Member-Only and the Distance-Priority-Based algorithm. A multicast session m= {*source*: 2 | *members*: 2~6} is required. Initially, source node 2 is in the subtree T. At each step, the nearest destination is added. With the Member-Only algorithm the light-tree in *Figure 8(a)* can be produced. It is interesting to note that when adding node 4, it could be connected via either the source 2 or node 5. If we use *destination priority*, it should be connected to the source as shown in *Figure 8(b)*. Still with regard to this graph, if node 4 (1 hop to the source in NSFNET) is added to T earlier than node 5 (2 hops to the source in NSFNET), then node 5 could be connected to node 4 as shown in *Figure 8(c)*, which corresponds to *connect node priority*. We compare the average delay and the diameter of the light-tree resulting from these three results in *Table 1*. It is apparent that the reduction in the delay and the diameter is significant while the link stress remains the same.

Virtual-Source-Capability-Based Algorithm

This algorithm can be viewed as an enhancement of the Member-Only algorithm. The enhancement derives from two heuristics, namely Spawn-from-VS and Capability-based-Priority. The network is assumed to have nodes with different capabilities, namely splitting (*MC*), wavelength conversion (*WC*), Tap-and-continue (*TaC*) and splitting plus wavelength conversion (*VS*) nodes. A priority is assigned to the nodes depending on their capabilities in the following descending order: *VS*, *MC*, *WC* and *TaC*. The node with the highest priority is used when a destination needs to be included in the tree and is equally distant to more than

Figure 8. (a) Member-Only. (b) Destination Priority. (c)Connect Node & Destination Priority

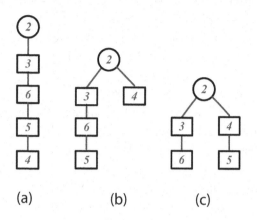

(a)   (b)   (c)

one node in the *MC_SET* for the current tree. In addition, if no constraints-satisfied shortest path from a destination *d* to the members in *MC_SET* could be found, the algorithm tries to find the nearest *VS* node $z \in VS\_SET$ for the current tree. If $dist(d,z) < dist(d,s)$, then *d* is joined to the current tree on another wavelength via *z*. Otherwise, a new tree is needed. This algorithm reduces the number of wavelength channels required and improves network resource utilization by using wavelength converters.

Power Aware Multicast Routing

As mentioned in the previous section, a light signal suffers power loss due to light splitting. Moreover, light attenuation along the long route from the source to the multicast members may not be negligible. Consequently, routing schemes should be carefully designed to guarantee signal delivery to multicast members, thus the multicast

routing problem in a sparse splitting network with power considerations is an important topic. Next, we present two power aware multicast routing algorithms: Centralized-Splitting algorithm (Wu, Wu, & Yang, 2001) and Balanced-Light-Tree (Xin & Rouskas, 2004).

Centralized-Splitting Algorithm

This algorithm aims to build a Steiner-based tree that achieves an efficient utilization of network resources while producing low power loss in order that the transmitted light signal is maintained above the signal sensitivity threshold. Initially, a multicast diffusion tree is constructed by applying the Member-Only algorithm without any consideration of the power-level impairment. Then some adjustments are made in the tree produced according to the following guidelines. Firstly, if there are more than two successive *MC* nodes in a subtree, they will produce a cascade effect on power loss (as indicated in *Figure 9(a)*). Hence, it is better to replace the successive *MC* nodes by a single *MC* node. Secondly, although a power splitter located near to the source can balance the power loss on each subtree, the effect of the power loss will be propagated to all children nodes located within its subtree. In order to reduce power loss, the algorithm assigns the splitting capability to the node furthest from the source node whenever possible. For instance in *Figure 9(b)*, the light splitting happens in the last level of the tree; hence the power loss decreases to $2e_0/3$ compared to the cascade splitter situation with $3e_0/4$ in *Figure 9(a)*. Thirdly, when the number of splittings at a node increases, the incremental power loss caused by each additional splitting decreases. As a result,

Table 1. Comparison of Light-trees in Figure 8

	Member-Only	Destination Priority	Two Priorities
Link Stress	1	1	1
Diameter	4	3	2
Average Delay	2.5	1.75	1.5

Figure 9. (a) Cascade Power Loss. (b) Splitting far from the Source

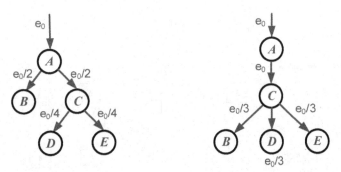

if a node is chosen to be a branching node in the multicast tree it is desirable to assign as many splittings as possible to this node.

Balanced-Light-Tree Algorithm

In the Balanced-Light-Tree (*BLT*) algorithm, it is assumed that signal attenuation is negligible while power loss due to light splitting is the dominant factor. Hence, the power loss imposes an upper bound on the splitting ratio on the path to each destination node. Furthermore, the splitting ratios of any two paths from the source to two destination nodes of the same multicast group should be within a tight range of each other. In other words, the multicast trees must be as balanced as possible. This is because an unbalanced tree results in two important disadvantages. Firstly, it is unfair for certain destination nodes, since the destination node at a smaller depth receives a better quality signal than the one at a large depth. Secondly, it is not scalable, since it may introduce excessive losses that make it impossible to deliver a light signal to a large destination set. Similarly to the Centralized-Splitting algorithm, an initial multicast tree spanning all multicast members is built by any existing algorithm such as Member-Only. Then, the balancing procedure is performed on the tree to check the splitting ratio of the nodes. Consider an intermediate multicast tree T, and let u (respectively v) denote the leaf node with maximum (respectively minimum) splitting ratio. The main idea behind the *BLT* algorithm is to delete

node u from T and then add it back to the tree by connecting it to some node y in the path from source s to v. This procedure reduces the splitting ratio of v, though it increases the splitting ratio of all nodes below y in the tree. Thus, it is desirable to perform this pair of delete/add operations as long as it does not increase the splitting ratio of any node beyond node u. It is worth noting that the difference between the maximum and minimum splitting ratio values decreases after the balancing operation.

On the Optimality of Multicast Routes in WDM Networks

In this section, we examine optimal routing structures for multicast communications under splitting constraints in meshed WDM networks. For source based multicast routing current practice is to propose light-trees. We shall see next that optimal multicast routing structures do not always correspond to trees. Let us suppose that the links can be used in both directions and the topology of the optical network is given by an undirected graph $G=(V, E)$. The multicast group is given by a source node s and a set D of destination nodes. A multicast route, a directed sub graph spanning the source and the set of the destinations is required. Remember that MC_SET contains the splitting capable nodes (accordingly, the nodes in $MI\_SET = V\backslash MC\_SET$ cannot duplicate the light). So, only the nodes in MC_SET can have a degree

Figure 10. Light-Tree and Light-Hierarchies

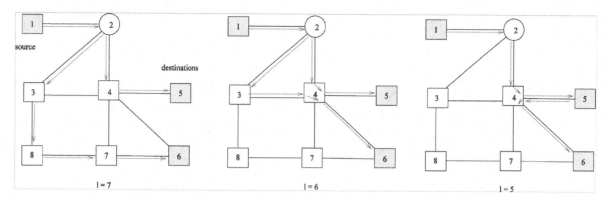

greater than two in the multicast routing structure. Wavelength constraint in the fiber implies that two arcs of the minimal length structure cannot use the same edge of the topology graph in the same direction. But in general the same switch can be used twice (or more) because the switch architecture offers several disjoint lightpaths between its interfaces. The optimal solution must have the minimal length of all sub graphs spanning $s \cup D$ and must satisfy the constraints. This connected and minimal length solution is not always a tree or a forest: the same switch can belong several times to the optimum structure which is called a hierarchy (Molnár, 2008). To illustrate this, let us consider the example in *Figure 10*. In this example the source node is node 1 and the destinations are the nodes 5 and 6. In the given network only node 2 can duplicate light and can be selected as a branching node. The light-tree with minimal length is shown in *Figure 10(a)*. The length of this tree is equal to 7. More advantageous spanning structures can be obtained by relaxing the tree construction constraint. *Figure 10(b)* illustrates an optimal hierarchy which uses node 4 twice when the wavelength is unique in each link. If the links can be used in both directions (there is bi-directional fiber or two fibers between the switches), the minimal length hierarchy corresponds to a light-trail illustrated in *Figure 10(c)*.

CONCLUSION

This chapter focused on multicast routing over an optical access network. Our first point was intended to demonstrate that in the future many optical access networks will have meshed topology, and will require automatic and smart management. Indeed, mesh topologies have inherently good properties: flexibility and dependability. A mesh topology can freely evolve with technology and with users' requirements and may be as redundant as required. Dependability becomes a very important factor in access networks because of the increasing variety of services which must be reliable.

Our second point was that data broadcasting to a specific set of end users over a certain network domain will increase with the development of new multimedia distributed applications, and the use of multicasting can lead to huge savings in network resources. However, due to the specific physical constraints which can be found in optical networks, the computation of efficient light trees is not a trivial task. For instance, some optical switches have to be selected as branching nodes and have to split the light and the power of the transmitted light has to be intelligently controlled to ensure sufficient signal level at the final receivers. Thus the relevant multicast routing algorithms which can be applied to meshed optical access networks under the physical constraints presented by the network were described.

One surprising concluding point is that the optimal multicast structure is not always a tree or a forest. Indeed the analysis shows that if you try to find the most efficient multicast structure which spans all desired destinations, taking into account the numerous optical constraints, it will lead to a hierarchical structure. In this hierarchy some switches may be used several times to transmit light to the destinations. So, the tree search constraint in the different route computation algorithms can be relaxed and more efficient hierarchies can be found for multicast routing.

Our last concluding remark is the following. To provide strong dependability, a high level of flexibility and to be as efficient as possible in network resource utilization, we forecast that some integration between access and core optical networks will be required. That will necessitate some network management and control coordination. For instance, some of the solutions developed for burst or packet optical networks could be reused in optical access networks.

REFERENCES

Ali, M., & Deogun, J. S. (2000). Cost-Effective Implementation of Multicasting in Wavelength-Routed Networks. *IEEE Journal of Lightwave Technology, 18*(12), 1628–1638. doi:10.1109/50.908667

Bianco, A., Galante, G., Leonardi, E., Neri, F., & Nucci, A. (2003). Scheduling Algorithms for Multicast Traffic in TDM/WDM Networks with Arbitrary Tuning Latencies. *Computer Networks, 41*(6), 727–742. doi:10.1016/S1389-1286(02)00436-X

Bogineni, K., Sivalingam, K. M., & Dowd, P. W. (1993). Low-Complexity Multiple Access Protocols for Wavelength-Division Multiplexed Photonic Networks. *IEEE Journal on Selected Areas in Communications, 11*(4), 590–604. doi:10.1109/49.221206

Borella, M. S., & Mukherjee, B. (1996). Efficient Scheduling of Nonuniform Packet Traffic in a WDM/TDM Local Lightwave Network with Arbitrary Transceiver Tuning Latencies. *IEEE Journal on Selected Areas in Communications, 14*(5), 923–934. doi:10.1109/49.510916

Desurvire, E. (1991). *Erbium-Doped Fiber Amplifiers: Principles and Applications.* New York, NY, USA: Wiley.

EICTA. *(2008).* Position on Next Generation Networks (NGN) & Next Generation Access (NGA):"Moving Towards a Very High Speed Europe

Elmirghani, J. M. H., & Mouftah, H. T. (2000). All-Optical Wavelength Conversion: Technologies and Applications in DWDM Networks. *IEEE Communications Magazine, 38*(3), 86–92. doi:10.1109/35.825645

Hamad, A., Wu, T., Kamal, A. E., & Somani, A. K. (2006). On Multicasting in Wavelength-Routing Mesh Networks. *Computer Networks, 50*(1), 3105–3164. doi:10.1016/j.comnet.2005.12.012

Han, K. H., Kim, H., & Chung, Y. C. (2001). Active Multi-Purpose Fiber-Optic Access Network. In *Asia-Pacific Optical and Wireless Communications Conference* (pp. 31-37). Beijing, China.

He, J.Y., Gary, Chan, S.H., & Danny Tsang, H.K. (2002). Multicasting in WDM Networks. *IEEE Communications Surveys & Tutorials, 4*(1), 2-20.

Hossain, A. D., Dorsinville, R., Ali, M. A., Shami, A., & Assi, C. (2006). Ring-based Local Access PON Architecture for Supporting Private Networking Capability. *Journal of Optical Networking, 5*, 26–39. doi:10.1364/JON.5.000026

Hossain, A. D., Erkan, H., Hadjiantonis, A., Dorsinville, R., Ellinas, G., & Ali, M. A. (2008). Survivable Broadband Local Access PON Architecture: A New Direction for Supporting Simple and Efficient Resilience Capabilities. *Journal of Optical Networking, 7*, 645–661. doi:10.1364/JON.7.000645

Jue, J. P., & Mukherjee, B. (1997). The Advantages of Partitioning Multicast Transmissions in a Single-Hop Optical WDM Network. In *ICC '97, 1*, 427-431.

Kim, J., Choi, J. Y., Im, J., Kang, M., & Kevin Rhee, J. K. (2007). Novel Passive Optical Switching Using Shared Electrical Buffer and Wavelength Converter. In *ONDM* (pp. 101-106).

Kramer, G., Mukherjee, B., & Pessavento, G. (2001). Ethernet PON (epon): Design and Analysis of an Optical Access Network. *Photonic Network Communications, 3*(3), 307–319. doi:10.1023/A:1011463617631

Kuznetsov, M., Froberg, N. M., Henion, S. R., Rao, H. G., Korn, J., & Rauschenbach, K. A. (2000). A Next-Generation Optical Regional Access Network. *IEEE Communications Magazine, 38*, 66–72. doi:10.1109/35.815454

Malli, R., Zhang, X. J., & Qiao, C. M. (1998). Benefits of Multicasting in All-optical Networks. In *SPIE Proceedings of All Optical Networking* (pp. 209-220).

Mayhew, A. J., Page, S. J., Walker, A. M., & Fisher, S. I. (2002). Fiber to the Home Infrastructure Deployment Issues. *BT Technology Journal, 20*(4), 91–103. doi:10.1023/A:1021374532690

Modiano, E. (1999). Random Algorithms for Scheduling Multicast Traffic in WDM Broadcast-and-Select Networks. *IEEE/ACM Transaction on Networking, 7*(3), 425-434.

Molnár, M. (2008). *Hierarchies for Constrained Partial Spanning Problems in Graphs* (Tech. Rep. No. 1900). Rennes, France: IRISA.

Mukherjee, B. (1992a). WDM-Based Local Lightwave Networks. I. Single Hop Systems. *IEEE Network, 6*(3), 12–27. doi:10.1109/65.139139

Mukherjee, B. (1992b). WDM-Based Local Lightwave Networks. II. Multihop Systems. *IEEE Network, 6*(4), 20–32. doi:10.1109/65.145161

Mukherjee, B. (1997). *Optical Communication Networks*. New York, NY, USA: McGraw-Hill.

Mukherjee, B. (2000). WDM Optical Communication Networks: Progress and Challenges. *IEEE Journal on Selected Areas in Communications, 18*(10), 1810–1824. doi:10.1109/49.887904

Papadimitriou, G. I., & Pomportsis, A. S. (1999). Centralized Wavelength Conversion Protocols for WDM Broadcast-and-Select Star Networks. In *Proceedings of the 7th IEEE International Conference on Networks* (pp. 11-18). Washington: IEEE Computer Society.

Qiao, C. M., & Yoo, M. (1999). Optical Burst Switching (OBS) - A New Paradigm for an Optical Internet. *Journal of High Speed Network, 8*(1), 69–84.

Ram Murthy, C. S., & Gurusamy, M. (2002). *WDM Optical Networks: Concepts, Design, and Algorithms*. USA: Prentice-Hall, Inc.

Sahasrabuddhe, L. H., & Mukherjee, B. (1999). Light-trees: Optical Multicasting for Improved Performance in Wavelength-Routed Networks. *IEEE Communications Magazine, 37*(1), 67–73. doi:10.1109/35.747251

Sheu, S. T., & Huang, C. P. (1997). An Efficient Multicast Protocol for WDM Star-coupler Networks. In *the 2nd Symposium on Computers and Communications* (p. 579). IEEE Computer Society.

Sivalingam, K. M., Bogineni, K., & Dowd, P. W. (1992). Pre-allocation Media Access Control Protocols for Multiple Access WDM Photonic Networks. *ACM Sigcomm Computer Communication Review, 22*(4), 235–246. doi:10.1145/144191.144289

Sreenath, N., Satheesh, K., Mohan, G., & Siva Ram Murthy, C. (2001). Virtual Source Based Multicast Routing in WDM Optical Networks. *Photonic Network Communications, 3*(3), 213–226. doi:10.1023/A:1011443013088

Takahashi, H., & Matsuyama, A. (1980). An Approximate Solution for the Steiner Problem in Graphs. *Mathematica Japonica, 24*, 573–577.

Tridandapani, S., Meditch, J. S., & Somani, A. K. (1994). The Matpi Protocol: Asking Uning Times Through Ipelining in WDM Optical Networks. In *INFOCOM'94* (pp. 1528-1535).

Wu, H. T., Ke, K. W., & Huang, S. Y. (2007). A Novel Multicast Mechanism for Optical Local Area Networks. *Computers & Electrical Engineering, 33*(2), 94–108. doi:10.1016/j.compeleceng.2006.08.002

Wu, K. D., Wu, J. C., & Yang, C. S. (2001). Multicast Routing with Power Consideration in Sparse Splitting WDM Networks. In *ICC2001, 2,* 513-517. Helsinki, Finland: IEEE Communication Society.

Xin, Y. F., & Rouskas, G. N. (2004). Multicast Routing Under Optical Layer Constraints. In *INFOCOM2004, 4,* 2731-2742). IEEE Communication Society.

Yan, S. G., Deogun, J. S., & Ali, M. (2003). Routing in Sparse Splitting Optical Networks with Multicast Traffic. *Computer Networks, 41*(1), 89–113. doi:10.1016/S1389-1286(02)00345-6

Yeh, C. H., Lee, C. S., & Chi, S. (2008). Simply Self-Restored Ring-Based Time-Division-Multiplexed Passive Optical Network. *Journal of Optical Networking, 7,* 288–293. doi:10.1364/JON.7.000288

Zhang, X. J., Wei, J., & Qiao, C. M. (2000). Constrained Multicast Routing in WDM Networks with Sparse Light Splitting. *IEEE Journal of Lightwave Technology, 18*(12), 1917–2000. doi:10.1109/50.908787

Zheng, S. Q., & Gumaste, A. (2006). Smart: An Optical Infrastructure for Future Internet. *The 3rd International Conference on Broadband Communications, Networks and Systems* (pp. 1-12).

Zhou, F., Molnár, M., & Cousin, B. (2008a). Avoidance of Multicast Incapable Branching Nodes for Multicast Routing in WDM Networks. In *the 33rd IEEE Conference on Local Computer Networks* (pp. 336-344). Montreal, Canada: IEEE Computer Society.

Zhou, F., Molnár, M., & Cousin, B. (2008b). Distance Priority Based Multicast Routing in WDM Networks Considering Sparse Light Splitting. In *the 11th IEEE Conference on Communication Systems*. Guangzhou, China: IEEE Communication Society.

Chapter 9

The Vertical–Cavity Surface–Emitting Laser
A Key Component in Future Optical Access Networks

Angélique Rissons
Université de Toulouse, France

Jean-Claude Mollier
Université de Toulouse, France

ABSTRACT

The proposal chapter aims at highlighting the tremendous emergence of the Vertical-Cavity Surface-Emitting Laser (VCSEL) in the FTTX systems. The VCSEL is probably one of the most important and promising components of the "last-leg" Optical Access Networks. To satisfy the bandwidth rise as well as the inexpensive design constraints, the VCSEL has found its place between the Light-Emitting-Diode (LED) and the Edge-Emitting-Laser (EEL) such as the DFB (Distributed-Feedback) laser. Hence, the authors dedicate a chapter to the promising VCSEL technology that aims to give an overview of the advances, the physical behavior, and the various structures regarding VCSELs. They discuss the VCSEL features and performance to weigh up the specific advantages and the weaknesses of the existing technology. Finally, diverse potentials of Optical Access Network architectures are discussed.

INTRODUCTION

For more than ten years, the access network market is attracted by the Vertical-Cavity Surface-Emitting Lasers well known as the acronym VCSEL.

Indeed, the numerous advantages of this component make the VCSEL technology suitable for access networks and today several studies have demonstrated the feasibility of various configura-

tions of Optical Access Network (OAN) using the VCSEL Technology.

Above all, the VCSEL has been designed to achieve the need of the optoelectronic circuits planarization. Since its invention in 1977 by Prof. K. Iga (Iga, Koyama, Kinoshita, 1988), the VCSEL structure is in a state of constant progress (Iga, 2000; Koyama, 2006; Chow, Choquette, Crawford, Lear, Hadley, 2006) Today, a wide wavelength emission range (from the Green-blue band up to the infrared)

DOI: 10.4018/978-1-60566-707-2.ch009

is covered that enables the usage of these components in various applications, not only in the field of digital datacommunications (Gigabit Ethernet) but also in consumer applications; such as optical switching, laser printers, laser mice, high-density optical disks, display systems, etc. Thereof the VCSEL market is composed of various fields of application: automotive, computer, consumer, industrial, military/aerospace, telecom, biomedical where computer and consumers are major fields in the VCSEL market (Szweda, 2006). As we focus on the OANs, we will study the VCSELs emitting at 850, 1310 and 1550nm.

Even though these three varieties of VCSELs have really different structures and performance, the following features are distinctive to the VCSEL-technology:

- The vertical laser emission is perpendicular to the layers which makes easier the one and two-dimensional integration (1D and 2D VCSEL-array) according to the electrical packaging constraints.
- The VCSEL is the smallest commercial available semiconductor laser diode (LD) type.
- By combining the small size and the perpendicular to the structure emission, this component responds to the criterion of planarization that makes the VCSEL, the LD with the higher integration level
- The AL low volume (Quantum Well) involves sub-milliampere threshold current and low electrical power consumption.
- The small cavity size allows for a singlemode longitudinal emission.
- The VCSEL presents low thermal variation close to the room temperature (wavelength, threshold current).
- The serial fabrication reduces the cost and allows on-wafer testing.
- Due to its cylindrical geometry, the light-beam cross-section is circular.

All these reasons have led to the VCSEL growth market in a wide application range.

Firstly, we present chronologically the VCSEL emergence through the pioneer structures, the salient features which allowed the improvement of the VCSEL operation. That led to the presentation of the main 850nm VCSEL and the emerging 1300 and 1550nm VCSELs. This section also gives the main characteristics of these three types of components.

Secondly, the second section of this chapter introduces the VCSEL modeling based on coupled carriers and photons rate equations and the main characterization needed to be known before integrating the VCSEL into a system. This knowledge is important to avoid a drawback that could be encountered by the inadequate utilization of the device, notably due to the frequency increase in the optoelectronics circuits and electrical mismatch in integrated circuits.

The last section discusses the VCSEL utilization in diverse system configurations to generate a signal for the OAN such as VCSEL-Based Oscillator, Optical injection-locking, etc. Some network architectures are also presented: Radio-Over-Fiber Distributed Antenna Systems, Hybrid OAN.

PRESENTATION OF THE VCSEL

Even if the first VCSEL structure emitted at long wavelength (LW), it's the 850nm GaAs-VCSELs which became the emerging technology in 2000. Today the 850nm VCSEL technology is the most competitive and reliable one (Ulm photonics company, 2008). All these advantages are due to the maturity of the AlGaAs/GaAs technology and the good performance of the structure: a Quantum-Well (QW) active layer (AL) between two Distributed Bragg Reflectors (DBR). Thus, due to the cost effectiveness linked to the massive VCSEL production, this device finds quickly its application in the 850nm short-distance Multimode fiber links giving an alternative to the LED.

Figure 1. Typical Edge Emitter Structure

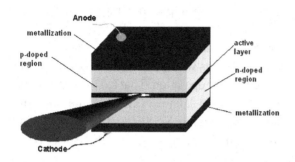

Recent progress in semiconductor device fabrication gives the possibility to find reliable 1550 (Vertilas, 2008) and 1310nm (Beam Express, 2008) VCSELs. That gives opportunities in single-mode fiber metropolitan area and wide area networks. Various materials allow reaching these two spectrum ranges. A review of the commercialized structures is given with emphasis on the component performance and reliability.

Brief Reminder Regarding Laser Diodes

Before presenting the VCSEL, it is important to retrieve the LD principle. Firstly a laser is composed of an active medium which ensures the capacity to produce the well known photon-electronic interactions: the absorption, the spontaneous emission, the stimulated emission which at the base of the laser effect and the non-radiative recombination.

Three kind of non-radiative recombination can be find in a laser. A conduction band electron can recombine with a valence band hole or with a recombination center (material defects) without photon emission (the energy is dissipated as heat in the crystal). A conduction band electron can also, by the recombination with a valence-band hole, transfer its energy to another electron or hole as kinetic energy. After its lifetime, the electron or hole gives its energy to the crystal lattice. This phenomenon is called the Auger effect. It occurs

laser diode with an emission wavelength $>1 \mu m$.

A typical structure of an EEL is described on the Figure 1: the electrodes are on the top and bottom surfaces and the light is emitted by the edge.

The VCSEL Technologies

The first part of this session aims to introduce the VCSEL emergence through a chronology of the diverse fabrication stages.

Since **1975**, many Surface-Emitting-Laser (SEL) structures were proposed, all based on the beam deflection from an EEL (Figure 2) (Iga, Koyama, Kinoshita, 1988). The large surface area provides a beam with low divergence but the size does not allow the 2-Dimensional array application and the integration in optoelectronic circuits.

In 1977, Prof. K. Iga gave the first suggestion of Vertical-Cavity SEL with a GaInAsP/InP and AlGaAs/GaAs active region for the optical fiber communication and for the optical disk, optical sensing and optical parallel processing.

The underlying motivation for the VCSEL design was the probe testing before separating into chips and monolithic integration into an optical circuit. Two years later, he obtained the first lasing operation of a GaInAsP/InP SEL structure with a 900mA threshold current (I_{th}) under a pulsed regime at 77K. In **1983,** a first lasing pulsed operation at room temperature (RT) with a GaAlAs/GaAs active region and I_{th}=1.2A (at 77K, I_{th}=350mA).

At that time, I_{th} was high in comparison with the EELs, due to structural differences. To obtain a RT Continuous-Wave (CW) operation with a narrower cavity, it was essential to increase the mirror reflectivity (>95%) and to improve the confinement current. (Figure 3)

The increase of the mirror reflectivity was obtained by introducing an Au/SiO2 mirror or a dielectric multilayer reflector.

The current confinement was enhanced by introducing a circular buried-heterostructure.

In **1987,** by using the Metal Organic Chemical

Figure 2. First structure proposal of Surface Emitting Laser

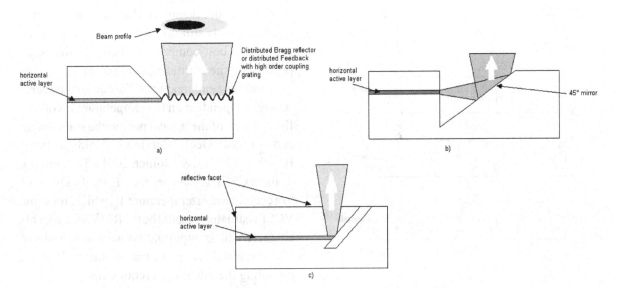

Vapor Deposition (MOCVD) growth technique, and by introducing an AlGaAs/GaAs circular buried-heterostructure, a 50mA-I_{th} in pulsed operation was obtained at 293K.

In **1989,** the first CW operation at RT with a GaAs structure was achieved by Prof. K. Iga. At the same time, Jewell published the characteristics of the first QW GaAs-based Vertical-Cavity Surface Emitting-Laser (Jewell, Scherer, McCall, Lee, Walker, Harbison, & Florez, 1989); Ibariki (Ibaraki, Kawashima, Furusawa, Ishikawa, Yamaguchi, & Niina, 1989) introduced the DBR in the VCSEL fabrication process.

This breakthrough accelerated the VCSEL technology advance. Indeed, the QW active region involved a strong decrease of I_{th} and the DBR introduction enabled to improve the mirror reflection coefficient.

Another important point concerning the VCSEL performance enhancement is the growth of the entire structure by the Molecular Beam Epitaxy (MBE) or MOCVD.

The MBE has dramatically reduced the fabrication cost and led to a broad-based production especially for the AlGaAs/GaAs structure (MBE unsuitable to grow the phosphorous-containing

materials).

However, the DBR utilization increased notably the series resistance. So the novel challenge was to reduce this resistance while keeping the optical performance intact.

Numerous research works have been carried-out to improve the doping profile (Kopf, Schubert, Downey, & Emerson, 1992; Wilmsen, Temkin, & Coldren, 1999).

As shown in Figure 4, several doping profiles were possible. The first generation of VCSELs had an abrupt doping profile allowing good reflectivity but involving a high resistance (>100Ohm). The alternative was to modify the doping profile

Figure 3. First VCSEL structure

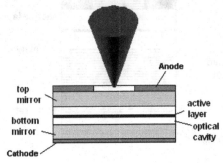

Figure 4. Different doping profile of $Al_xGa_{1-x}As$

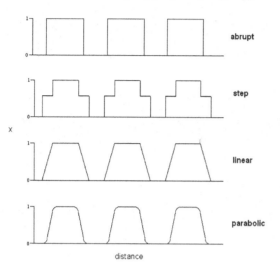

for each DBR interface. Better performance has been obtained with parabolic profiles.

850nm VCSELs

Today, many GaAs-VCSEL structures are commercialized, but we assert that the most competitive VCSELs are the oxide-confined VCSELs presented on the Figure 5 (right side).

But the simultaneous electrical and optical confinement is still unresolved. There is a need to find the best compromise between the beam profile and the emitted optical power.

As in any semiconductor, the carrier number is strongly dependent on the temperature involving fluctuations of the optical power, the wavelength and I_{th} (Scott, Geels, Corzine, & Coldren, 1993; Boucher, Rissons, & Mollier, 2001). The earmark of the VCSEL is the parabolic I_{th} evolution close to a characteristic temperature T_0, which, for some VCSEL structures, could be the RT (Figure 6). This characteristic is important because in non severe environment, it let to eliminate the thermal control regarding the integration requirement.

However, if the thermal effect does not degrade I_{th}, it influences on the power evolution especially due to Joule heating in the DBRs. Indeed, the carrier transport through the DBRs modifies the refractive index of each layer generating multiple transverse modes. Furthermore, the imperfect carrier confinement leads to their inhomogeneous spatial distribution in the AL, which is responsible for a strong Spatial Hole Burning (SHB). The

Figure 5. Main VCSEL structures

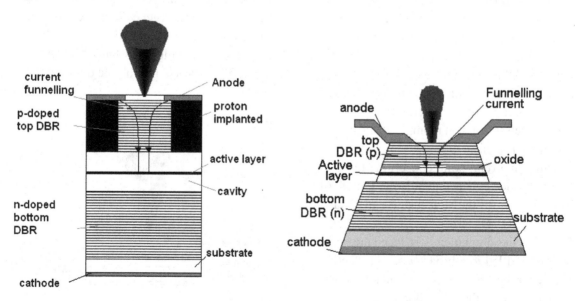

Figure 6. Temperature dependence of the threshold current

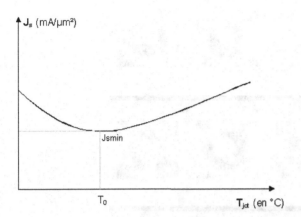

Figure 7. Kink on the Light-Current curve due to the multiple transverse mode influence

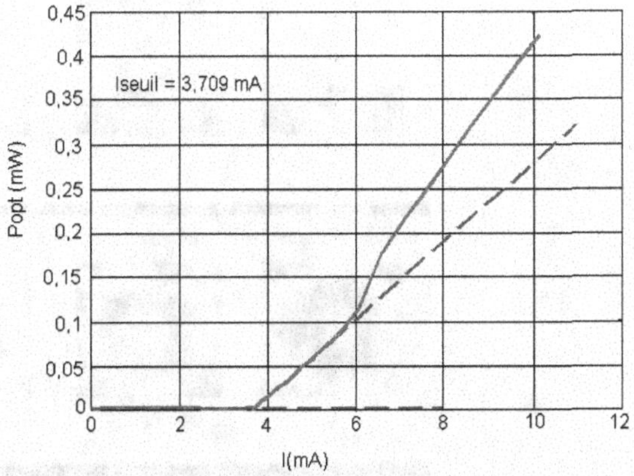

carrier accumulation in some zones implies a heating entailing a modification of the optical cavity volume and the refractive index (called thermal lens effect). This phenomenon implies a multiple transverse mode behavior that can be observed as kink on the Light-Current (L-I) curve (Figure 7). The beam profile and the optical spectrum are also affected by this effect: Figure 8 shows the evolution of the resulting annular beam profile for different bias current, and Figure 9 presents the multiple transverse modes in the optical spectrum.

The solution to reduce these effects is the diminution of the oxide-aperture diameter. Thus, for AlOx-diameter lower than 5µm, the VCSEL can provide a single-mode operation. As we can see in Figure 10, the optical spectrum presents a single-mode emitting and Figure 11 reveals an L-I curve devoid of kink and a Gaussian beam profile. In counterpart, the higher DBR resistivity is increased reducing the carrier injection efficiency in the AL consequently reducing the optical power (<1mW).

The typical performance of the different 850nm VCSEL is summarized in the Table 1. According to this table, the 850nm VCSEL users can find a good compromise between the different characteristics.

Long Wavelength VCSEL

Contrary to the 850nm VCSEL, the fabrication of LW-VCSELs has been curbed by the difficulty in growing InP-based compound. Another issue is to keep the advantage of the GaAs-VCSEL structures, especially in terms of low-I_{th} and high integration-level. Indeed, in InP-based structures, for emission wavelength higher than 1µm, due to the small bandgap, the large Auger recombination coefficient and other non-radiative recombinations raise I_{th}. Furthermore, the lattice mismatching between DBR and the InP AL, the small index contrast in the DBR structure adapted for these wavelengths and the low thermal resistance (decreasing the thermal range operation) require finding another mirror configuration resulting in the VCSEL size increasing.

Before heading farther in the description of the LW-VCSEL structures, it is important to remind the emission wavelengths associated to the materials. The various materials utilized in the VCSEL AL fabrication are summarized in Table 2.

Regarding the mirrors, several configurations have been proposed by the VCSEL designers (Chang-Hasnain, 2003). This includes metamorphic DBR (Boucart, Starck, Gaborit, Plais, Bouche, Derouin, Remy, Bonnet-Gamard,

Figure 8. Beam profile of large oxide-aperture-VCSEL for various bias current (5mA up to 25mA)

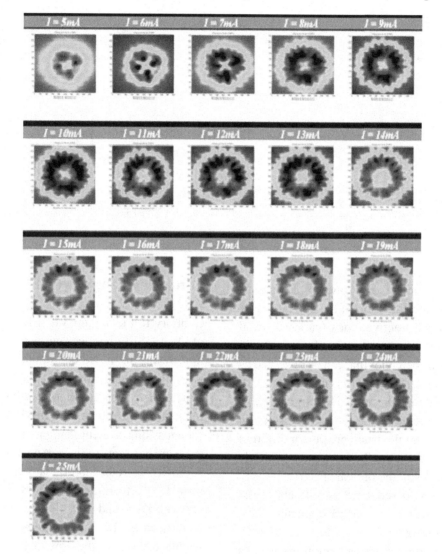

Goldstein, Fortin, Carpentier, Salet, Brillouet, & Jacquet, 1999), dielectric mirror (Shau, Ortsiefer, Rosskopf, Boehm, Lauer, Maute, & Amann, 2004), wafer-fused AlGaAs/GaAs DBR, InP Air-gap DBR.

Metamorphic DBR

The metamorphic DBR design consists of using metamorphic materials to relax the constraints imposed by lattice matching. Moreover this kind of structure permits to introduce an oxide aperture for the current confinement.

Dielectric Mirror

As it has been presented before, one of the first generation VCSELs used dielectric mirrors. Thanks to the progress in technology, the operation has been considerably improved resulting in the introduction of a buried n^+-p^+-p tunnel junction (BTJ) on the top of the AL. This ensures the current

Figure 9. Multiple transverse mode of an 850nm oxide-confined-VCSEL with a 12μm oxide-aperture diameter

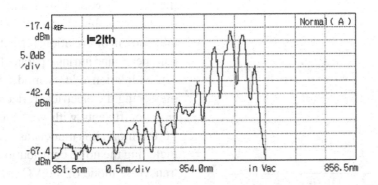

injection with an efficient carrier confinement.

Wafer-Fused AlGaAs/GaAs DBR

The wafer fusion or wafer bonding consists in growing the AL and AlGaAs/GaAs DBR on two different substrates. This technology provides an InGaAlAs AL with the thermal advantage of GaAs/AlGaAs DBRs. By introducing intracavity contact to ensure the current injection and a BTJ just above the AL, this structure gives the best performance in terms of singlemode optical power emitting at 1.3μm and 1.55μm (Syrbu, 2005). In 2008, VCSELs designed by Beam Express attain an optical power close to 6mW for the 1.3μm VCSEL and 4mW for the 1.55μm VCSEL

VCSEL MODELING AND CHARACTERIZATION FOR THE FUTURE OPTICAL LINKS

Whatever the application, the simulation of an optical link for all applications is of great interest. Each component has to be perfectly modeled before implementation in an optical sub-assembly (Toffano, Pez, Le Brun, Desgreys, Hervé, Mollier, Brabary, Charlot, Constant, Destrez, Karray, Marec, Rissons, & Snaidero, 2003). Moreover, it is also important to achieve a perfect experimental characterization of these components with emphasis on LD modeling which leads to the link behavior (Tucker, 1985). As the structure of the VCSEL is very different from that on the EEL one, a specific model has to be established.

Figure 10. Optical spectrum of a singlemode 850nm oxide-confined-VCSEL with a 5μm oxide aperture diameter

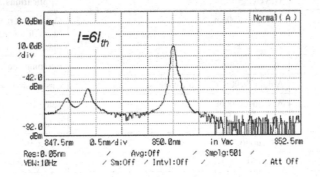

Figure 11. L-I curve and intensity profile of singlemode VCSEL

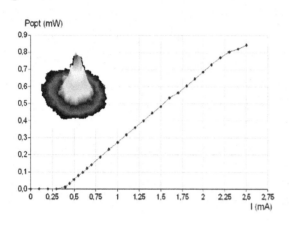

(Rissons, Perchoux, Mollier, & Grabherr, 2004) based on a small-signal equivalent-circuit including the DBR contribution and the rate equations resolution is thus presented. That allows the establishment of a relationship between measurable parameters and intrinsic parameters of the VCSEL. The validation of this model was achieved by measuring the electrical reflection and transmission coefficients with vector network analyzer. Furthermore, this model has been generalized by including the noise contribution and the multiple transverse mode of the VCSEL (Perchoux, Rissons, & Mollier, 2007)

Table 1. Typical values of main electro-optical characteristics of oxide-confined-VCSELs versus the oxide aperture diameter (Ulm-photonics company, 2008)

VCSEL type	P_{opt} output optical power (mW)	I_{th} (mA)	η_s slope efficiency	R_s Differential series resistance (Ω)
<5µm oxide aperture	1	0.2	0.3	<100
Close to 12µm oxide aperture	1.5	1.5	0.4	50
>25µm oxide aperture	24	3	0.8	20

In this way, a VCSEL model is presented.

The first step of the modeling is the description of carrier and photon exchange through the rate equations . We are interested in the dynamic behavior of this LD. Thus an electrical model

The modeling and characterization have been extended to the 850, 1310 and 1550nm VCSEL which could be integrated in OAN.

Table 2. Wavelength and applications versus materials of the AL

Compound AL	Wavelength	Applications
AlGaAs/ GaAs	680 up to 890nm	Optical interconnection Short distance fiber link Plastic optical fiber link
GaInAs/GaAs	Around 980nm	Pumping source, gas sensing
GaAsSb/GaAs	Around 1200nm (up to 1300nm)	Gas sensing
GaInNAs/GaAs	Around 1300nm	Telecom
AlGaInAs/InP	1300 up to 1600nm	Telecom
GaInAsP/InP	1200 up to 1600nm	Telecom
InGaAs/InP	Above 1600nm	Sensing water and CO2

VCSEL Rate Equation Model and Steady State Resolution

The laser rate equations are based on two coupled rate equations in which the electron and photon interactions are converted into mathematic form. These equations are established from the VCSEL structure. That is why a general simplified form could be adapted to the various VCSEL configurations (Rissons, Mollier, Toffano, Destrez, Pez, 2003).

The carrier rate equation is the difference between the carrier injection with the recombination rates and the photon rate equation is the difference between the photon generated which can be amplified by stimulated emission and the lost photons. These equations can be written in the following form:

$$\frac{dN}{dt} = \frac{\eta_i \cdot I}{q \cdot N_w} - R(N) - G(N,P) \cdot P \qquad (1)$$

$$\frac{dP}{dt} = R_{sp}(N) + N_w \cdot G(N,P) \cdot P - \frac{P}{\tau_p} \qquad (2)$$

where N is the electron numbers in one QW, P, the photon number in the cavity, N_w, the QW number, η_i, the internal quantum efficiency.

Furthermore, $\frac{\eta_i \cdot I}{q \cdot N_w}$ represents the population inversion in each QW.

$R(N)$ and $R_{sp}(N)$ are the carrier recombination rate and the spontaneous emission rate respectively. The $R(N)$ term may be written as: $R(N) = (\tau_e)^{-1} = A + B.N + C.N^2$.

A is the nonradiative recombination contribution, B, the spontaneous emission coefficient, and the C, the Auger recombination coefficient which can be neglected at 850nm wavelength.

$G(N, P)$, the modal gain, is obtained from the relationship $G(N,P) = g_0 \frac{N - N_{tr}}{1 + \varepsilon \cdot P}$ where N_{tr} is

the transparency electron number, ε is the nonlinear gain coefficient. $g_0 = v_{gr} \cdot \Gamma \cdot \frac{a}{V_{act}}$, is the modal gain coefficient, where the a, is the differential gain coefficient, V_{act}, the AL volume, v_{gr}, the group velocity and Γ the confinement factor.

R_{sp} is defined as $(N_w.\Gamma.\beta.N)/\tau_e$ where $\Gamma.\beta$ corresponds to the photons generated by spontaneous emission contributing to the laser emission and τ_p is the photon lifetime.

These rate equations are adapted to a QW Laser by the presence of the terms η_i and N_w. The Γ parameter takes into account the vertical light emission and possibly the DBR contribution. Moreover, the value of each parameter depends upon the VCSEL structure.

The hereafter table gives an example of 850nm oxide-confined VCSEL parameters range of value. (Table 3)

When the steady-state is reached, the rate equations are equal to 0, such as:

$$0 = N_w \cdot \Gamma \cdot \beta \cdot B \cdot N^2 + N_w \cdot g_0 \cdot \frac{N - N_{tr}}{1 + \varepsilon \cdot P} \cdot P - \frac{P}{\tau_p}$$

$$(3)$$

$$0 = \frac{\eta_i \cdot I}{q \cdot N_w} - \frac{N}{\tau_e} - g_0 \cdot \frac{N - N_{tr}}{1 + \varepsilon \cdot P} \cdot P \qquad (4)$$

The evolution of the carrier and photon number against bias current is given in Figure 12.

Below I_{th}, N increases linearly. Above I_{th}, N becomes constant and keeps the value of N_{th}. We can obtain the value of N_{th} when $I = I_{th}$ and $P \cong 0$ thus:

$$N_{th} = \tau_e \frac{\eta_i I_{th}}{q N_w} \qquad (5)$$

Event if τ_e is strongly depending on N, it can be considered as a constant because the minor fluctuations of N around the threshold value are too small as compared to N_{th}.

Table 3. Range of intrinsic parameter value for 850nm VCSEL

Parameters	Unit	Values
A	s^{-1}	[10^8 ; 1.3 10^8]
B	s^{-1}	[0.7 10^{-16}; 1.8 10^{-16}] x V$_{act}$
N$_{tr}$	-	[0.83 10^{24} ; 4.40 10^{24}] x V$_{act}$
τ_p	ps	[1 ; 6]
a	m^2	[0.2 10^{-20} ; 3.7 10^{-20}]
v$_{gr}$	m.s^{-1}	[8.33 10^7, 8.60 10^7]
Γ	-	[0.045 ; 0.06]
β	-	[10^{-5} ; 10^{-4}]
η_i	-	[0.6 ; 0.86]

If the bias current increases above the threshold, P is greater than 0. In that case, we can calculate the value of P in steady-state regime according to the bias current value I.

From the equations (4) and (5), the following relationship is written:

$$G \cdot P = \frac{\eta_i \cdot \left(I - I_s\right)}{q \cdot N_w} \quad (6)$$

By injecting this term in the equation (3), we obtain an analytic expression for P:

Figure 12. Variation of the carrier and the photon numbers versus the bias current

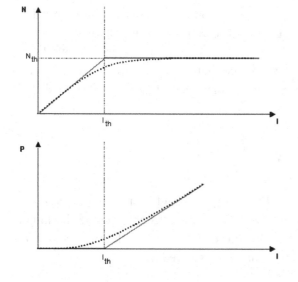

$$P = \tau_p \cdot \left(N_w \cdot \Gamma \cdot \beta \cdot B \cdot N_s^{\,2} + N_w \cdot \frac{\eta_i \cdot \left(I - I_s\right)}{q \cdot N_w} \right) \quad (7)$$

While the Equations (1) and (2) give a general and simplified representation of P-N interaction in any VCSEL structure, this model needs to be adapted for particular structure.

For example, the multimode behavior of 850nm VCSEL can be introduced in the rate equations.

The multiple transverse mode contributions can be introduced in the gain term of the rate equations by the mode competition.

Thereby, the gain term becomes:

$$G_m = g_0 \frac{N - N_{tr}}{1 + \varepsilon P_m + \sum_{m \neq n}\left(\varepsilon_{mn} P_n\right)} \quad (8)$$

where ε_{mn} is the gain compression factor of mode 'n' due to the mode 'm'. It is given by the following relationship:

$$\varepsilon_{mn} = \varepsilon \frac{1 + \alpha \tau_p \Omega_{mn}}{1 + \left(\tau_p \Omega_{mn}\right)^2} \quad (9)$$

α is the Henry Factor or wavelength enhancement factor and $\Omega_{mn} = |\omega_m - \omega_n|$ is the frequency detuning between the 'm' and 'n' modes that represents the gain degradation.

The stochastic evolution of N and P generates noise. Indeed, the operation of the LD is disturbed by several noise sources whose influence varies according to the different regimes. For targeted applications, the preponderant noise source is the spontaneous emission. The randomness of the spontaneous emission generates an amplitude and phase fluctuations of the total optical field. Moreover, these photons which are produced in the laser cavity follow the feedback of the stimulated photons and interact with them. By taking into account the wave-corpuscle duality of the light, a quantum approach is better suited to describe the emission noise generation including the photon-electron interactions: each state of a photon or electron is associated to a noise pulse. For the purposes of noise generation quantification, recombination and absorption rates in the cavity allow the utilization of the electron and photon Langevin forces to give a mathematical representation of the optical emission noise.

By going back on Equation (1) and Equation (2) and by adding the noise contribution, a new rate-equation couple can be written as follows:

$$\frac{dN}{dt} = \frac{\eta_i \cdot I}{q \cdot N_w} - R(N) - G(N,P) \cdot P + F_N(t)$$

(10)

$$\frac{dP}{dt} = R_{sp}(N) + N_w \cdot G(N,P) \cdot P - \frac{P}{\tau_p} + F_P(t)$$

(11)

where $F_N(t)$ and $F_P(t)$ are the Langevin functions representing electrons and photons fluctuations respectively.

By using the EEL model, the spectral densities and the cross-correlation function can be expressed using the intrinsic parameters of the VCSELs:

$$\langle F_N(t)^2 \rangle = \sum r_N^+ + \sum r_N^-$$

(12)

$$\langle F_P(t)^2 \rangle = \sum r_P^+ + \sum r_P^-$$

(13)

$$\langle F_N(t) \cdot F_P(t) \rangle = -\left[\sum r_{NP} + \sum r_{PN} \right]$$

(14)

r_N^- and r_N^+, r_P^+ and r_N^- r_N^+ r_P^+ r_P^- and r_N^+, r_P^+ and r_P^- are the rate of increasing and decreasing N and P,

r_{NP} and r_{NP} r_{PN} and r_{PN} are the rates of exchange between the photons and electrons reservoirs.

By using the rate-equations, a simplified form of the spectral densities and cross-correlation function is derived as:

$$\langle F_N(t)^2 \rangle = 2g_0 \frac{N_{tr}}{1 + \varepsilon \bar{P}} \bar{P} + \frac{\eta_i \bar{I}}{q N_w}$$

(15)

$$\langle F_P(t)^2 \rangle = 2N_w g_0 \frac{N_{tr}}{1 + \varepsilon \bar{P}} \bar{P} + \frac{2\bar{P}}{\tau_p}$$

(16)

$$\langle F_N(t) \cdot F_P(t) \rangle = -\left[2g_0 \frac{N_{tr}}{1 + \varepsilon \bar{P}} \bar{P} + \frac{\bar{P}}{\tau_p} \right]$$

(17)

These expressions are currently used to model the VCSEL Relative Intensity Noise (RIN) which will be presented next.

Furthermore, the mode competition in the multiple transverse-mode VCSELs strongly affects the noise. Therefore, the previous relation must be formulated for a multimode case. As for the singlemode case, the Langevin functions are added to the rate equations:

$$\frac{dN}{dt} = \frac{\eta_i \cdot I}{q \cdot N_w} - R(N) - \sum_m G_m P_m + F_N(t)$$

(18)

$$\frac{dP_m}{dt} = R_{sp}(N) + N_w \cdot G_m \cdot P_m - \frac{P_m}{\tau_p} + F_m(t)$$

(19)

Where $F_m(t)$ is the Langevin function of one mode.

Thus, new expressions for the spectral densities and the cross correlation of the Langevin functions can be written as:

$$\left\langle F_N(t)^2 \right\rangle = \sum_m 2g_{0m} N_{tr} \bar{P}_m + \frac{\eta_i \bar{I}}{qN_w}$$

(20)

$$\left\langle F_P(t)^2 \right\rangle = 2N_w g_{0m} N_{tr} \bar{P}_m + \frac{2\bar{P}_m}{\tau_p}$$

(21)

$$\left\langle F_N(t) \cdot F_P(t) \right\rangle = -\left[2\bar{g}_{0m} \frac{N_{tr}}{1+\varepsilon\bar{P}} \bar{P}_m + \frac{\bar{P}_m}{\tau_p} \right]$$

(22)

The equations (18) and (19) give a complete description of photon-carrier exchanges in a multimode VCSEL. Other physical phenomena, such as carrier diffusion and optical polarization could be included but in our experience this pair of rate-equations is complete enough to accurately simulate the VCSEL behavior. Indeed, the main effects are presented in a simplified rate equation model and the inclusion of each physical contribution would complicate the equation making the simulation software implementation more difficult.

Dynamic Behavior: Modeling and Characterization

This session presents the dynamic behavior of the VCSEL through rate-equation modeling and the associated experiments for the validation. This approach is interesting for several reasons: it gives an overview of the most useful dynamic characteristics in the VCSEL-optical-fiber-link and the model is simplified and easily implement-

able in simulation tools.

Small Signal Characterization and Modeling

This model presents the small signal VCSEL behavior via an electrical equivalent circuit. This method, well-known in electronics, allows the creation of a clear and original model. Such models are helpful for the designers of the optical transceivers especially for impedance-matching the electronic circuit driving the laser chip.

The procedure of this modeling is decomposed into several steps presented in Figure 13. To get a relationship between the intrinsic parameters, the photon and carrier rate equations along with Kirchhoff circuit equations have been used.

The first step is a qualitative approach which consists in describing the physical phenomena that occur into the VCSEL structure using lumped components.

In this section, we present the particular case of an oxide-aperture 850nm VCSEL. This example is of great interest because it allows the introduction of the DBR contribution during electronic injection.

The funneled current, crossing the VCSEL structure involves electronic phenomena not only in the AL but also in the DBRs. On the one hand the required population inversion in the QWs induces resistive and capacitive effects while on the other hand the crossing by the carriers of different heterojunctions constituting the DBRs implies the introduction of resistances and capacitances at each interface.

We have to adapt the electrical model of the optical cavity of an EEL diode to the VCSEL (Figure 14). The electronic mechanism in the multiple-QW AL can be represented by the resistance, R_j, and the capacitance, C_j. The photon behavior into the cavity is symbolized by following equivalent electrical elements: the photon storage corresponds to the inductance L_p and the resonance damping to the resistance R_s.

Figure 13. Data processing and parameters extraction

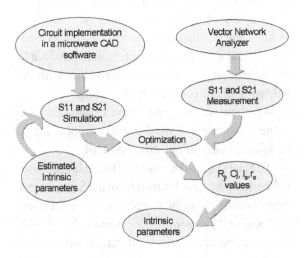

Figure 14. Behavioral electrical equivalent circuit of oxide confined VCSEL

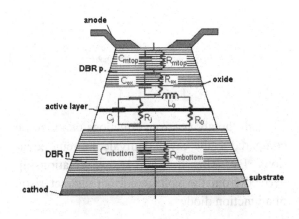

of relationships between the VCSEL intrinsic parameters and the lumped components of the cavity equivalent circuit.

Thereby, to associate the equivalent circuit and the rate-equations, the equations (1) and (2) have to be linearized under the small-signal approximation. In this way, the carrier number $N(t)$ and the photon number $P(t)$ can be defined in terms of $I(t)$.

$$N(t) = N_0 + \Delta N(t), \ P(t) = P_0 + \Delta P(t)$$
with $\Delta N << N_0$ and $N(t) = N_0 + \Delta N(t)$
$P(t) = P_0 + \Delta P(t) \ \Delta N << N_0 \ \Delta P << P_0,$
$P(t) = P_0 + \Delta P(t)$ with $\Delta N << N_0$ and
$\Delta P << P_0$. So we get the linearized form of the rate equation as follows:

The DBR being periodical growth of two material layers is represented by distributed RC-cells which can be simplified into an equivalent resistance in parallel with a capacitance.

The top and the bottom DBRs having different doping and layer pair numbers, the equivalent resistances and capacitances are thus different for the both mirrors: C_{mtop} and R_{mtop} and $C_{mbottom}$ and $R_{mbottom}$ refer to the top and bottom DBR resistances and capacitances respectively.

Considering the small-signal regime, the injected current is the superimposition of the bias current I_0 and a sinusoidal current $\Delta I(t)$ so that $\Delta I(t) << I_0$. In this way, the Kirchhoff equations of the cavity circuit can be expressed according to the convention given in Figure 15 where ΔV and ΔI are the input voltage and current respectively, and i_L is the equivalent current of the photon flow variation.

$$\frac{d\Delta V}{dt} = \frac{\Delta I}{C_j} - \frac{\Delta V}{R_j C_j} - \frac{i_L}{C_j} \quad (23)$$

$$\Delta V = L_0 \frac{di}{dt} + R_0 i_L \quad (24)$$

These equations will allow the establishment

Figure 15. Small-signal driving of a VCSEL cavity equivalent circuit

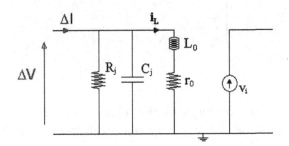

$$\frac{d\Delta N}{dt} = \frac{\eta_i}{q \cdot N_w} \cdot \Delta I - g_0 \frac{N_0 - N_{tr}}{(1 + \varepsilon P_0)^2} \cdot \Delta P - \left(g_0 \cdot \frac{P_o}{(1 + \varepsilon P_0)} + A + 2B \cdot N_0 \right) \cdot \Delta N$$

$$(25)$$

$$\frac{d\Delta P}{dt} = N_w \left(\Gamma \cdot 2 \cdot \beta \cdot B \cdot N_0 + g_0 \cdot \frac{P_o}{1 + \varepsilon P_0} \right) \cdot \Delta N + \left(N_w g_0 \cdot \frac{N_0 - N_{tr}}{(1 + \varepsilon P_0)^2} - \frac{1}{\tau_P} \right) \cdot \Delta P$$

$$(26)$$

Both the above-given equations can then be compared to the cavity equivalent circuit equations. By using the well-known relationship derived from the voltage-current characteristic of a junction diode:

$$\frac{\Delta V}{V_T} = \frac{\Delta I}{I_0} = \frac{\Delta N}{N_0} \tag{27}$$

We can derive various relations between the intrinsic parameters and the circuit elements by comparison of equations (23) and (24) with (25) and (26), as follows:

$$C_j = \frac{N_o \cdot q \cdot N_w}{V_T \cdot \eta_i} \tag{28}$$

$$R_j = \frac{V_T}{N_0} \cdot \frac{1}{\left(\dfrac{I_0 - I_{th}}{N_0 - N_{tr}} + \dfrac{q \cdot N_w}{\eta_i} \cdot (A + 2BN_0) \right)} \tag{29}$$

$$L_0 = \frac{V_T}{N_0} \cdot \frac{q}{\eta_i} \cdot \frac{P_0^2 \cdot g_0 \cdot (N_0 - N_{tr})}{(I_0 - I_{th})^2} \cdot \frac{1}{\Gamma \cdot 2 \cdot \beta \cdot B \cdot N_0 + \dfrac{\eta_i}{q \cdot N_w} \dfrac{I_0 - I_{th}}{N_0 - N_{tr}}} \tag{30}$$

$$R_0 = L_0 \cdot \left(\frac{1}{\tau_p} - \frac{\eta_i^2 \cdot (I_0 - I_{th})^2}{q^2 \cdot N_w \cdot P_0^2 \cdot g_0 \cdot (N_0 - N_{tr})} \right) \tag{31}$$

The modeling of the electronic or optoelectronic devices is always confronted with the problem of intrinsic parameter estimations. The values of the parameters implied in the equation are not provided by the VCSEL manufacturer. Nevertheless, by knowing the materials constituting the VCSEL and several extrinsic parameters, it is possible to estimate a range of the parameter values with the help of data collected in the literature.

All these values enable the estimation of a range of values for the lumped-components to be injected into an optimization tool.

The last stage of this model is the experimental validation which is obtained from the S-parameter measurement. This measurement is of great interest because it gives information about the VCSEL frequency response and the electrical impedance-matching.

To obtain the S11 and S21 parameters, the measurements can be achieved with the experimental setup presented in Figure 16. Once again, these measurements are realized on an 850nm oxide-confined VCSEL since the topology of the chip allows a high-frequency measurement without the packaging influence. Indeed, when a VCSEL is mounted on a TO Package (for example), the frequency response shows a degradation of the signal due to the package cut-off frequency having a significant impact on the transmission link. Moreover in optical subassemblies integrating VCSEL arrays, an electrical access mismatch can generate electrical crosstalk between adjacent lasers. For this reason, it is important to match the VCSEL to its drive-assembly in order to preserve the VCSEL advantages. The electrical access is thus determined form the dimensions and the material properties and is included in the equivalent circuit (Figure 17).

In order to extract the VCSEL response from the measurement of S11 and S21, the equivalent circuit is implemented in the RF simulation tool ADS and an optimization of simulations is achieved to extract the values of the lumped components.

As shown in Figure 18, the measurements are very consistent with the simulations. Consequent-

Figure 16. Experimental setup for S11 and S21 measurement

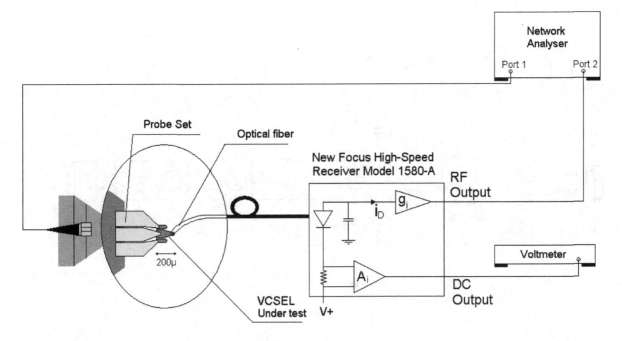

ly, this model gives a reliable simulation of the electrical access via the S11 and S21 parameters and the VCSEL frequency response. Moreover, by applying the optimization results to extract lumped component values it is possible to get values of intrinsic parameters by using the relationships (28), (29), (30) and (31) according to the range of values taken from the literature.

Relative Intensity Noise (RIN)

The RIN is the quantification of the photon number fluctuation in a LD as compared to the total photonic emission i.e. the ratio between the spectral density of the optical power, $S_{Pe}(f)$ to the square of the mean power $\bar{P}_e$ emitted by the LD. The optical power being directly proportional to the photon number, the RIN could be expressed as the spectral density of the photon number fluctuation over the square of the photon number mean value.

$$RIN = \frac{S_{Pe}(f)}{\bar{P}_e^2} = \frac{\langle \delta P^2 \rangle}{\bar{P}^2} \tag{32}$$

The RIN is widely given in dB such as:

$$RIN_{dB} = 10 \cdot \log_{10}\left(\frac{S_{Pe}(f)}{\bar{P}_e^2}\right) = 10 \cdot \log_{10}\left(\frac{\langle \delta P^2 \rangle}{\bar{P}^2}\right) \tag{33}$$

An electrical equivalent model allows the RIN simulation by using the values of intrinsic parameters.

In **Figure 19**, we can show see that the simulation is consistent with the experimental RIN extraction.

Digital Link Modeling

As mentioned previously, the VCSEL is a ubiquitous component for the Gigabit Ethernet transmission modules employing multimode optical fiber. This is due to several reasons the most important of

Figure 17. Small signal equivalent circuit of the VCSEL and its electrical access

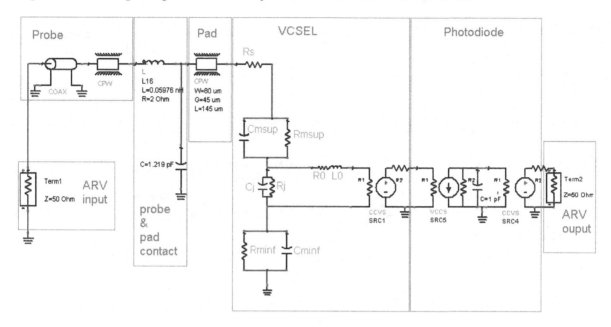

which being the cost effectiveness of the VCSEL technology while keeping good performance for short distance optical links.

In that context, this section gives an overview of the main characteristics of the optical link.

Once again, a rate equation based model is proposed in order to simulate the behavior of a VCSEL digital link (Ly, Rissons, Gambardella, & Mollier, 2008). The main source of signal degradation in this type of link is the jitter effect. So, this section focuses on the jitter modeling using the turn-on delay of the LD. Moreover, an estimation of these delays is important to predict the maximum deliverable data rate. We will see that the emission delay is strongly dependant to the lower level of current (I_{off}).

When a LD is directly modulated by a current pulse train from $I_{OFF} < I_{th}$ up to I_{ON}, the photon response is not instantaneous. These delays appear before the steady state of the stimulated emission (Figure 20) is reached. This delay occurs with damped ringing at the relaxation resonance frequency f_R yielding upper limit for the achievable modulation frequency.

In order to quantify the effects occurring during the turn-on event, the dynamics of the photon number P(t) is separated into a stochastic regime ($t < t_c$) and a deterministic regime ($t > t_c$). There is a delay time t_d between the applied bias and the pulse rise time to the ON level, equal to the sum of the carrier density rise time to the lasing threshold τ_D and the optical switch-on time to the optical ON level t_{on}: $t_d = \tau_D + t_{on}$. Resolution of the two-coupled equations (1) and (2) under large signal direct modulation leads to these turn-on delay times τ_D and t_{on} in the stochastic and deterministic regimes respectively.

The turn-on delay τ_D is defined as the time elapsed between the application of an ideal step current and the lasing threshold and can be determined from the carrier equation (1). According to Dixon, & Joyce (1979), we can express the turn-on delay τ_D as follows (when the OFF state is slightly below the threshold):

$$\tau_D = \tau_e' \cdot \frac{I_{ON} - I_{OFF}}{I_{ON} - I_{TH}} \tag{34}$$

Figure 18. Comparison between simulated and measured S11 Module and Phase and S21 module

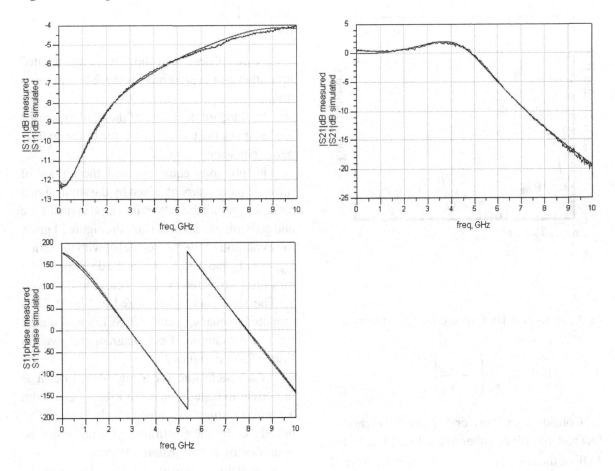

Where $\tau_e^{'} = \dfrac{d\tau_e}{dN}$ is defined as the differential

Figure 19. Measurement and simulation of 850nm multimode VCSEL RIN for three bias currents, $1.5I_{th}$, $3I_{th}$, $6I_{th}$

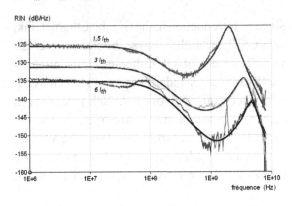

lifetime which is assumed to be constant in the entire range of integration and equal to its threshold value: $\tau_e^{'} = (A + 2B \cdot N_{th} + 3CN_{th}^{2})^{-1}$. When the bias level is far below threshold, the expression of the turn-on delay becomes:

$$\tau_D = \tau_e \cdot \frac{I_{ON} - I_{OFF}}{I_{ON} - I_{TH}} \qquad (35)$$

Modulating the laser under pseudorandom bit sequences (PRBS) leads to fluctuations of this delay τ_D due to bit-pattern effect, represented by the crossing time t_c. The statistics of the turn-on delay time τ_D allow the derivation of the turn-on jitter. Previous studies have led to the expression of the following probability density function (PDF)

Figure 20. Turn on event from Ioff=0 to Ion>Ith

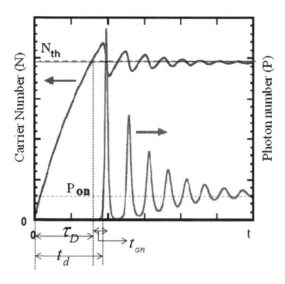

of the turn-on delay t_c where D is the data rate:

$$p(t_c) = \ln 2 \cdot B \cdot \frac{\tau_e}{\tau_D}\left(1 - \frac{t_c}{\tau_D}\right)^{\ln 2 \cdot D \cdot \tau_e - 1} \quad (36)$$

Considering the two orthogonal polarizations that both contribute to the emitted VCSEL light, the PDF of the rise time $t_{on} = t_d - t_c$ can be derived from the photon rate equation(2) as:

$$p(t_{on}) = 4 \cdot \omega_R^2 \cdot \frac{P_m^2}{\langle P_{off}^2 \rangle} \cdot t_{on} \cdot \exp[-(\omega_R \cdot t_{on})^2] \cdot \exp\left[-2 \cdot \frac{P_{on}}{\langle P_{off} \rangle} \cdot \exp\left(\frac{-(\omega_R \cdot t_{on})^2}{2}\right)\right] \quad (37)$$

where $\langle P_{off} \rangle$ represents the mean value of the photon number at the turn-on delay t_c and f_R refers to the resonance frequency at the ON state. The jitter is defined as the standard deviation of the PDF. Since those two processes are not correlated, the PDF accounting for both bit pattern effect and spontaneous emission is obtained by the convolution of expressions (5) and (6), leading to:

$$p(t_d) = \int_0^{\tau_D} p(t_c) \cdot p(t_{on}) \cdot dt_c \quad (38)$$

The model described above was implemented in Matlab and is presented in the following section.

The standard deviation of the PDF expressed in (36) gives the turn-on jitter due to the spontaneous emission.

The previous equations and the values of intrinsic parameters obtained by the small signal modeling permit the delay simulation in stochastic and determinist regimes. Thus, the Figure 21 gives the evolution of the turn-on delay with the ration I_{OFF} over I_{th} and clearly shows the decrease of the turn-on delay when the I_{OFF} is close to I_{th}.

The test-bench schematized in Figure 22 is mounted to characterize a VCSEL-based digital link. Two example of eye diagram are given in Figure 23 and Figure 24.

As a conclusion, for a high data rate, it is important to superimpose the PRBS signal with a prebias current in order to reduce the delay introduced by the electrons (τ_D) which could be amplified by the bit pattern effect.

Even if this section presented modeling results of 850nm VCSELs, the methodology is the same for longwavelength VCSELs but needs to be adapted to the VCSEL intrinsic structure.

OPTICAL ACCESS NETWORK ARCHITECTURES USING VCSELS

Since its introduction in the Gigabit Ethernet link, the VCSEL technology has seen a steady progress. VCSEL-based 10 GB-Ethernet conforming to the IEEE 802.3 standard has already been available since 2001 (Simoneaux, 2001). Moreover, numerous research groups are trying to use the VCSELs for RF transmission. Thus, the VCSEL can be used in Radio-over-Fiber (RoF) Systems for Distributed Antenna Systems (Lethien, Loyez, &

Figure 21. Turn-on delay τ_D for $I_{ON}=12*I_{TH}$

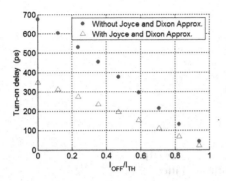

Figure 22. Test setup of a VCSEL based digital link

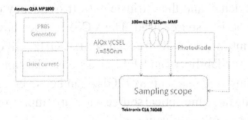

Vilcot, 2005; Sauer, Kobyakov, & George, 2007; Chia, Luo, Yee, & Hao, 2003). In this way, some systems have been demonstrated to improve the RF performance of the VCSEL such as:

- Optoelectronic oscillators to generate spectrally pure microwave signals (Varon, Le Kernec, & Mollier, 2007; Hayat, Varon, Bacou, Rissons, & Mollier, 2008).
- Optically injection-locked VCSELs to improve the cut-off frequency, to narrow the linewidth, to reduce the laser RIN and the laser frequency chirping (Hayat, Bacou, Rissons, Mollier, Iakovlev, & Sirbu, 2008; Lu, Tzeng, Chuang, Chi, & Liao, 2007).

Recently, VCSEL-based hybrid base-station architecture has been proposed (Bakaul, Nadarjah, Nirmalathas, & Wong, 2008). This architecture is able to deliver wireless RF Signal, and Wired Ethernet signal simultaneously.

Finally, the in-system performance of each application is discussed.

The Optoelectronic Oscillators (OEO), are frequently used to generate spectrally pure microwave signals for quite some time now. The optical signal was generally provided by a DFB LD. However, the configuration was not adapted to low cost and integrated systems. Hence, the solution is to utilize a directly modulated VCSEL, in lieu of DFB laser. The architecture of a VCSEL-Based Oscillator, such is presented in the Figure 25, is composed by a loop containing a LD, an optical fiber as a delay line, a photodiode, an RF filter centered on the frequency f_0, an RF amplifier

Figure 23. Eye diagram of a 100m fiber link at 1,25Gbps for $I_{off}<I_{th}$

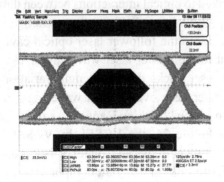

Figure 24. Eye diagram of a 100m fiber link for $I_{off}>I_{th}$

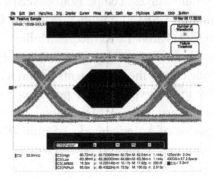

Figure 25. VCSEL Based Oscillator test bench

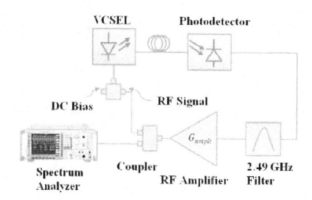

with a gain G_{ampli}.

This OEO allows converting the VCSEL lightwave into a spectrally pure RF signal whose frequency is selected by the RF filter of the loop. The noise generated in the loop, containing all frequency components, directly modulates the VCSEL. Out of these diverse modulating spectra only the frequency components satisfying the loop phase and gain conditions are amplified and emerge from the noise floor. These components are then filtered by the RF filter and amplified by the RF amplifier. This amplified signal becomes the modulating signal for the VCSEL. The resulting optical output from the VCSEL is a modulated optical carrier and the spectrally pure microwave signal. This signal is then detected by a photodetector, filtered and injected again in the VCSEL which involves a continuous production of high spectral purity oscillations at the desired microwave frequency.

The Table 4 gives a comparison between 850nm multimode VCSEL, 1.55 μm VCSEL and DFB-based OEO (Varon, Le Kernec, & Mollier, 2007; Hayat, Varon, Bacou, Rissons, & Mollier, 2008). It is evident from the experiments performed that in order to achieve a similar performance, the electrical consumption of a DFB laser is more than 8 times that of a single-mode 1.55μm VCSEL. The multimode 850nm VCSEL consumes about twice as much as current than the 1.55μm single

mode VCSEL. The oscillator frequency is limited by the availability of RF filter at a certain desired frequency and the intrinsic cut-off frequency of the optical source used. The VCSELs used have an intrinsic cut-off frequency of 12GHz. Given the availability of an RF filter, a sinusoidal signal of any frequency up to 12GHz can be generated using the same optical source without compromising any of the advantages.

In addition, the VCSEL RF performance can be enhanced by the Optical Injection-Locking (OIL). In effect, this technique gives the possibility to increase the VCSEL cut-off frequency. As technology of LW VCSELs is getting more mature and ready for practical applications, the potentiality of VCSEL-by-VCSEL OIL bring a relevant prospect for the high speed optical source. The OIL method consists in the injection of a light emitted by a master laser into a follower laser operating at the same wavelength. The dynamics of this technique are linked to optical interactions between the two lasers. Coherent light from the master laser is injected into the follower laser optical cavity to lock it to the master laser wavelength. The follower laser tracks the wavelength variations of the master laser over a short wavelength range known as the "Locking Range". As it presented in the Figure 26, the VCSEL-by-VCSEL OIL architecture employs an isolator, a 4-port optical coupler to link-up different components used to

Table 4. Comparison of optoelectronic oscillator characteristics using different optical source optical sources

	DFB SM (1.55µm)	VCSEL SM (1.55µm)	VCSEL MM (850nm)
I_{th}(mA)	13	1.5	3
Operation current (mA)	50	6	12.5
Phase noise @10kHz (dBc/Hz)	-108	-95	-100
Oscillator linewidth (Hz)	10	16	10
Fiber loop length (m)	100	100	120
Oscillator frequency (MHz)	900	2490	900
Temperature Control	Yes	No	No

conceive the test-bench. The utilization of two identical VCSEL significantly simplifies the IOL experimental setup because it eliminates the all types of polarization maintaining equipment.

The Figure 27 presents the 2-fold increase in the 3-dB cut-off frequency by VCSEL-by-VCSEL OIL technique. Moreover several studies in DFB IOL demonstrate a chirp and RIN reduction. Therefore, improving and integrating this system should give the possibility to improve the RF performance of the light source.

Another method to improve the modulation bandwidth is to use the non-linearity of the light-current curve to realize a microwave mixing and thus up-conversion or a data signal. This technique is frequently use for baseband digital signal or data signal centered on a low intermediate frequency (Constant, Le Guennec, Maury, Corrao, & Cabon, 2008). When the VCSEL is directly modulated in a non linear region of the L-I curve, the centre frequency of the signal is shifted to a higher frequency. The experiment consists in directly modulating a 1550nm VCSEL by the combination of a local oscillator (LO) CW-signal with an RF carrier or an RF carrier modulating by a digital signal. The CW mixing was obtained with an RF frequency f_{RF}= 1.1GHz and an LO frequency

f_{LO}=2GHz. An optimized conversion gain (ratio of the electrical power photodetected at the output of the link at the mixing product $f_{RF} + f_{LO}$ over the electrical power injected at f_{RF}) have been obtained by adjusting the bias current I_{bias} in the range 1.4 to 2mA and the input power (P_{RF} and P_{LO}) from -15 dBm to -10dBm according to a modulation in the non linear region.

This technique is applied to the RoF applications by modulating the RF frequency by a digital signal. The authors present two digital up-conversion results, the first case is a 10Msym/s quaternary phase-shift-keying (QPSK) signal and the second case is a multiamplitude, multicarrier wireless signal (WLAN IEEE 802.11a standard) at the bit rate 54Mb/s.

The up-conversion of the QPSK signal was achieved with I_{bias} = 1.4 mA according to a maximum conversion gain and by adjusting P_{RF} to find the best error vector magnitude (EVM). An optimum value of 3.9% EVM was obtained with P_{RF}=-23dBm and P_{LO} = -10dBm.

The up-conversion of a complex WLAN 802.11a digital signal is more critical. The same value of f_{RF}, f_{LO}, P_{LO} and I_{bias} than previously and P_{RF} = -27dBm provide a minimal EVM of 8.9% which is above the maximum value (5%) of the

Figure 26. VCSEL-by-VCSEL optical injection-locking experimental setup and spectrum of the optically injection-locked VCSEL

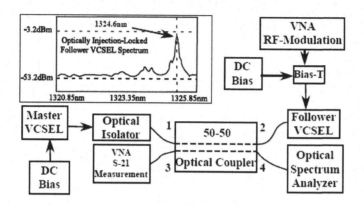

IEEE 802.11a standard.

Even if the results are not in good agreement with the WLAN standard, this technique should be improved for two reasons. Firstly, this frequency mixing is very interesting in the radio signals over optical fiber distribution allowing to achieve a low-cost up-conversion of RF digital signals. Secondly, the technology of the VCSEL utilized is not mature, the future generations of 1550nm VCSEL would provide upper optical power with better reliabilities.

In parallel to the improvement of the VCSEL properties, several network architecture using VCSEL have been demonstrated.

For example, Bakaul, Nadarjah, Nirmalathas, & Wong, (2008) proposed a VCSEL-Based hybrid base station (H-BS) able to provide simultaneously wireless and wired services (Figure 28). Previously, this scheme utilized a reflective semiconductor optical amplifier (RSOA). This technique was limited by the narrow modulation bandwidth of the RSOA, which need to be ther-

Figure 27. Comparison of the S21 spectra of free-running and optically injection-locked single-mode VCSELs

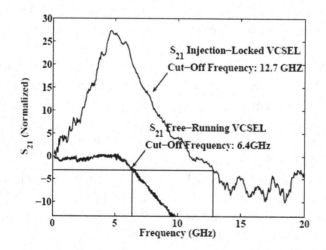

mally controlled, and also by the high cost. The utilization of the VCSEL was suggested by the author to provide a cost effective solution while allowing the possibility of integration and a good thermal stability. Thus, at the output of the remote access node a VCSEL H-BS provide wireless and wired services according to the scheme introduced by Figure 29. A three port optical oscillator is linked to the star coupler of the Remote Access Node (RAN). The downlink signal is sent to the photodetector via the port 2 of the optical circulator. The detected signal is divided into two parts by a 3dB electrical splitter to be sent to the RF antenna and the Ethernet Interface. The RF signal and the Ethernet signal are separated by a band pass filter to provide the 2,5GHz LO Signal with 155Mb/s data to the RF antenna and a low pass filter to provide the 1,25Gb/s Ethernet Signal to the Ethernet hub. The VCSEL is directly modulated by the output signal of an RF combiner which is the mixing of the wireless signal (provided by an RF Antenna) and the Ethernet signal. Thereby, the VCSEL emits an optical uplink signal which is sent to the RAN via the port-1 of the optical circulator.

A mock-up has been realized to evaluate the H-BS concept. A measurement of the recovered RF spectra demonstrates the feasibility of the H-BS. Furthermore, the BER of the wired Ethernet and wireless data have been measured after 5.6 and 10km single mode fiber length such as BER<10^{-9} for received optical power close to -22.5dBm for the 1.25Gb/s Ethernet signal and close to -21dBm for the 155Mb/s data at 2.5GHz. Thus, the concept of the VCSEL-based hybrid base station has been demonstrated by the authors. With an optimization of the mock-up, the architecture could be easily enhanced while given a new potential in the multipoint-to-multipoint communications.

Furthermore, Sauer, Kobyakov, & George (2007) proposed the introduction of the VCSEL in Picocellular Networks. They suggested a direct modulation of 850nm and 1310nm VCSEL for radio-over-fiber link on multimode and singlem-

ode fiber as it shown in Figure 30. Experiments have been realized in the 2.4 and 5GHz bands to achieve record fiber transmission over 1.14km with uncooled 1310nm VCSEL on MMF and 30km on SMF. A maximum EVM of 4.8%rms has been obtained. A transponder for a dual-band WLAN fiber radio picocellular network have been designed according to the system requirement in terms of sensitivity (<-90dBm) and a spurious free dynamic range (SFRD>95dB. $Hz^{2/3}$). It was demonstrated that the RF signal provided to cells is available to the propagation by implementing this scheme in a 14-cell demonstrator system and widely proofs the interest of VCSEL in this network. Thus, the authors have presented a reliable transponder for full WLAN throughput at both frequency ranges.

Even if this list of application is non-exhaustive, it gives up to date overview VCSEL utilization in the OAN.

CONCLUSION

For all these reasons, we could foresee that the VCSEL could become a versatile device in the OANs. Thanks to its rapid evolution and cost-effectiveness, the present performance makes the VCSEL one of the most used Lasers in wide application fields.

A presentation of the component, its structure, its operation and its model has been introduced to give a versatile knowledge of the VCSEL in order to allow its usage in existing systems.

Moreover, the last section gave some techniques to improve the RF performance of the VCSEL such as the VCSEL-Based Oscillator, the Optically Injection Locked VCSEL and the modulation in non-linear region of the Light-Current curve. After optimization depending on the future generation VCSEL performance and integration of mock-up, all these techniques could be used in Optical-Access-Network scheme.

Finally two kind of OAN architecture based

Figure 28. Wired and wireless access network architecture

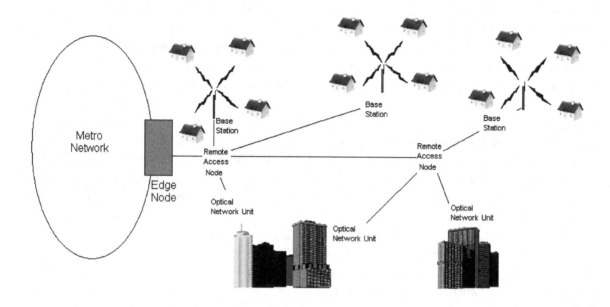

on VCSEL have been presented. The first one was a VCSEL based Hybrid-Base Station to provide wired and wireless signal. The concept has been proved and the advantage of the VCSEL use has been demonstrated, but the performance needs to be improved before the implantation. The second one was a radio transmission link based on directly modulated VCSEL. The results obtained were very promising because a reliable transponder has been developed.

Thus, with the current progress of the VCSEL technology, we can guess that the VCSEL will

Figure 29. H-BS: Optical Network Unit /Base Station

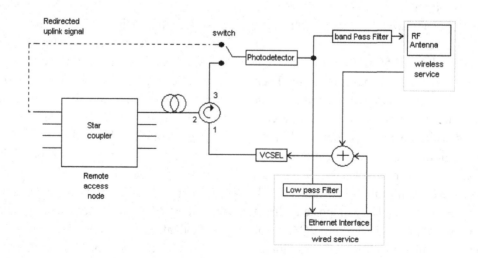

Figure 30. Architecture for radio transmission based on directly modulated VCSEL

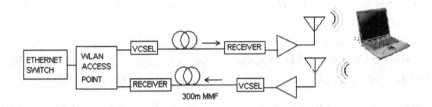

become a component of choice for the novel Optical Access Network.

REFERENCES

Bakaul, M., Nadarjah, N., Nirmalathas, A., & Wong, A. (2008). Internetworking VCSEL-Based Hybrid Base Station towards Simultaneous wireless and wired Transport for Converged Access Network. *IEEE Photonics Technology Letters*, *20*(8), 569–571. doi:10.1109/LPT.2008.918875

Beam Express. (2008). *VCSEL, Long Wavelength VCSEL*. Retrieved in August 2008, http://www.beamexpress.com/

Boucart, J., Starck, C., Gaborit, F., Plais, A., Bouche, N., & Derouin, E. (1999). Metamorphic DBR and tunnel-junction injection. A CW RT monolithic longwavelength VCSEL. *IEEE Journal on Selected Topics in Quantum Electronics*, *5*(3), 520–529. doi:10.1109/2944.788414

Boucher, Y., Rissons, A., & Mollier, J. C. (2001, April). *Temperature dependence of the near-threshold emission wavelength a linewidth in a Vertical-Cavity Surface-Emitting Laser (VCSEL)*. Paper presented at the conference SIOE01, Cardiff, UK.

Chang-Hasnain, C. J. (2003). Progress and prospects of long-wavelength VCSEL. *IEEE Optical Communications*, *41*(2), 530–534.

Chia, M. Y. W., Luo, B., Yee, M. L., & Hao, E. J. Z. (2003). Radio over multimode fiber transmission for Wireless LAN using VCSELs. *Electronics Letters*, *39*(15), 1143–1144. doi:10.1049/el:20030724

Chow, W., Choquette, K. D., Crawford, M., Lear, K., & Hadley, G. (1997). Design, Fabrication, and Performance of infrared and Visible Vertical-Cavity Surface-Emitting Lasers. *IEEE Journal of Quantum Electronics*, *33*(10), 1810–1824. doi:10.1109/3.631287

Constant, S. B., Le Guennec, Y., Maury, G., Corrao, N., & Cabon, B. (2008). Low-Cost All-Optical Up-Conversion of Digital Radio Signals Using a Directly modulated 1550-nm Emitting VCSEL. *IEEE Photonics Technology Letters*, *20*(2), 120–122. doi:10.1109/LPT.2007.912543

Dixon, R. W., & Joyce, W. B. (1979). Generalized Expressions for the turn-on delay in semiconductor lasers. *Journal of Applied Physics*, *50*(7), 4591–4595. doi:10.1063/1.326566

Hayat, A., Bacou, A., Rissons, A., Mollier, J.-C., Iakovlev, V., & Sirbu, A. (2008, November). *1.3 μm Single-Mode VCSEL-by-VCSEL Optically Injection-Locking for Enhanced Microwave Performance*. Paper Presented at IEEE LEOS Conference, Newport, CA.

Hayat, A., Varon, M., Bacou, A., Rissons, A., & Mollier, J.-C. (2008, September). *2.49 GHz Low Phase-Noise Optoelectronic Oscillator using 1.55μm VCSEL for Avionics and Aerospace Application*. Presented at the Conference Microwave Photonics 08, Gold Coast, Australia.

Ibaraki, A., Kawashima, K., Furusawa, K., Ishikawa, T., Yamaguchi, T., & Niina, T. (1989). Buried Heterostructure GaAs/GaAlAs Distributed Bragg Reflector Surface Emitting Laser with Very Low Threshold (5.2 mA) under Room Temperature CW Conditions. *Japanese Journal of Applied Physics*, *28*(4), 667–668. doi:10.1143/JJAP.28.L667

Iga, K. (2000). Surface-Emitting Laser – Its Birth and Generation of new Optoelectronics Field. *IEEE Journal on Selected Topics in Quantum Electronics*, *6*(6), 1201–1215. doi:10.1109/2944.902168

Iga, K., Koyama, F., & Kinoshita, S. (1988). Surface Emitting Semiconductor Lasers. *IEEE journals of Quantum Electronics, 24*(9), 1845-1855.

Jewell, J. L., Scherer, A., McCall, S. L., Lee, Y. H., Walker, S., Harbison, J. P., & Florez, L. T. (1989). Low threshold electrically pumped vertical-cavity surface-emitting microlaser. *Electronics Letters*, *25*(17), 1123–1124. doi:10.1049/el:19890754

Kopf, R. F., Schubert, E. F., Downey, S. W., & Emerson, A. B. (1992). N- and P- type doping profiles in Distributed Bragg Reflector Structures and their effect on resistance. *Applied Physics Letters*, *61*(15), 1820–1822. doi:10.1063/1.108385

Koyama, F. (2006). Recent advances of VCSEL Photonics. *IEEE Journal Of Lightwave Technology*, *24*(12), 4502–4513. doi:10.1109/JLT.2006.886064

Lethien, C., Loyez, C., & Vilcot, J.-P. (2005). Potentials of Radio over Multimode Fiber Systems for the In-Buildings Coverage of Mobile and Wireless LAN Applications. *IEEE Photonics Technology Letters*, *17*(12), 2793–2795. doi:10.1109/LPT.2005.859533

Lu, H.-H., Tzeng, S.-J., Chuang, Y.-W., Chi, Y.-C., & Liao, C.-W. (2007). Bidirectional Radio-over DWDM Transport Systems Based on Injection-Locked VCSELs and Optoelectronic Feedback Techniques. *IEEE Photonics Technology Letters*, *19*(5), 315–317. doi:10.1109/LPT.2007.891627

Ly, K., Rissons, A., Gambardella, E., & Mollier, J.-C. (2008, April). Optimization of An Avionic VCSEL-Based Optical Link Through a Large Signal Characterization. In Proceeding of SPIE, *Photonics Europ Symposium*: Vol.6997.

Rissons, A., Mollier, J.-C., Toffano, Z., Destrez, A., & Pez, M. (2003). Thermal and Optoelectronic Model of VCSEL Arrays for Short Range Communications. In *Proceedings of SPIE, Vertical-Cavity Surface-Emitting Lasers VII, 4994*, 100-111.

Rissons, A., Perchoux, J., Mollier, J. C., & Grabherr, M. (2004). Noise and Signal modeling of various VCSEL structures. In . *Proceedings of SPIE Vertical-Cavity Surface-Emitting Lasers VIII, 5364*, 81–91.

Sauer, M., Kobyakov, A., & George, J. (2007). Radio Over Fiber for Picocellular Network Architectures. *Journal of Lightwave Technology, 25*(11), 3301–3320. doi:10.1109/JLT.2007.906822

Scott, J. W., Geels, R. S., Corzine, S. W., & Coldren, L. A. (1993). Modeling Temperature effects And Spatial Hole Burning To Optimize Vertical-Cavity Surface-Emitting Laser Performance. *IEEE Journal of Quantum Electronics, 29*(5), 1295–1308. doi:10.1109/3.236145

Shau, R., Ortsiefer, M., Rosskopf, J., Boehm, G., Lauer, C., Maute, M., & Amann, M. (2004). Longwavelength InP-Based VCSELs with buried tunnel Junction: properties and applications. In . *Proceedings of SPIE Vertical-Cavity Surface-Emitting Lasers VIII, 5364,* 1–15.

Simoneaux, C. (2001). Optical Communications: VCSELs propel 10-Gbit Ethernet. *EETimes*. Retrieved August 2001, from http://www.eetimes.com/story/OEG20010806S0059.

Syrbu, A. (2005). Wafer fused long-wavelength VCSELs with InP-Based active Cavities. In IEEE, *International Conference on Indium Phosphide and Related Materials* (pp. 670-674).

Szweda, R. (2006). VCSEL applications diversify as technology matures. *III-V Review, 19*(1), 34–38. doi:10.1016/S0961-1290(06)71477-6

Toffano, Z., Pez, M., Le Brun, C., Desgreys, P., Hervé, Y., & Mollier, J.-C. (2003). Multilevel Behavioral Simulation of VCSEL based Optoelectronic Modules. *IEEE Journal on Selected Topics in Quantum Electronics, 9*(3), 949–960. doi:10.1109/JSTQE.2003.818348

Tucker, R. (1985). High-Speed Modulation of Semiconductor lasers. *IEEE Transactions on Electron Devices, ED-32*(12), 1180–1192.

Ulm photonics company (2008). VCSEL technology, Retrieved in August 2008, from http://www.ulm-photonics.de/.

Varon, M., Le Kernec, A., & Mollier, J.-C. (2007). Opto-Microwave source using a harmonic frequency generator driven by a VCSEL-Based Ring Oscillator. *Proceeding of the European Microwave Association, 3*(3).

Vertilas (2008), Technology, Retrieved in August 2008, from http://www.vertilas.com/

Wilmsen, C. W., Temkin, H., & Coldren, L. (1999). *Vertical-Cavity Surface-Emitting Lasers, Design, Fabrication, Characterization and applications*. Cambridge University press.

Section 4
Business Models and Techno–Economic Evaluations

Chapter 10
Business Models for Municipal Metro Networks
Theoretical and Financial Analysis

Vagia Kyriakidou
University of Athens, Greece

Aristidis Chipouras
University of Athens, Greece

Dimitris Katsianis
University of Athens, Greece

Thomas Sphicopoulos
University of Athens, Greece

ABSTRACT

In Europe initiatives towards the development of optical networks infrastructures' have undertaken in order to address the need of faster and more telecommunications services. Consequently, the implementation of appropriate business model seems to be meaningful tool for infrastructures as it could ensure commercial viability and limit investment's risks. Although, the variety of forms characterizes a business model, there are some aspects that networks owners or decision makers should consider in each case. The purpose of this book chapter is to highlight the main issues that should be considered from the main actors involved in the business models like telecom operators, fiber optical constructors as well as municipalities that own the main part of the infrastructure.

INTRODUCTION

From dial-up connections to xDSL Internet users demand continuously higher bit rate in order to enjoy new offered services like Video on Demand, Voice over IP and more speed stream. Nowadays, people enjoy broadband communications through xDSL, FTTx and other technologies. High speed stream Internet turn to be a necessity for modern people that causes a continuous increase in the demand for higher bandwidth (Zukerman, Mammadov, Tan, Ouveysi, & Andrew, 2008).

Fiber optics networks seem to be the solution in this problem. The transmission capacity of optical

DOI: 10.4018/978-1-60566-707-2.ch010

fibers depends on the fiber length, type and transmitter technology (Cloetens L., 2001). However, optical fibers can achieve data rates up to 10 Gbps or more. Without this capacity we couldn't have real broadband services.

There are many areas where fiber optics networks have constructed, as such kind of networks are considered as the most future-proof access technology (Green P.E., 2004). Fiber optics networks seem to be a reliable solution for broadband access as well as for real broadband services. These services have a variety of applications in all aspects of daily life. They can be offered to business and residential costumers. There are applications like e-education, e-health, e-games etc, providing many options to all subscribers.

These infrastructures should be operated in an effective way so that business modeling must be applied. Talking about business models, we must make clear that we are going to analyze the operation of fiber optics networks, which players appear, the obligations they have and the relationships between them (Parr Rud O., 2001).

Differentiations among societies don't allow a strict deployment for optical networks. There is a variety of political, technological and economic drivers behind the development of this kind of networks. A deep study of these drivers could give answers and lead to the best practice for each case. Engineers and decision makers should keep in mind that each network is a different case (Michalakelis, Varoutas, & Sphicopoulos, 2008).

However there could be a basic standardization in networks' deployment. The collection of data from many cases and classification would lead to these standards.

The rest of this chapter is structured in the following way: In "Case Studies" optical MANs (Metropolitan Area Networks) from all over the world are discussed in order to overview practices that have already employed. In the section "Business Models' analysis", the proposed models are presented based on SWOT analysis. Next section "Technoeconomic analysis for Business

Model "Private Initiative in Active Layer"", deals with the financial results and the viability of the networks based on a specific business scenario. Finally, in the last section, conclusions based on the main results reached from previous sections are presented.

Case Studies

This section presents and validates the existing deployments in optical infrastructures from Europe as well as worldwide. Many countries invest on fiber optics networks and so far many towns all over the world enjoy broadband services through that kind of networks. Through this study, useful information according to the international practices can be acquired and based on them benchmarking can be employed in areas where such kinds of initiatives start. The following cases have some similarities, although, their differences are more interesting for this chapter. The most important information is the ownership of the network, the decision makers who administrate it, and the population of considered Municipalities. Population is a substantial factor regarding to that decision, because, combined with broadband penetration, determines the potential users.

Sweden

It is widely known that Scandinavia countries are pioneers on technological issues. Firstly, we present some initiatives that have taken place in Sweden.

The city of Stockholm could not ignore the challenge of broadband development in country. The capital city of Sweden contains the largest Municipality in the country with almost 800.000 inhabitants. Municipality decided to found municipal company Stokab in order to promote initiatives at regard of economic growth through IT development. Stokab build a fiber optic network and is responsible for its operation and maintenance. The company owns passive layer and active equipment and leases them to Services Providers. The network is open access for all players on equal terms and company has found the way to operate it with the most effective way. Nowadays, Stokab cooperates with many ISPs and end users can enjoy a variety of broadband services as they have to choose among many ISPs (AB Stokab, 2006a, 2006b). Stockholm is the economic centre of Sweden and this network gives an additional advantage for further development in the area. Many sectors have already benefited such as business, research, public domain etc.

Sollentuna and Hudiksvall are two Swedish cities that own fiber optic networks as well (Sollentuna Energi, 2002). Through municipal companies both cities decided to cooperate with Ericsson and invested in FTTH (Ericsson AB, 2002a). They developed a three layer network which is open access to all Service Providers (Ericsson AB, 2002b; Hudiksvallsbostäder, 2007). The construction of these networks was partly ready in 1999 and in the next years networks were extended in order to provide sufficient coverage in the Municipalities. Both cases have similarities with the previous case of Stockholm as far as the ownership and the administration of the networks. Although, it must be mentioned, that the total population of these cities is 60.000 and 38.000 inhabitants respectively. In both cases, end users can choose freely which SP satisfies their needs and they can enjoy high speed broadband connectivity (up to 100Mbps) for a low monthly cost. It seems that these projects were a very optimistic vision of Municipal decision makers, but prove that such kind of infrastructures should be developed not only in vast urban areas.

Ireland

Ireland is also a brilliant example of a country that faced the challenge of broadband reality as a national issue. It was decided to develop a large optical network that is to be installed in towns from Co. Donegal to Cork under the so-called 19

towns' project. The project is targeting in three main issues, which are the financial support for the development of broadband infrastructures, the transformation of legal and regulatory environment and, finally, e-Business, Lifelong learning and e-Inclusion. Irish Government estimated that these infrastructures will enhance the effectiveness of public sector, will encourage private investments and will improve the quality of daily life for all citizens.

Network will provide always-on Internet Services that can be used for many applications in education, health, industry, business and research area. The network is partly completed by now and gives a great advantage to all country and an opportunity for economic growth. The project will be funded by European and National sources while municipalities will give 10% of the total cost (Ireland's NDP, 2006). According to the "NDP 2007-2013" the initial plan was extended and included, among others, in national investment priority (Ireland's NDP, 2007).

Netherlands

Netherlands is a country that has developed broadband infrastructures in many areas. We present some of these examples in the passages below.

The city of Amsterdam is a typical example of a great initiative. The project that will be completed in 2013 called CityNet and is the largest in Europe. The project is an optical network that could connect 420,000 costumers – citizens and businesses –. The first face is already begun, and there are 40,000 fiber-to-the-home (FTTH) connections. The network is open access as it helps competition among ISPs. Municipality owns the passive layer while BBned that belongs to Telecom Italia operates active network. Customers in Amsterdam can choose among a variety of offered services and can enjoy them for a low monthly cost. All project is funded by municipal sources at a cost of € 300 M, however municipality believes that the network will give the economy a boost which will cover the investment cost (Bbned, 2007; Citynet Amsterdam, 2006; Damien Chew, 2006).

Almere is a Netherlands city that since 2004 tries to play a crucial role in global broadband map. Through the Almere Fiber Project the municipal company UNET cooperated with commercial parties such as Cisco Systems, IBM, SARA Computing etc, and founded a Private Public Partnership (PPP). This Partnership decided the construction of a fiber optic network that is open access to all players on equal terms. There are many ISPs which set out in Almere and they provide a diversity of broadband services on competitive prices (Cisco Systems, 2004; Sara, 2005).

The city of Rotterdam followed the same plan as Amsterdam. The local real estate company, Stadswonen decided to invest in fiber-to-the-home and gain its optic network. The project was decided in 2006 and it is to be completed in 2008. According to Stadswonen's forecasts, network could offer broadband services to 5,500 customers. In order the project is realized the same players as in Amsterdam will cooperate. BBned will operate active layer and deal with ISPs. Stadswonen figures on offering cheap services to citizens and local businesses, while it plans to offer free Internet access to all students in town (City of Rotterdam, 2007; SURFnet, 2007).

In the city of Hillegom, Lijbrandt Telecom that is a private company, decided to invest in fiber optic network. This company belongs to Dik Wessels who owns Volker Wessels that takes part in Amsterdam and Rotterdam projects for network construction. Lijbrandt Telecom has the know-how and by now it has over 7,500 connections. Company believes that can gain 70% of total demand in the town. It has also decided to expand its network to neighbor town Lisse, which has 10,000 citizens and in all neighbor area. The aim of this plan is to reach 100,000 connections in next years. The case of Hillegom shows that private segment could invest in optic networks as they seem to be a profitable business (Lijbrandt telecom, 2007; VolkerWessels, 2007).

Iceland

Iceland is a country with high rates of development. The city of Reykjavik has many times awarded of innovative projects. The municipal company Reykjavik Energi started to build an open access optical network in 1999. In 2004, network served 4,000 subscribers while 2005 the number of subscribers was 15,000. The company believes that will gain 65,000 connections by the end of 2008, when total population of the city is about 120.0000 inhabitants. Reykjavik Energi holds passive layer and active layer and leases the infrastructure to ISPs. The company operates also as an ISP and all subscribers should pay a monthly fee to the company (Gislason, 2004; Rosen, 2004). Reykjavik Energi succeeded to run a profitable and viable optical network, by gaining vertical convergence from passive layer to broadband services.

USA

The most representative sample of a regional broadband initiative in USA is the project UTOPIA. It has taken place in Utah region and initial included 11 towns from the wide area. The first part of the construction has completed in 2007 and the expected penetration was calculated about 55% of potential demand. From the beginning of construction, the whole issue was faced as regional project and five more cities were jointed to this project. In addition, it was claimed that the total demand in an area will motivate more ISPs and so competition will be developed. The total population of this area is estimated at more than 700.000 inhabitants. Network is open access for all players on equal terms (Marinkovich & Sybrowsky, 2003; UTOPIA, 2006). UTOPIA project was developed through a very analytical financial analysis, based on the perspective of share costs that gives economic advantages. The developed network is also a three layer network, in which Municipalities own, operate and maintain passive and active layer.

BUSINESS MODELS' ANALYSIS

In this section the proposed Business Models are presented, together with a SWOT analysis for each one. Based on this analysis the evaluation of the models can be done. It is assumed that the considered candidate models are the most appropriate for the deployment of fiber optical networks in emerging markets. The adoption of an existing model, even if it was a successful one, cannot ensure positive results in another country or region. According to the analysis, proposed models could be the best practices in countries or regions where there is no competition in infrastructures, e.g. Hillegom, and Municipalities are the only network providers. In addition, the following solutions are suitable when there is no previous experience in the administration of such kinds of networks and Municipalities don't want to gain the required know how.

Although, the diffusion of broadband services and the necessity of broadband infrastructures are in political agendas of many countries, there is a lack in bibliography according to the basic concept of building and operating broadband networks (Owen & Raj, 2003).

Apart from the huge investments needed for a FTTH network there are some questions with equal importance than financial issues (Aagedal et al., 2002). Next paragraphs present:

1. Modified Business models for fiber optical networks
2. SWOT analysis for each Business model

The presentation of many case studies gives the dominant mood according to the deployment of optical infrastructures. The classification of these cases could be a useful object for future decisions.

The analysis of Modified Business models

for fiber optical networks analyzes the network's architecture, the players who participate and the relationships among them. Taking into account the discrete layers of the networks, answers are given to the following questions:

a. Who owns each layer
b. Who administrates each layer

The administration of these networks is a very important factor as the effective operation, which could ensure networks' viability, depends on administrator(s) decisions.

Needless to say, SWOT analysis is a very useful tool for marketers to focus on key issues. Strengths and Weaknesses are internal factors while Opportunities and Threats are external ones (Adkins I., 2001).

In Figure 1 proposed Business Models are presented. As it comes along from the analysis of existing cases, an optical network is consisted of three layers, which are Passive, Active and Service Layer. Passive or Physical Layer includes all the necessary pipes, ducts, fibers etc in order a complete optical fiber network to be constructed. Active Layer consists of all the elements needed to transmit, forward and route packets of information. Finally, Service Layer referrers to the Services that Service Providers (SPs) will offer to the end users. The discrimination among the proposed Models is depended on ownership of each Layer. According to this analysis, there are five possible participants in each Business Model, Figure 2.

In the Service Layer, SPs are going to set out and trying to gain their market share. Public Sector in this analysis is identified with Local Municipalities. As long as Private Sector, it could be consisted of any kind of firm or company, (national or international), that would like to participate in the initiative of constructing and operating optical MANs, even if they don't have previous experience. According to the previous presentation of international cases or projects there are real-estate, housing associations etc

Figure 1. Proposed Business Models

Network Layers	BUSINESS MODELS			
	PPP in Active Layer	Private Initiative in Active Layer	Public Ownership-Outsource Administration	PPP in Infrastructure
Services	▯▯▯	▯▯▯	▯▯▯	▯▯▯
Administration	◧☐	☐☐⬚	⬚	
Active Layer	◧	◧☐		◧
Passive Layer	▬	▬	▬	

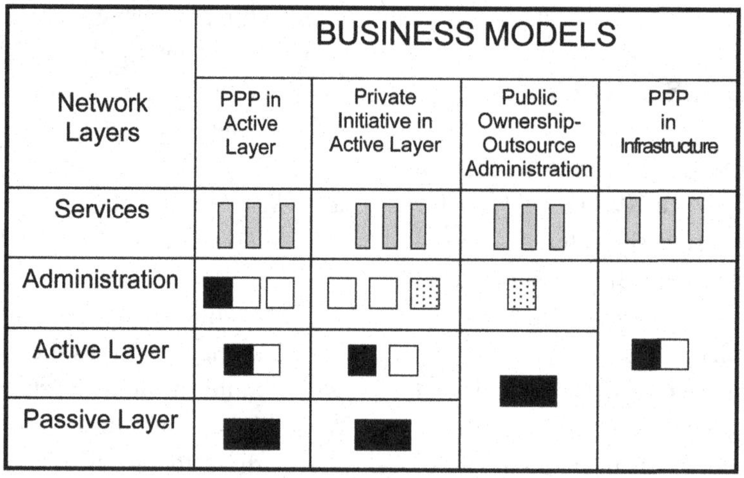

companies which participate in such kind of initiatives. Public Private Partnership (PPP) could be established in order to assume the responsibility to operate and develop the initial infrastructures. If Municipalities don't have the required know how to operate optical MANs, an Outsource Operator can be awarded (after an open contest) by them and ensure networks' operation.

Apart from the three layer structure that is adopted, the joint action of Municipalities is also proposed such as UTOPIA project. In that case, economies of scales and cost limitation are expected. One more similarity is that the proposed models refer to regions with medium or low population, e.g. Sollentuna and Hudiksvall cases, as real broadband infrastructures have become a necessity not only for the vast urban areas. Moreover, the establishment of a PPP and the collaboration with private sector is also proposed as an alternative, e.g. cases of Amsterdam,

Rotterdam and Almere. We also believe that, mainly in developing countries, subsidization of the construction of fiber optical networks will be a common practice, as countries include such kind of initiatives in their development plans, e.g. Ireland, and Municipalities should take the advantage to invest in optical networks.

However, there are some crucial differences from the existing cases that presented in the previous section. Based on the characteristics of the regions in which proposed models are deployed, existing business models were modified in order to fit better and become best practices for specific cases.

Hence, some practices were rejected, for example, in the considered case of vertical convergence, it seems that this prospect creates high complexity to the Municipalities. In addition, in some Municipalities operate optical networks on their own e.g. Stockholm, Reykjavik etc.

Figure 2. Participants in proposed Business Models

Service Provider	Public Private Partnership (PPP)	Public Sector	Private Sector	Outsource Operator
▯	◧	▬	☐	⬚

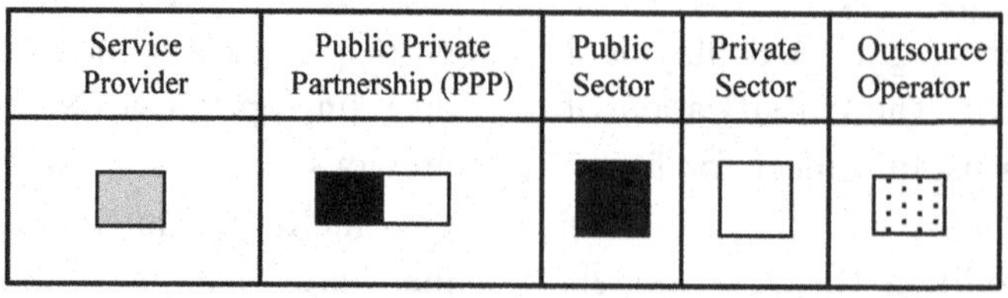

Table 1. SWOT analysis for the Business Model "PPP in Active Layer"

Strengths	Weaknesses
Lower expenses for the extensions of Active equipment – share with Private Partner Municipalities own 100% of Passive Layer and receive all the profits Municipalities can benefit from other municipal excavation projects and extend the initial infrastructure ISPs can get into the market with low budget	Municipalities must finance the cost of all the extensions of the Passive Layer Great complexity due to the participation of many participants Private Sector's participation will probably lead to higher prices of offered services The procedure of PPP's establishment will delay the effectuation of the model
Opportunities	**Threats**
Faster extension of Active Layer Participation of the Public sector provide a level of safety to the Private sector Joint administration in the region enhances joint development of the infrastructures and decreases joint expenses due to the economies of scale End users' utility will grow up as penetration rate will increase The initial interconnection of all Public Departments in the Region encourages the use of broadband services by the Public sector and improves the quality of offered services Citizens will benefit from the better operation of the Public sector in their Region	ISPs should offer their services through the existent equipment Private sector could invest in particular equipment and create lock-in problems Low penetration rate in the Region Low penetration means low profits The lack of the necessary know how lead to the need of an external administrator which means more expenses The compulsory replacement of the administrator of the Passive Layer and the possible cost should be considered

We believe that, at the beginning of the roll out, administration responsibilities should be charged either to private domain -in case of collaboration with the public sector- or outsource.

PPP in Active Layer

In this business model Municipalities will own Passive layer. In the active layer a Public Private Partnership is proposed to be established. This Partnership will take the responsibilities to transact with ISPs which in turn will offer final services to end users. According to administration PPP have two alternatives, either experts to employ or to award the administration to a private operator (outsourcing). In Table 1 SWOT analysis for this Business Model is presented.

Private Initiative in Active Layer

In this business model Municipalities will own Passive layer and give the operation of the networks to an operating company. Active layer is proposed to be developed by Private Sector. Private Sec-

tor will take the responsibilities to transact with ISPs which will offer final services to end users. Municipalities probably would own some Active equipment as well, although it would not be for commercial reasons. Outsource administrator will transact with Private sector or ISPs and will take all responsibilities for networks operation. SWOT analysis for the Business Model "Private initiative in Active Layer" is illustrated in Table 2.

Public Ownership-Outsource Administration

In this business model Municipalities will own Passive and Active layer as well. They will appoint the operation of the infrastructures to an operating company (outsourcing). This operating company will take the responsibilities to transact with ISPs which in turn will offer final services to end users. In Table 3, SWOT analysis for the Business Model "Public Ownership-Outsource Administration" is presented.

Table 2. SWOT analysis for the Business Model "Private initiative in Active Layer"

Strengths	Weaknesses
Municipalities focus on the extension of the Passive Layer Municipalities decide exclusively for networks' extensions and can take into account social issues Municipalities own 100% of Passive Layer and receive all the profits Municipalities can benefit from other municipal excavation projects and extend the initial infrastructure ISPs can get into the market with low budget Low complexity	Municipalities must finance the cost of all the extensions of the Passive Layer The lack of the necessary know how lead to the need of an external administrator which means more expenses Municipalities should convince administrator to take into account social issues regarding to decisions
Opportunities	**Threats**
Competition can be easily developed in Active Layer Private sector in Active Layer accelerates extension's procedure Joint administration in the region enhances joint development of the infrastructures and decreases joint expenses due to the economies of scale Higher penetration rate is expected as Private sector will invest in equipment in order to serve more end users End users' utility will grow up as penetration rate will increase The initial interconnection of all Public Departments in the Region encourages the use of broadband services by the Public sector and improves the quality of offered services Citizens will benefit from the better operation of the Public sector in their Region	Competition in Active Layer would cause an extra cost to the operators Private sector will probably invest in different equipment in Active Layer and it would be very difficult for that layer to be administrated Low penetration rate in the Region Low penetration means low profits The compulsory replacement of the administrator of the Passive Layer and the possible cost should be considered

Table 3. SWOT analysis for the Business. Model "Public Ownership-Outsource Administration"

Strengths	Weaknesses
Ownership of Passive and Active Layer leads to higher quality in offered services Economies of scale in both Layers (Passive and Active) Public ownership is expected to lead to lower prices due to the cost oriented approach Offered services would be common to all Region Low complexity The appropriate Administrator will ensure the efficient operation Municipalities decide exclusively for infrastructure' extensions and can take into account social issues Municipalities own 100% of the infrastructure and receive all the profits Municipalities can benefit from other municipal excavation projects and extend the initial infrastructures ISPs can get into the market with low budget	Municipalities must finance the cost of all the extensions of the infrastructures The lack of the necessary know how lead to the need of an external administrator which means more expenses Municipalities should convince administrator to take into account social issues regarding to decisions ISPs can not have access to the infrastructure Limited space for service differentiation from ISPs Different priorities by the participants could decrease efficient operation Competition can be developed only in Service Layer
Opportunities	**Threats**
Vertical integration in infrastructure attracts more ISPs Faster procedure for ISPs to get into the market Joint administration in the region enhances joint development of the infrastructures and decreases joint expenses due to the economies of scale End users' utility will grow up as penetration rate will increase The initial interconnection of all Public Departments in the Region encourages the use of broadband services by the Public sector and improves the quality of offered services Citizens will benefit from the better operation of the Public sector in their Region	Other Telecom operators could develop their own infrastructures Low penetration rate in the Region Low penetration means low profits The compulsory replacement of the administrator of the Passive Layer and the possible cost should be considered

Table 4. SWOT analysis for the Business Model "PPP in infrastructure"

Strengths	Weaknesses
Expenses for extensions would be shared between Public and Private sector ISPs can get into the market with low budget Ownership of Passive and Active Layer leads to higher quality in offered services Economies of scale in both Layers (Passive and Active) Offered services would be common to all Region Municipalities can benefit from other municipal excavation projects and extend the initial infrastructures	Great complexity due to the participation of many participants Municipalities should convince Private participants to take into account social issues regarding to decisions Competition can be developed only in Service Layer ISPs can not have access to the infrastructure Limited space for service differentiation from ISPs Different priorities by the participants could decrease efficient operation
Opportunities	Threats
Joint administration in the region enhances joint development of the infrastructures and decreases joint expenses due to the economies of scale Higher penetration rate is expected as Private sector will invest in equipment in order to serve more end users The initial interconnection of all Public Departments in the Region encourages the use of broadband services by the Public sector and improves the quality of offered services Vertical integration in infrastructure attracts more ISPs Faster procedure for ISPs to get into the market End users' utility will grow up as penetration rate will increase Citizens will benefit from the better operation of the Public sector in their Region	Low penetration rate in the Region Low penetration means low profits Private participation in administration will influence networks' development Private sector will probably insist on higher margin revenues and that could affect prices Other Telecom operators could develop their own infrastructures

PPP in infrastructure

In this business model, Municipalities will participate with Private Sector and a PPP will be established. This PPP will own Passive and Active layer as well. Moreover, PPP will take the responsibilities to operate infrastructures and transact with ISPs which will offer final services to end users. Public sector will have a consulting role and due to the lack of previous experience, Private sector will be responsible for decision making. SWOT analysis for Business Model "PPP in infrastructure" is presented in Table 4.

QUALITY ANALYSIS

The evaluation of the proposed Business models could be done through a quality analysis (Luna-Reyes Luis F. & Andersen D. F., 2003) which is based on 10 important factors. These factors are easy decision making, complexity, transpar-ency, total value of networks, risk and required investments, time scaling, expected performance, competition, functionality and break-through participation. According to the study of international cases, considered factors were the most popular among cases. Furthermore, they deal with all stages of networks' development. If t_0 is the time when networks are ready and services are offered to end users, some of these factors are referred to time before t_0 and some after that.

We assume that active municipalities' participation will be embarrassed by the lack of previous experience. Complexity has to do with the number of players in a business model and with the difficulty to reach an agreement. According to time scaling, there must be a public control so that delays are avoided. The total value of networks is similar in all cases, but where municipalities take more active role, the value is not just commercial but also social. Outsource operation seems to be more risky for operating company where there is no previous application.

According to required investments, outsource operation has also unfavourable prospects as operating company should defray them on its own. Time scaling is defined always from the beginning but in outsource operation model, operating company could be more efficient in construction and at the offset. Expected performance and competition will be the same regardless to business model, as we suppose that in all cases players will have common targets, which will be to perform efficiently and to develop competition. Although, it doesn't mean that the evaluation of each model will be the same, on the contrary, it means that the effort from all involved player would be. The evaluation results depend on the aggregate influence of each model to the effective operation of the Municipal networks.

Functionality will be depended on operation planning where private sector can be more effective. An initiative like the construction, operation and development of an optical network requires a detailed plan which should at least includes a technical, financial and operational analysis. Private domain traditionally has more experience with commercial issues and its participation in the process of the development of such a plan could be beneficial for the network' operation.

Fiber optics networks are considered as breakthrough infrastructures. The participation to their operation is a challenge for private sector but it is also a bigger one for municipalities. In addition, the participation of the public sector in operation of the developed networks could give guarantees to potential investors.

In Table 5 all the above factors and the proposed business models' evaluation are presented. The attribute of all factors can be either positive or negative and the scale ranges from three minus to three plus. Each considered factor, positive or negative, is evaluated based on three basic measures that are High (three plus or minus), Medium (two plus or minus) and Low (one plus or minus). This general normalization provides information according to the expected dynamics

from the employment of each business model. The advantage of a model to ensure competition could be limited if high complexity is required.

TECHNOECONOMIC ANALYSIS FOR BUSINESS MODEL "PRIVATE INITIATIVE IN ACTIVE LAYER" AND VIABILITY

Technoeconomic analysis is based on the assumption that Municipalities choose to own, operate and develop only Passive Layer, living Active and Service Layer to private sector (selected Business Model: "Private Initiative in Active Layer"). The following analysis is based on data from a typical emerging market. Greece is in the first stages of broadband evolution and therefore is chosen as suitable case study for running a technoeconomic analysis.

Initially, Capital Expenses (CapEx) and Operational Expenses (OpEx) that come along with the operation cost of networks are determined and calculated by using technoeconomic tools developed in (Ims, 1998; Katsianis et al., 2001; Olsen et al., 2006). CapEx consist of office requirements, such as software licenses, and coverage cost, which based on available data is estimated at 48€ per meter. On the other hand, OpEx include employees, training, space rental, maintenance for Passive Network etc.

Apart from these business modeling assumptions the analysis includes the detailed examination of three main scenarios in three different Regions. Each one of these Regions has a different number of inhabitants. There is a small (S), a medium (M) and a large (L) sized Region in terms of population density as well as required network length. The density of the Large sized region will improve the profitability since the required network length per inhabitant will be lower. It is assumed that Municipalities decide to act jointly in order to gain economies of scale. As a result, Municipalities for the same Region are going to

Table 5. Quality analysis of proposed Business Models

	PPP in Active Layer	Private initiative in Active Layer	Public Ownership-Outsource Administration	PPP in Infra structure
Easy Decision Making	+	++	+++	++
Complexity	--	-	-	--
Transparency	+++	+++	+	++
Total Value of Networks	++	+++	+++	++
Risk & Required Investments	--	-	--	--
Time Scaling	+	++	+++	++
Expected Performance	+	++	+++	++
Competition	+	+++	+++	++
Functionality	+	++	+++	++
Break-through Participation	+++	+++	+	++
TOTAL	13+, 4-	20+, 2-	20+, 3-	16+, 4-

corporate and operate their infrastructures as a joint network. Initially, optical networks provide about 35% of coverage in each municipality of the Region, according to the deployment of optical Metropolitan networks within the frame of Operational Program "Information Society" through "Invitation call 93" (InfoSoc, 2005), in remote and less developed areas in Greece. However, private funds will be needed in any case as infrastructures must be realistic and competitive and therefore networks should be extended in order to achieve 100% coverage in the next 2-3 years. In all scenarios the costs have been discounted at an interest rate of 10%. A 6 year time frame, spanning the period 2008 to 2013 has been chosen since nowadays most of the telecommunications project aim to be analyzed in a period less that 7-10 years (Blum, 2005).

According to the model "Private Initiative in Active Network", Service Providers will offer services and invest with their own private funds in active equipment. Municipalities will expand the network availability (passive equipment) in order to reach 100% coverage.

There is also a sensitivity analysis deployed, based on the number of Service Providers which are going to offer services to end users. In this analysis fiber is always a pair of fibers, up-link down link. The rental price for fiber depends on the number of Service Providers since the total cost should be covered by them, as Figure 3 and Figure 4 are shown.

Prices will depend also on the existence or not of funding from Europe or National sources. The mean value per year is estimated based on annual discounted expenses that should be diminished to zero (cost oriented approach) so that Municipality avoids losses and moreover adds a fair profit margin. The total Fiber length of the network is based on the number of covered curbs multiply by the ring length plus the extensions to the curbs. The municipalities can offer the access to all the curbs (or point of interest) by using this simple rule (e.g. 2 rings 5 km, 20 curbs per ring, Total

Figure 3. Annual Rental cost of fiber per meter without funding and for different number of SPs

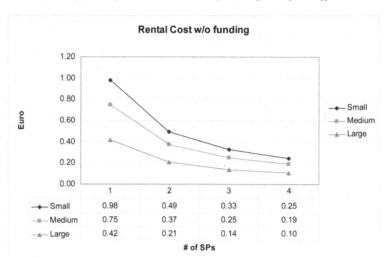

fiber length aprox= 2x5x20=200 km, with no extentions)

In order to present reliable results, data from the majority of the Greek regions are taken and analyzed (NSSG, 2008). Based on the results, demand appears to play a crucial role to the profitability of these networks. In this study, the case where two services providers offer broadband services via the municipalities' network is assumed. In the Table 6, general characteristics and first results

are presented.

Total Network length (curbs) in meter, includes all the rings and network extensions for covering more curbs (almost 100% coverage of the towns in 3 years). Cost without Funding refers to operational cost, space rental for the new company, training, salaries, maintenance etc. The differentiation between with and without funding refers to the intention of the EU to cover the physical extensions of these network (passive

Figure 4. Annual Rental cost of fiber per meter with funding and for different number of SPs

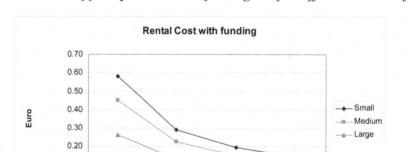

Table 6. General Information about Regions – Investment Cost

Regions	S	M	L
Total Network length (curbs) (m)	94,413	166,636	200,830
Initial network (m) – (2008)	46,890	80,192	90,300
Curbs (m)	14,700	30,310	47,320
Cost w/o funding (€)	4,285,729	7,467,738	9,059,150
Cost with funding (€)	2,710,225	4,773,286	6,025,070
SP cost w/o funding (€)	6,825,725	12,854,174	19,787,247
SP cost with funding (€)	5,839,578	11,168,343	17,888,249
Population	119,641	247,788	388,788
# of Municipalities in the Region	5	8	8
Network (m) per Inhabitant	0.79	0.67	0.52
SP Rental Cost per year from 2008 up to 2013 w/o funding	138K € (in 2008) 485K € (in 2013)	242K € (in 2008) 847K € (in 2013)	290K € (in 2008) 1,030K € (in 2013)
SP Rental Cost per year from 2008 up to 2013 with funding	82K € (in 2008) 287K € (in 2013)	145K € (in 2008) 507K € (in 2013)	182K € (in 2008) 648K € (in 2013)
Connections (end of the Study period 2013)	6,568	12,787	22,538
Demand (end of the Study period 2013)	5.49%	5.16%	5.80%
Cost w/o funding per connection (€)	652.52	584.01	401.95
Cost with funding per connection (€)	412.64	373.29	267.33
SP cost w/o funding per connection (€)	1,039.24	1,005.25	877.95
SP cost with funding per connection (€)	889.10	873.41	793.63

infrastructure), as it has subsidize the initially infrastructures.

Service Provider (SP) cost without funding covers main Categories of cost such as Rental of fiber, DSLAMs, interconnection circuits, regional circuits etc. This is the cost per SP which rents the network and offers services to end users in the Region. In the case where the Municipalities companies will be benefitted from the EU, the rental cost for the SPs will be significant lower as it is illustrated in SP cost with funding row. Connections (and Demand) represent the 20% of the total market since we assume that at least 1 of 5 of the possible customers will be connected to the new metropolitan network (market share).

SP should pay a rental cost per year which is

going to increase as network will be extended. For the first year of operation, in 2008, and for 35% coverage, the rental cost without funding in a Small Region is around 138K €, in Medium Region 242K € and in Large Region 290.249€. From 2010 up to 2013, network will provide 100% coverage and rental cost for SP will be 485K €, 847K € and 1.0030K € for Small, Medium and Large Region respectively. In the case of funding rental cost will vary from 82K € to 287K € in Small Region, from 145K € to 507K € in Medium Region and from 182K € to 648K € in Large Region, for the years 2008 and 2010-2013 correspondingly.

The cost per potential customer is almost 1.692€ (652,52+1.039,24) or 1.302€ (412,64+889,10) for the small areas without and with funding respec-

Figure 5. Cash Flows without funding up to 2013

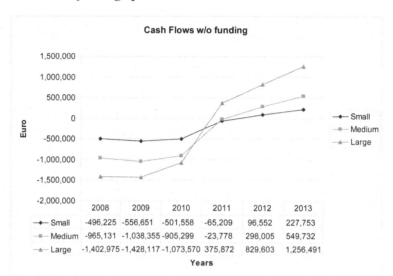

Cash Flows w/o funding

	2008	2009	2010	2011	2012	2013
Small	-496,225	-556,651	-501,558	-65,209	96,552	227,753
Medium	-965,131	-1,038,355	-905,299	-23,778	298,005	549,732
Large	-1,402,975	-1,428,117	-1,073,570	375,872	829,603	1,256,491

Years

tively. All the areas are in the same range for the investment cost but in the large areas the cost per potential subscriber is significant lower which is 1.279,90€ (401,96+793,63) without funding and 1.061,02€ (267,33+793,63) with funding. This cost consists of passive and active equipment and operational cost as well. Similar to other Broadband economics studies (OECD, 2008a), all this cost is heavily effected by the cost for civil work (assuming around 50€ per meter in this study).

The total estimated cost per customer per region negatively affect the fiber optic deployment in the small areas and European Union funding is necessary for attracting Private funds. Similar conclusions can be derived for the Net Present Value evaluation.

For the time period from 2008 to 2013 Net Present Value (NPV) is calculated for each Region for the SP. According to the results, all three Regions have a negative NPV for the SP, in both cases, with or without funding. However, NPV for a wider time period could be opposite, as profit is continuously increasing and losses decreasing. The investment after 2-3 years will turn to be profitable (likely after 2015).

Even if the NPV for the large areas is more

negative than in the smaller areas the potential profits (cash flows) as it is illustrate in Figure 5 and Figure 6 are quite promising. The large areas are by far engaged with huge investments point out at the profits after 2013.

For each examined Regions Cash Balance curve is calculated, with and without funding, as Figure 7 and Figure 8 illustrate. Plots are shown the required investments from each Region. It is obvious that in large Region SPs are going to invest the biggest amount. The deployment cost is estimated 3M euro or 4M euro with and without funding respectively based on the considered data. This numbers are comparable to other economic studies (OECD, 2008b) in which the deployment cost without operational expenses per costumer reach up to 370 euro, for VDSL solution, and up to 1.150 euro, for FTTH solution. These investments are mainly targeting to the development and operation of the networks. SPs should plan and assure financial sources, as in long term these investments are going to ensure networks viability. However, in the large Region, cash balance curves indicate a steeper slope after third year, in both cases (with or w/o funding), yielding to profits after 2013 (Figure 8).

Figure 6. Cash Flows with funding up to 2013

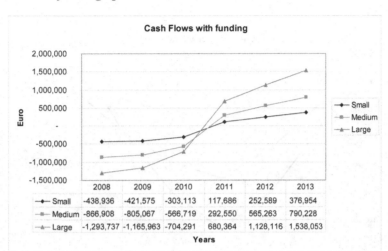

CONCLUSION

As stated in the introduction, the demand for high-speed broadband services is continuously increasing. The necessity of new infrastructures that could cover this demand is a high priority concern in many countries. Fiber optical networks seem to be the key issue, as they provide high-speed transfer rate to the end user. EU and countries independently are co-operating in order to develop optical fiber networks by funding such kind of initiatives. Many of them are focusing in less developed areas, where competition and broadband diffusion are in low levels. There are a lot of Municipalities that have comprehend the seriousness of this situation and decided to develop their own optical MANs.

Due to the huge required investments, the proper business model should accompany the deployment of an optical network. There must

Figure 7. Cash Balance without funding up to 2013

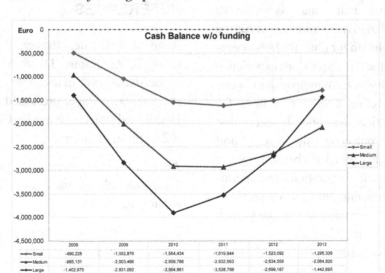

Figure 8. Cash Balance with funding up to 2013

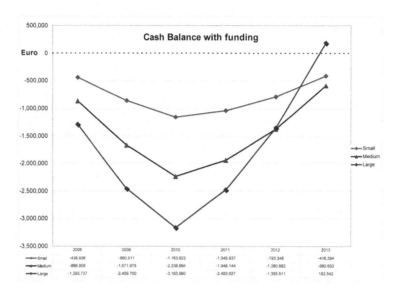

be a number of factors such as economic development, number of businesses and households, broadband diffusion and population that should be taken into account in order to reach a conclusion about the proper one. The creation of a success business model could include all the above factors and much more. The road mapping of broadband networks' deployment could differ according to each case.

In this chapter the most suitable Business models for developing countries are presented. Through SWOT and quality analysis we arrive to the conclusion that Private initiative in Active Layer is the best solution for optical MAN operation in cases where there is no previous experience of such kind of networks. Municipalities choose to corporate with other neighboring ones for deploying fiber optical networks. It seems that joint actions give strong economies of scales and ensure profitability in a long term basis. They can either allow an outsource operation to an operating company or cooperate with private sector in the frame of a PPP in infrastructure due to lack of experience and required knowledge.

According to financial analysis, which is based on real data from Greece, funding could be the

key that will ensure profitability and viability for the optical MANs in shorter time period for non competitive areas. Municipalities should support such kind of initiatives even if more investments are required and losses are expected within. In the long term the deployment of such networks will be beneficial for public and private sector, and people will raise all the advantages that such initiatives guarantee.

REFERENCES

Aagedal, J. O., den Braber, F., Dimitrakos, T., Gran, B. A., Raptis, D., & Stolen, K. (2002). *Model-based risk assessment to improve enterprise security.* Paper presented at the Enterprise Distributed Object Computing Conference EDOC '02 Sixth International

Adkins, I. (2001). *Strategies for Utilities in the European Telecommunications Market*. London: Reuters & Mason.

Bbned. (2007). *Subsidiary company of Telecom Italia*. from http://www.bbned.nl/content/english.shtml

Blum, S. (2005). In F. t. t. H. Council (Ed.), *Financial Analysis of FTTH System Proposals: An Operations- Based Approach.* Tellus Venture Associates.

Cisco Systems. (2004). *Almere looks to a thriving digital future with real broadband.*

City of Rotterdam. (2007). from http://www.rotterdam.com/

Citynet Amsterdam. (2006). *European municipal Fiber and Fiber backbone projects.* from http://www.citynet.nl/

Cloetens, L. (2001). *Broadband access: the last mile.* Paper presented at the IEEE International Solid-State Circuits Conference.

Damien Chew, C. (2006). ING wholesale banking. *European telecoms: Citynet Amsterdam: Fiber-to-the-home is becoming a reality.*

Ericsson AB. (2002a). Case Study. *Sollentuna Energi: A Broadband Pioneer.*

Ericsson AB. (2002b). Case Study. *Hudiksvalls-bostäder: We got fiber all the way.*

Gislason, H. (2004). Reykjavik: Fiber to Every Home. *BYTE.com.* United Businee Media.

Green, P. E. (2004). Fiber to the home: the next big broadband thing. *IEEE Communications Magazine, 42*(9), 100–106. doi:10.1109/MCOM.2004.1336726

Hudiksvallsbostäder. (2007). from http://www.hudiksvallsbostader.se/

Ims, L. A. (1998). *Broadband Access Networks Introduction strategies and techno-economic evaluation, Telecommunications Technology and Applications Series.* Chapman & Hall.

InfoSoc. (2005). *Information Society.* from http://europa.eu/pol/infso/index_el.htm

Ireland's NDP. (2006). Ireland's National Development Plan. *European Union Structural Funds.* from http://www.ndp.ie/viewdoc.asp?fn=/documents/homepage.asp

Ireland's NDP. (2007). *Irelands' Broadband Strategy.* from http://www.ndp.ie/viewdoc.asp?fn=%2Fdocuments%2FNDP2007-2013%2Foverview.htm

Katsianis, D., Welling, I., Ylonen, M., Varoutas, D., Sphicopoulos, T., & Elnegaard, N. K. (2001). The financial perspective of the mobile networks in Europe. *IEEE Pers. Comm. Mag., 8*(6), 58–64. doi:10.1109/98.972169

Lijbrandt telecom. (2007). from http://www.lijbrandt-telecom.nl/

Luna-Reyes Luis, F., & Andersen, D. F. (2003). Collecting and analyzing qualitative data for systems dynamics: methods and models. *System Dynamics Review, 19,* 271–296. doi:10.1002/sdr.280

Marinkovich, M., & Sybrowsky, J. (2003). *UTOPIA: A Public Network based on FTTP, Layer 2 Ethernet Access and the "OSPN" Model.* Converge Network Digest.

Michalakelis, C., Varoutas, D., & Sphicopoulos, T. (2008). Diffusion models of mobile telephony in Greece. *Telecommunications Policy, 32*(3-4), 234–245. doi:10.1016/j.telpol.2008.01.004

NSSG. (2008). General Secretariat of National Statistical Service of Greece (Publication.: www.statistics.gr

OECD. (2008a). *Working Party on Communication Infrastructures and Services Policy.* from http://ec.europa.eu/information_society/index_en.htm

OECD (Ed.). (2008b). *Broadband Growth and Policies in OECD Countries*: An OECD Browse it Edition.

Olsen, B. T., Katsianis, D., Varoutas, D., Stordahl, K., Harno, J., & Elnegaard, N. K. (2006). Technoeconomic evaluation of the major telecommunication investment options for European players. *IEEE Network*, *20*(4), 6–15. doi:10.1109/MNET.2006.1668398

Owen, M., & Raj, J. (2003). An Introduction to the New Business Process Modeling Standard. *BPMN and Business Process Management*. Popkin Software.

Parr Rud, O. (2001). *Data Mining Cookbook: Modeling Data for Marketing, Risk, and Customer Relationship Management*. Wiley Computer Publishing.

Rosen, E. (2004). Reykjavik leads the way on fiber to the curb. *Network World*.

Sara. (2005). *Computing & Networking services*. from http://www.sara.nl/news/newsletters/20050105/news_lett_20050105_eng.html

Sollentuna Energi. (2002). from www.sollentunaenergi.se

Stokab A. B. (2006a). from http://www.stokab.se/templates/StandardPage.aspx?id=306

Stokab, A. B. (2006b). *Stokab Annual Report*. from http://www.stokab.se/upload/Ladda%20ner/dokument/Stokab%20Annual%20Report_05_ENG_I.pdf

SURFnet. (2007). from http://www.surfnet.nl/info/home.jsp

UTOPIA. (2006). *Connecting Communities*. from http://www.utopianet.org/

VolkerWessels. (2007). from http://www.volkerwessels.com/corporate/bin/en.jsp?enDispWhat=Zone&enPage=HomePage&enDisplay=view&

Zukerman, M., Mammadov, M., Tan, L., Ouveysi, I., & Andrew, L. L. H. (2008). To be fair or efficient or a bit of both. *Computers & Operations Research*, *35*(12), 3787–3806.

Chapter 11
Modeling and Techno–Economic Evaluations of WDM–PONs

Jürgen Schussmann
Carinthia University of Applied Sciences, Austria

Thomas Schirl
Carinthia University of Applied Sciences, Austria

ABSTRACT

In the near future, broadband access networks will be required with data rates of over 1Gbit/s per customer. Currently, time-division multiple access passive optical networks (TDMA-PONs) are deployed. However, TDMA-PONs cannot keep up with the requirements for the broadcasting of a great number of HDTV channels and the unicasting of several triple-play services (voice, data and video). In contrast, wavelength-division-multiplexed PONs (WDM PONs) will be able to provide these required high data rates per user causing higher costs than with TDM-PONs. The introduced paradigm shift, at least one wavelength per service and user, leads to the introduction of new aspects in the design of future WDM PON access networks. In techno-economic evaluations, new network architectures with the highest potential concerning economic considerations have been identified. Access to these newly identified network architectures will prompt market introduction as well as market penetration helping Fiber-to-the-Home (FTTH) to become reality.

INTRODUCTION

Due to the enormous and still rocketing bandwidth demand, there is no doubt that the age of Fiber-to-the-Home (FTTH) has arrived. The landscape of telecommunications, especially in the field of access networks, is now undergoing a major change towards FTTH. This major change is accepted worldwide and is driven by cutting-edge countries like Sweden, Korea, California, Japan, and projects like the Utah Telecommunications Open Infrastructure Agency (UTOPIA) in the USA or the Multi-Service Access Everywhere (MUSE) in the Netherlands and many other projects (Maeda, 2004; Lin, 2006).

Today, most Passive Optical Networks (PONs) and Point-to-Point (P2P) networks are being deployed only on a limited scale. Mass deployment of PONs was in its early stages. In Japan, Korea,

DOI: 10.4018/978-1-60566-707-2.ch011

Sweden, Denmark and Norway there were high levels of activities, while in the US and in other European countries the penetration and growth was more modest (Figure 1 a)). The countries in Figure 1 a) are sorted in ascending order by the total of broadband subscribers. In this context it is important to define the terms DSL, cable and fibre. In the terminology of the OECD (OECD, 2008), DSL includes all DSL lines offering Internet connectivity which is capable of download speeds of at least 256kbit/s. Cable includes all subscribers with a download speed greater than 256kbit/s. Fiber includes all fiber-to-the-premises subscribers at download speeds greater than 256kbit/s. In summary, only 1.1 broadband subscribers per 100 inhabitants over all OECD countries use fiber connections with at least 256kbit/s.

In Figure 1 b), the average advertised download speed sorted by the same countries as in Figure 1 a) is shown. The arithmetic mean value of 13.7Mbit/s over all countries shows the poor bitrate compared with high speed passive optical access networks. Unfortunately, only a few exceptions like Japan, France, Korea and Sweden show significantly higher download speeds.

Figure 1 shows the significant lack of high-speed optical access networks in the world.

Bitrate Per User Requirements

Today, the major challenge is the bitrate bottleneck in access networks due to the ever increasing bitrate requirement per user. In the past, many projections of future bitrate requirements have been made (Coffman & Odlyzko, 2001). I refrained from such forecasts, as they have all proven to be too optimistic. However, it is clear that bitrates will, in fact, continue to grow faster than the world population grows. Due to the dropping prices for Internet access (and computers) and as the Internet begins to serve more data-intensive applications, it is also clear that the bitrate requirements per user will grow. Furthermore, the unshakeable demand for bitrates caused by the growth of high bitrate

applications and broadband systems leads to higher bitrate requirements per user (Figure 2).

In a deployment, not only one service but a service mixture is used and leads to a further increase of the bitrate requirements. (Lund, 2001) points out that a future application mixture leads to a bitrate requirement of approximately 40Mbit/s per user. In a residential area with 3,500 supported users, the total bitrate required is roughly 140Gbit/s. (Lund, 2001) also shows that a 2m node rack can service these 3,500 users. Further information about the cable plant design can be found in (Mahlke & Gössing, 2001).

The development of a killer application, for which many customers would be prepared to pay a premium price, could generate a compelling commercial argument for deploying fiber-to-the-home (FTTH) more quickly (Wolf & Zee, 2000; Mayhew et al., 2002). A possible killer application could be the video transmission in an unforeseen quality. Symmetry in terms of downstream and upstream bitrates is an important parameter in access networks. The symmetry is defined as the ratio between downstream and upstream bitrates. It was reported that the current symmetry by means of current traffic is approximately 1.4 and is further tending downwards in the future (Reed, 2003). New and emerging applications such as video conferencing or data file repositories require symmetric bandwidth in both directions. In the future, the access network must therefore be capable of transporting symmetrical traffic. WDM PON solutions presented in this chapter are highly symmetric.

Bitrate Limitation

The big question is: does it make sense to ever increase the bitrate in telecommunication networks to reduce the transmission time? The transmission time depends on two parameters. The first one is the bitrate, which is limited by the transmission channel (Shannon, 1948). It can be easily calculated by the relation

Figure 1. a) Broadband subscribers per 100 inhabitants, by technology, December 2007; b) average advertised download speeds, October 2007; (OECD, 2008)

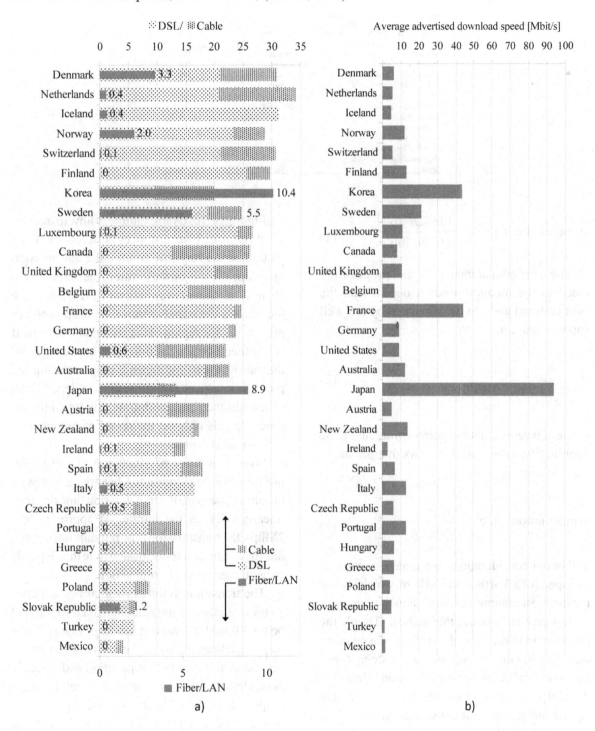

Figure 2. Typical bitrate requirements of different current and future applications (Hecht, 2005; Lund, 2001; Simcoe, 2002; Biraghi et al., 2002; Weldon & Zane, 2003; Ramsey, 2005)

Application		Bitrate [Mbit/s]
Video conference (uncompressed)	4k (4096 x 2048 pixel)	6000
Video conference (uncompressed)	1600 x 1200 pixel, 30fps	1200
Video conference (JPEG 2000)	4k (4096 x 2048 pixel)	400
Storage Area Network	SAN Service	100
LAN to LAN Connectivity	VPN IP (L3) - Transp. LAN (L2)	100
HDTV (1080i, compressed)	1920 x 1080 pixel, 30fps	28
HDTV (720p, compressed)	1280 x 720 pixel, 60fps	28
Video (uncompressed)	320 x 240 pixel, 15fps	10
PAL (compressed)	720 x 576 pixel, 25fps	4
Internet gaming		2
CD quality audio		1.2
Broadband web surfing		0.3
High-quality audio sessions		0.125
Voice	POTS / ISDN	0.064

$$\text{transmission time} = \frac{\text{file size [bit]}}{\text{bitrate [bit/s]}}. \quad (1)$$

The second parameter is the propagation speed in the medium used. It depends on the used material and can be calculated by the well known equation

$$v = \frac{1}{\sqrt{\mu\varepsilon}} \quad (2)$$

whereas μ represents the permeability and ε the permittivity. From this it follows that the transmission time is

$$\text{transmission time} = \frac{\text{d [m]}}{\text{v [m/s]}}. \quad (3)$$

For our considerations, we assume a propagation speed of 2/3 of the speed of light, which is valid for the most common transmission mediums.

It must be mentioned that in the real world the transmission time depends on many other parameters like the line coding, protocols, delay times due to the packet switching and so on. These additional parameters enhance the transmission time significantly. Thus, our consideration is reliable, and the published data represents theoretical limits. This means that the transmission time cannot fall below these theoretical limits.

Figure 3 gives us an example of how to transfer a 10kByte file over distances between 1 km and 40,000km. At very low bitrates, the transmission time depends only on the bitrate of the transmission channel, as shown in equation (1). If we increase the bitrate, the transmission time drops and approximates the region of the propagation speed. If we further increase the bitrate, at a specific point the transmission time depends only on the limited propagation speed (equation (3)). In dependence of the distance and the bitrate, the transmission time depends only on the speed of propagation and not on the bitrate.

Over a distance of 10,000km, for example, the transmission time goes linearly down with the bitrate, because the transmission time depends linearly only on the bitrate. At approximately 2Mbps, the transmission time remains unchanged due to the constant transmission time that depends only on the speed of propagation.

The transition to the horizontal line depends on the distance and the amount of data which will be transferred. As we can see from Figure 3, the horizontal lines show that the transmission time is limited by the speed of propagation and varies in dependence of the distance (and therefore of the propagation delay). The slanting line depends only on the file size. For larger file sizes, the slanting line moves parallel to greater transmission times and vice versa.

As a consequence, we can see an interesting

Figure 3. Transmission time to transmit and acknowledge a 10kByte file over distances between 1 and 40,000km. The transmission time declines linearly to a specific point. At this point, the transmission time is constant because it depends only on the velocity of propagation.

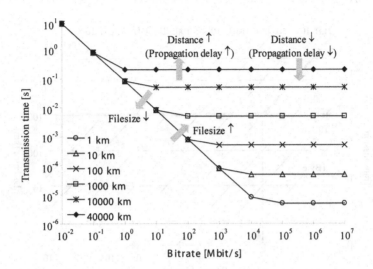

property: for smaller file sizes, the transmission time depends on the speed of propagation for lower bitrates. That implies that only for larger file sizes is it reasonable to increase the bitrate to reduce the transmission time.

Figure 3 shows the transmission time in dependence on the bitrate and different distances. This diagram can be generalized and is shown in Figure 4 for distances between 1km and 40,000km and for file sizes between 1Byte and 100GByte. With this diagram, we are able to determine the theoretical limitation of the transmission time.

In Figure 4 an example is given for a file size of 1kByte over a distance of 1,000km. We can see that for bitrates of greater than 2Mbit/s the transmission time is constant and cannot underrun the 4ms limitation. For this example, it would be pointless to increase the bitrate over 2Mbit/s. As mentioned before, it must be noted that the transmission time depends on many other factors and is larger in the real world.

This subsection identifies the influencing factors and show the implications on the transmission time. It could be shown that the transmission time

depends on the file size, the distance between the communication nodes and the bitrate of the communication channel (additional parameters like protocols, delay times due to the packet switching, ... are unaccounted). The importance of increasing the bitrates mainly because of the ever increasing data volume has been shown.

TDM PON

This subsection shortly describes three important TDM PON standards. Due to the highly comprehensive standards, only the main characteristics are stated.

One of the first papers describing PONs was published by British Telecom researchers in 1988 (Stern et al., 1988). Down to the present day, several alternative architectures for TDM PON-based access networks have been standardized (Prat, 2002). The main difference today is the choice of the bearer protocol (Sivalingam & Subramaniam, 2005). Currently, three different standardized specifications exist:

Figure 4. Generalized diagram to determine the transmission time in dependence on the bitrate, file size and distance. An example of how to transfer a file with a size of 1kByte over a distance of 1,000km is indicated. In this example, it is obvious that an increase of the bitrate beyond approximately 2Mbit/s can not reduce the transmission time. Therefore, the transmission time can not go below 4ms.

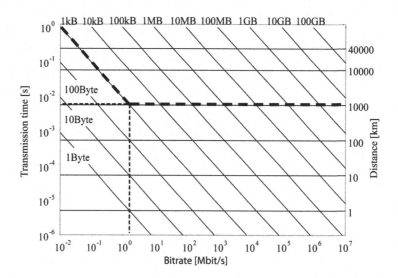

- ATM-based PON (APON) or broadband PON (BPON)
- Generic frame based PON (GPON)
- Ethernet based PON (EPON)

The following paragraphs give a brief overview of these standardized architectures.

A/BPON

ATM PON was formed by the Full Service Access Network (FSAN) initiative (FSAN, 2008) which is an incorporation of major network operators and many equipment vendors. The goal was to specify a specification that uses ATM as its layer-2 protocol (Killat, 1996; Dixit, 2003). Later, the name APON was replaced by broadband PON (BPON) which symbolizes the support of Ethernet services, video services and virtual private line services (Gillespie, 2001; Green, 2005). Since 1997, the ITU-T has published a series of A/B-PON-related recommendations (Nakanashi & Maeda, 2002). APONs are mainly used in North America (Kettler et al., 2000; Lin, 2006).

GPON

Because of the ever growing traffic volume, the FSAN group (FSAN, 2008) strove to standardize a new PON (Generalized PON) network with bitrates higher than 1Gbit/s and improved efficiency for data traffic (Green, 2005). To overcome the inefficiency in the ATM based PON, the generic framing procedure (GFP) (ITU-T G.7041, 2003) can be used, while allowing a combination of ATM cells and variable-size frames. GPON is standardized in the ITU-T standard G.984.1 (ITU-T G.984.1, 2003), G.984.2 (ITU-T G.984.2, 2003) and G.984.3 (ITU-T G.984.3, 2003).

In summary, GPON includes full service support like voice (POTS, ISDN), Ethernet (10/100 BaseT), ATM, leased lines and more. In principle, GPON supports various bitrate options. Downstream direction bitrates of 1.2 and 2.4Gbit/s and upstream direction bitrates of 155Mbit/s, 622Mbit/s, 1.2Gbit/s and 2.4Gbit/s are standardized.

On the basis of the power budget limit and dispersion limitations, the distance between the OLT and the ONU is standardized and cannot exceed the distance 10km or 20km (in dependence of the bitrate used). The transmission convergence (TC) layer permits split ratios up to 1:128.

The biggest advantage of the GPON is its multi-service support, especially the large amount of TDM services which can be supported efficiently. Thus, for Fiber-to-the-Office (FTTO), GPON is the preferred PON technology (Lin, 2006).

EPON

Ethernet PON (EPON) (also termed as Ethernet in the first mile (EFM)) is a PON-based network that carries data traffic encapsulated in Ethernet frames (Kramer & Pesavento, (2002); Dixit, 2003; Green, 2005; Kramer, 2005). In most countries, EPON is the dominant PON technology for homes and small offices (Lin, 2006).

In principle, two modes of operation are standardized. In the first one, it can be deployed over a shared medium using the carrier-sense multiple access with collision detection (CSMA/CD) protocol. In the second one, stations can be connected via full duplex point-to-point links. It uses a standard 8b/10b line coding and operates at standard Ethernet speed.

In the downstream direction, Ethernet frames transmitted by the OLT pass through a 1:N passive star coupler and therefore reach each ONU. N is standardized between 4 and 64. In the downstream direction, packets are broadcast by the OLT and extracted by their destination ONU based on the media-access control (MAC) address.

In the upstream direction, due to the directional properties of a passive star coupler, data frames from any ONU will reach only the OLT, and not other ONUs. If data frames from different ONUs are transmitted simultaneously, they may still collide. Thus, in the upstream direction, the ONUs needs to employ some arbitration mechanism to avoid data collisions and share the fiber-channel capacity fairly. All ONUs are synchronized to a common time reference and each ONU is allocated

to a time slot. An ONU should buffer frames received from a user until its timeslot arrives.

The possible time slot allocation schemes could range from a statistic allocation TDMA (time division multiple access) to a dynamically adapting scheme based on an instantaneous queue size in every ONU (statistical multiplexing scheme). More allocation schemes are possible, including schemes utilizing notions of traffic priority, quality of service (QoS), service-level-agreements (SLAs), oversubscription, and so on (G. Kramer et *al.*, 2001).

WDM PON

While a TDM PON uses a broadcast structure in which all users receive copies of the same optical signal, it has been realized that wavelength-division multiplexing (WDM) had certain advantages: increased capacity, security, privacy, upgrade-ability (Gumaste & Antony, 2002; Iannone et *al.*, 1998; Krishna & Subramaniam, 2000). Due to the collision-free nature of a WDM PON, it offers the n-fold bandwidth compared with TDM PONs implemented with a passive star coupler. This fact implies the evolution of FTTH solutions illustrated in Figure 5. It should be indicated that B-ISDN and xDSL are not pure FTTH solutions. But it illustrates the ever increasing fiber penetration and the fact that fibers are moving closer to the user.

A passive-star coupler represents a broadcast-and-select component in which each optical input signal is equally distributed to all output ports. This reality leads to optical power splitting losses and can be avoided by using wavelength router (WR). In principle, there are several possible implementations of WR depending on the physical concept (Agrawal, 2002; Keiser, 2003; Mestdagh, 1995; Wan, 2000).

The preferred embodiment of the wavelength router is the arrayed-waveguide-grating multiplexer (AWGM), also known as wavelength grating router (WGR or simply WR) or an optical phased array (PHASAR). Each wavelength can be simultaneously used on all input ports without channel collision, thus enabling wavelength reuse within the network (Barry & Humblet, 1993; Glance et al., 1994; Lin, 2004; Milorad, 2004; Sivalingam & Subramaniam, 2005; Smit & van Dam, 1996). The wavelengths are routed as virtual channels λ_i^k, where the subscripted index i denotes the physical wavelength and the superscripted index k represents the input port number.

Several applications can be imagined with such an AWGM device. The AWGM is mostly used as a wavelength mutliplexer (Figure 6 (a)) and a wavelength demultiplexer (Figure 6 (b)) (Takahashi et al., 1995). Both applications allow for setting up WDM networks. If we join the wavelength multiplexer and demultiplexer, we create a new component called the static wavelength router (see Figure 6 (d)). The fourth application is add-drop multiplexing which is essential to set up optical WDM rings (Takahashi et al., 1995).

As shown in Figure 6 (d), all available wavelengths at input port 1 are routed to port numbered 1, 2, 3 and 4 without having collisions with other wavelengths of the other input ports 2, 3, and 4. The wavelengths of the other input ports 2, 3, and 4 are similarly routed in a cyclic way.

Convenient property of the AWGM is the support of a high number of ports (e.g. 80 x 80 AWGMs are currently commercially available (Photeon, 2008)) and its excellent channel selection characteristics.

Compared to a passive-star coupler, in which a given wavelength may only be used on a single input port, the AWGM with M input and M output ports is capable of routing a maximum of M^2 connections, as opposed to a maximum of M connections in the passive-star coupler. Because the AWGM is an integrated device, it can be fabricated at low cost. The disadvantage of the AWGM is the fixed routing matrix which cannot be dynamically reconfigured. Despite its static nature, an AGWM router has many applications in WDM networks as stated before.

Figure 5. Evolution of wired access networks and FTTH solutions

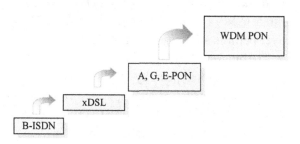

New Paradigm

In the past, for each service, a specialized telecommunication network was built. An example is the voice network, which was originally designed to transport only voice services. Each telecommunication network was optimized for a respective service. Thus, the utilization of the network with other services was undesirable. For the first time, with ISDN (Integrated Services Digital Network) (Kanbach & Körber, 1999), the paradigm "one service–one network" is discarded. The objective of the new paradigm is to transport all services in the past and future over a single telecommunication network. Today, we know that ISDN, xDSL and other up to date wired network technologies are due to the already enormous and still rocketing bandwidth demand overtaken by the internet.

In the near future, broadband access networks will be required with data rates over 1Gbit/s for each customer. Due to this aspect the widespread deployment of fiber-to-the-home (FTTH) technology is speeding up (Lin, 2006). Currently, time-division multiple access passive optical networks (TDMA-PONs) are deployed. However, TDMA-PONs, like ATM PON (APON), Broadband PON (BPON), Generalized PON (GPON) and Ethernet PON (EPON) cannot keep up with the requirements for the broadcasting of a great number of HDTV channels and the unicasting of several triple-play services (voice, data and video) (Söderlund et al., 2004).

TDM-PONs have limited bitrates per user.

Figure 6. Schematic illustration of the functionality of an AWGM device. Potential applications are the wavelength multiplexer (a), demultiplexer (b), the add-drop multiplexer (c) and the static wavelength router (d). In the case of the static wavelength router, 4 wavelengths per input port were cyclically routed to the 4 output ports without collisions. Therefore, wavelength reuse is feasible.

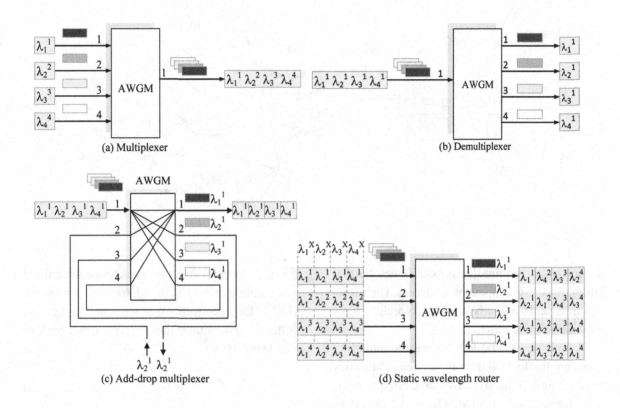

They are complicated to upgrade, and all ONUs must work at the aggregate bit-rate. Privacy is an important issue due to the broadcast of the downstream information. Network integrity is a field of study because one ONU can corrupt the entire upstream transmission. A delicate operational problem is how TDM PONs will be upgraded in the future in terms of higher bitrates per user.

In contrast, wavelength-division-multiplexed PONs (WDM PONs) will be able to provide these required high data rates for every user however with, the limitation of higher costs compared with TDM-PONs.

In contrast to cutting-edge scientific papers (An et *al.*, 2004; Banerjee et *al.*, 2005; Bock, et *al.*, 2004; Maier et *al.*, 2000;), the focus is to provide each user with at least one wavelength.

This paradigm can be formulated as follows

$$\text{\#user} \leq \text{\#service} \leq \text{\# } \lambda \qquad (4)$$

so that each user and each service obtain at least one wavelength (Figure 7). From another point of view, this work represents a generalization of the special case: one wavelength per user for all services.

However, there still are some drawbacks in WDM PON networks. WDM components such as multiplexers/demultiplexers, wavelength routers, multiwavelength sources are more expensive compared with TDM PON components. Yet the costs for WDM components are rapidly decreasing. Due to the cost sensitivity in access networks

Figure 7. The paradigm shift: at least one wavelength for each user. Each wavelength transports at least one service.

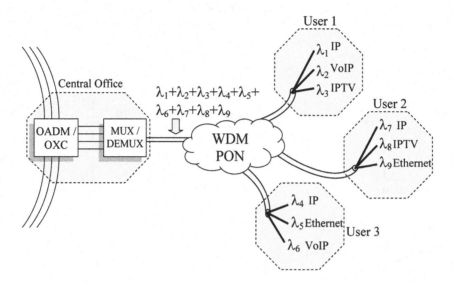

it is important to introduce new architectures that are able to share components and costs. Therefore, it is important to identify WDM PON architectures which minimize costs.

A topic in a WDM PON is the assignment of wavelengths to ONUs or to services (McGarry et al., 2006). This layer two task is part of the medium access control (MAC) protocol and does not affect our current work scope. We focus on practical WDM PON architectures which allow the transportation of many different wavelengths collision free from the OLT to each ONU.

SINGLE-STAGE NETWORKS

In order to be in a position to study and define the connectivity properties of the WR, it is useful to consider optical channels routed by a WR as belonging to a discrete bidimensional domain: space and wavelength. It must be noted that the WR is a passive device and is able to perform space permutations while it does not modify the wavelength of the signals.

The WR behaves like a periodical pass-band filter. Its transfer function peaks repeats at fixed wavelength intervals called free-spectral-range (FSR). Designing the WR we make use of the following equation (Glance et al., 1994; Smit & van Dam, 1996)

$$FSR = M \cdot \Delta\lambda \quad M \in \mathbb{N}. \qquad (5)$$

In the real world, the FSR has a more complicated expression. The FSR and the transfer function are not constant and vary over the λ-axis. The following discussion is based on an idealized WR model. The wavelength domain for the channels routed by the WR is a discrete set of wavelengths, numbered by an index k that can assume all the integer values from 1 to (theoretically) infinity (Figure 8). Within optical networks, the bandwidth is divided into frequency bands numbered by positive integers k. Wavelengths are spaced by a wavelength interval $\Delta\lambda$, whereas λ_0 correlates with k=1 and represents the lowest useable wavelength. Every wavelength used depends on the wavelength number k and the wavelength interval $\Delta\lambda$ and can be expressed by

Figure 8. Discrete wavelength domain in a WDM PON

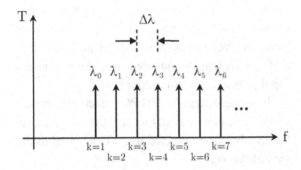

$$\lambda\big(k\big) = \lambda_0 + \big(k-1\big)\Delta\lambda \quad \forall k \in \big[1,\infty\big]. \quad (6)$$

A second parameter to characterize a WR which is called coarseness c and represents the number of adjacent wavelength channels belonging to the wavelength comb routable on the same output port can be introduced.

In summary, a WR can be specified by the size M and the coarseness c. The WR routing function is given by (Barry & Humblet, 1993)

$$o = 1 + \left(i-1+\left\lfloor\frac{k-1}{c}\right\rfloor\right) \bmod M \quad (7)$$

where $o,i \in [1,M]$ represents the output and input port and $k \in \big[1,\infty\big]$ represents the discrete wavelength number (Figure 8). In equation (7) the important modulo-M division is a consequence of the periodical behavior of the WR and is named "cyclic-routing". In the frequency domain, the cycling-routing property appears as the free spectral range (FSR).

On the basis of the routing properties of the WR, we are now going to define rules to evaluate the WR parameters in a WDM PON. The space demultiplexing function depends on the choice of the input occupancy pattern. Thus, we must define different input patterns for a WR. In the past, only designated input occupancy patterns have been used (Maier et *al.*, 2000). Thus, we define

a generic input pattern for a WR. The generically terminated input port $\tilde{i}$ in an input pattern can be described as

$$\tilde{i} = \left\lfloor\frac{\left|2\cdot I\cdot \bmod\big(2^o\big)\right|}{2^o}\right\rfloor \cdot o \qquad I \in \big[0,2^M-1\big], \quad o \in \big[1,M\big], \quad \tilde{i} \in \big[0,M\big]$$

$$(8)$$

where I is the terminated input pattern described as a decimal number. The expression in the round-off brackets results in two numbers: one and zero. The result is multiplied by the output port number o so that $\tilde{i} \in \big[0,M\big]$. The identifier $\tilde{i}$ depends on the two parameters input occupancy pattern and the number of the used output ports. Substitution (8) in (7) leads to

$$o = 1 + \left(\left(\left\lfloor\frac{\left|2\cdot I\cdot \bmod\big(2^o\big)\right|}{2^o}\right\rfloor \cdot o -1+\left\lfloor\frac{k-1}{c}\right\rfloor\right)\right) \bmod M.$$

$$(9)$$

The commonly known inverse function of a=b mod c is b=c·q+a where $q \in [0,\infty]$. By utilizing this correlation in equation (9), we obtain

$$Mq+o-1 = \left\lfloor\frac{\left|2\cdot I\cdot \bmod\big(2^o\big)\right|}{2^o}\right\rfloor \cdot o -1+\left\lfloor\frac{k-1}{c}\right\rfloor + z.$$

$$(10)$$

Each channel inside a consecutive wavelength comb is indexed by the number z in (10). Multiplying equation (10) with c and transformation to k leads to

$$k = Mqc + \big(o-1\big)c - \left(\left\lfloor\frac{\left|2\cdot I\cdot \bmod\big(2^o\big)\right|}{2^o}\right\rfloor \cdot o -1\right)c + 1 + z.$$

$$(11)$$

The result of equation (8) leads to figures

between zero and M. The parameter i in equation (7) is defined between one and M. Thus, equation (11) must be adjusted for the case $\tilde{i} = 0$. This can be done by multiplying (11) with an expression so that the result of (11) is zero for the case $\tilde{i} = 0$. The expression in the round-off brackets in (8) fulfils our requirements, and k is then determined by

$$k = \left\lfloor Mqc + (o-1)c - \left\lfloor \left\lfloor \frac{2 \cdot \mathrm{I} \cdot \bmod(2^o)}{2^o} \right\rfloor \cdot o - 1 \right\rfloor c + 1 + z \right\rfloor \cdot \left\lfloor \frac{2 \cdot \mathrm{I} \cdot \bmod(2^o)}{2^o} \right\rfloor. \tag{12}$$

With the equation (12), the set of all wavelengths that, being inserted into the WR from any of its used input port occupancy patterns, reach a given output port o can be calculated. It depends only on the coarseness c, the size M and the input occupancy pattern I of the WR.

Another possibility to fully calculate the WR is the Wavelength Transfer Matrix (WTM) (Barry & Humblet, 1993) and the generalized WTM for any coarseness factor and any number of free spectral ranges (FSR) used (Schussmann, 2008).

Network Design

In single stage WDM PON networks, there are three potentials (see equation (7)) to comply with the introduced paradigm shift

(1) the use of WR with a coarseness factor greater than one,
(2) the utilization of the cyclic routing property and
(3) the simultaneous usage of the possibility (1) and (2).

Starting from equation (7) and the inverse function of $a = b \bmod c$ is $b = c \cdot q + a$ whereby $q \in [0,\infty]$, we are able to convert equation (7) to k

$$k = Mqc + oc - ic + 1. \tag{13}$$

With (13) all the discrete wavelength indices k can be identified on an output o. It linearly depends on the size M, the coarseness c and the number of the considered input port i. The q factor results from the cycling routing property and describes all the wavelengths that are routed from a certain input port to a certain output port.

In a single stage WDM PON each output port of the WR is connected to an ONU. Therefore, the number of connected users U is equal to the size of the WR

$$U = M. \tag{14}$$

The free spectral range (FSR) can be calculated by means of equation (5). Because the variable $\Delta\lambda$ refers to the analog behavior of the WR, we introduce another expression to describe the FSR

$$FSR = M \cdot c. \tag{15}$$

In a WDM PON, only a fraction of the whole FSR can be utilized. It is clear that due to the cyclic routing property of the WR, more than one FSR can be used. To express this characteristic, the number of FSR is given by

$$\#FSR = \frac{n_\omega}{FSR} = \frac{n_\omega}{M \cdot c}, \tag{16}$$

where n_ω is the number of wavelengths per input fiber of the WR. It should be noted that the number of used FSR is the absolute value and can be a decimal number. When the used FSR is, for instance, half the whole FSR, then the variable #FSR is 0.5 (independent from a possible overlapping of more than one FSR). The number of wavelengths routed to the same output port (is equal to the number of wavelengths per user U_ω) leads to the following design rule

$$U_\omega = c \cdot \#FSR \cdot I \tag{17}$$

Figure 9. Maximum number of possible wavelengths N_ω and the network capacity limit B_N for different channel spacings Δf with respect to $\Delta F=200nm$, $M=80$ and $B_0=1Gbit/s$.

Δf [nm]	$N_\omega = \dfrac{\Delta F}{\Delta f} \cdot M$	B_N [Gbit/s]
20	800	800
1.6	10,000	10,000
0.8	20,000	20,000
0.4	40,000	40,000
0.2	80,000	80,000
0.1	160,000	160,000

where I represents the number of terminated input ports of the WR. Substituting (16) in (17) leads to

$$U_\omega = \frac{n_\omega I}{M}. \tag{18}$$

Equation (18) shows that the number of wavelengths per user is independent from the coarseness factor. However, the wavelength indices are dependent on the coarseness factor. Another point of view is that the ratio I/M is a decimal number between 0 and 1 and therefore U_ω can be maximal n_ω.

Capacity Limits

The capacity limit in a single-stage WDM PON is the product of the channel bitrate on each source or wavelength B_0 and the number of sources or wavelengths N_ω in the whole network. The number of wavelengths depends on the accessible optical spectrum ΔF and the channel spacing Δf (Figure 8). The capacity limit for a single-stage WDM PON is then given by

$$B_N = B_0 \cdot N_\omega = B_0 \cdot M \cdot \frac{\Delta F}{\Delta f} = B_0 \cdot M^2 \cdot U_\omega. \tag{19}$$

On the strength of the restricted optical spec-

trum of the used components, the accessible optical spectrum is limited. The channel spacing depends on the WDM technology. Typical channel spacings are 20nm (ITU-T G.694.2, 2002) for CWDM and 1.6nm, 0.8nm, 0.4nm, 0.2nm and 0.1nm for DWDM technology (ITU-T G.694.1, 2002). The optical spectral range ΔF is not limited in both ITU-T standards. If we assume an accessible optical spectrum of $\Delta F=200nm$ and a channel bitrate of 1Gbit/s, we achieve the number of wavelengths and the capacity limits can be seen in Figure 9.

Due to technology restrictions, the size of the WR and therefore the maximum number of users is limited. Today a maximum size of 80 inputs and outputs are realistic (Smit & van Dam, 1996; Photeon, 2008) and therefore a maximum number of users of 80.

According to the paradigm of at least one wavelength per service and user, the maximum possible number of users and therefore the maximum network structures is given. In other words, the multiplication of the number of users U with the number of wavelengths per user cannot exceed the number $\dfrac{\Delta F}{\Delta f} \cdot M$.

MULTISTAGE NETWORKS

The single-stage architecture suffers from multiple limitations. These include difficulties in scaling

Figure 10.

• n_ω	number of wavelengths in a fiber
• U_ω	number of wavelengths per user
• N_ω	total number of wavelengths in the entire network
• U	number of users connected to the WDM PON
• S	number of stages of the considered WDM PON
• $\vec{N} = \{N_1, N_2, ..., N_S\}$	vector containing the number of WR of all S stages
• $\vec{I} = \{I_1, I_2, ..., I_S\}$	vector containing the number of input ports used of a single WR of all S stages. This vector represents also the number of fibers per WR that connect stage s with stage s+1
• $\vec{c} = \{c_1, c_2, ..., c_S\}$	vector containing the coarseness factor of the WR of all S stages
• $\vec{M} = \{M_1, M_2, ..., M_S\}$	vector containing the size of the WR of all S stages
• a	interval between terminated input ports of a WR

the number of ONUs once the network is laid out. Another problem is the limited number of users due to the limitations on the WR size. Furthermore, there is a limited spectral range in the network and therefore a limited number of useable wavelengths. A multistage WDM PON uses more than one WR in serial. A similar concept based on passive-star couplers can be found in (Borella et *al.*, 1998, p. 275 et seq.). This concept is called a multilevel optical network (MON).

The implementation of the paradigm shift in (4) needs many different wavelengths. Due to the cost sensitivity of access networks, channel spacings of 20nm (ITU-T G.694.2, 2002) offer the most advantages for optical access networks to use uncooled DFB Lasers (Tanis & Eichenbaum, 2002). Thus, only a handful of wavelengths can be used simultaneously. In this case, we need architectures that allow for the reuse of wavelengths in the same WDM PON network.

For the description of the network topology, the following symbols similar to (Maier et al., 2000) are introduced in Figure 10.

To allow only regular and practicable topologies, the network structures considered in this work is based on the following design conventions:

• Each user could obtain at least one wavelength

$$U_\omega \geq 1. \tag{20}$$

• Due to the considered tree topology, the first stage consists of one WR device

$$N_1 = 1. \tag{21}$$

• Each output port of a stage s is either connected to an input port of the following stage s+1 or to an ONU; therefore, the number of connected users are

$$U = M_S \cdot N_S. \tag{22}$$

Compared with single-stage networks (equation (14)), the number of connected users is N_s times greater.

• The coarseness for the stage s is

$$c_s \geq 1. \tag{23}$$

It would be pointless to use a WR with a coarseness factor of 0.

• We observe only networks with a tree topology. The number of WR is increasing stage by stage, so

$$N_{s-1} < N_s \quad s \in [2, S]. \tag{24}$$

• To accomplish the design convention (III) and to permit only networks with WR that are connected (at least by one link) to a WR from the previous stage or to the OLT in the first stage, we define

$$N_{s-1} \cdot M_{s-1} < N_s \cdot M_s. \qquad (25)$$

- The size M and the coarseness factor c of the WR are constant for all WR in the same stage.

- For all WR, the number of input ports is equal to the number of output ports.

- All ONUs connected require the same number of wavelengths U_ω.

- In an increased coarseness (IC) WDM PON, the coarseness factor of the first WR is one ($c_1=1$) to receive coarseness factors in the last stage in the range of technology feasibility.

- Similarly as in design convention (X), the coarseness factor in a decreased coarseness (DC) WDM PON is one in the last stage ($c_S=1$).

- The coarseness factor in an IC WDM PON increases stage by stage or is equal to the previous stage

$$c_{s+1} \geq c_s. \qquad (26)$$

- In a DC WDM PON, the coarseness factor decreases stage by stage or is equal to the previous stage

$$c_s \geq c_{s+1}. \qquad (27)$$

Network Model

Figure 11 represents the architecture to model the multistage WDM PON similarly as in (Maier et al., 2000). The WDM PON network connects a number of ONUs to a single OLT, which is typically located in the CO. In the model, the WDM PON is partitioned in stages and interstages. The WR are placed at the defined stages where an interstage represents the fibers which connect the device in stage s with the device in stage s+1. The model consists of S+1 stages, which are connected with S+1 interstages. The constant interval between terminated input ports of a WR in stage s is the relation M_s/I_s and is denoted by a_s. Stages and interstages are numbered in ascending order beginning with zero in the case of the first stage and interstage. Both, stages and interstages end with the index S. The number of fibers in an interstage s is denoted by I_s. The number of WRs in stage 1 is always 1 and is increasing stage by stage to form a tree topology. The number of ONUs U, which can be connected to the WDM PON, is equal to the number of output ports in the last stage S and can be calculated by design convention (III).

Design Parameters

First of all, the maximum wavelength per fiber link must be calculated. For the time being, one wavelength per ONU is assumed. From there, the following statement can be expressed: at each stage of the network, the number of users U must be equal to the number of wavelengths in the whole network N_ω. At the interstage 0, all wavelengths N_ω are grouped in I_0 fibers and must feed all connected ONUs. Thus, the following condition holds (Maier et al., 2000)

$$I_0 = \frac{U}{n_\omega}. \qquad (28)$$

The total number of used wavelengths in the whole WDM PON can be calculated by

$$N_\omega = U \cdot U_\omega. \qquad (29)$$

To perform the paradigm shift, we assume the number of wavelengths per ONU U_ω according to the design convention (I) and (IX). This rule can be obtained by upgrading equation (28) to the following relationship

Figure 11. Reference architecture of the multistage WDM PON. It illustrates the stages and interstages which are used for designing the network.

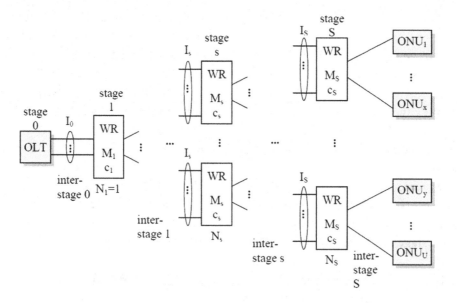

$$I_0 = \frac{N_\omega}{n_{\omega 0}} = \frac{U \cdot U_\omega}{n_{\omega 0}} \qquad (30)$$

in which each user obtains the same number of wavelengths. The next step is to determine the number of required WRs in stage s. The number of output ports in stage s-1 is $M_{s-1} \cdot N_{s-1}$. Because of the design conventions (III) and (VI), we must connect all the outputs from the stage s-1 to the stage s, which leads to the following relation $N_s \cdot I_s = M_{s-1} \cdot N_{s-1}$. Converting it to N_s leads to

$$N_s = \frac{M_{s-1} \cdot N_{s-1}}{I_s} \qquad s \in \left[2, S\right]. \qquad (31)$$

The general relationship for each stage can be obtained by recursion (Maier et al., 2000)

$$N_s = \frac{\prod_{x=1}^{s-1} M_x}{\prod_{x=2}^{s} I_x} = \frac{I_1 \prod_{x=1}^{s-1} M_x}{I_s \prod_{x=1}^{s-1} I_x} \qquad s \in \left[2, S\right]. \qquad (32)$$

Converting equation (31) to I_s leads to the required number of input ports on stage s

$$I_s = \frac{M_{s-1} \cdot N_{s-1}}{N_s} \qquad s \in \left[2, S\right]. \qquad (33)$$

IC WDM PON

In an IC WDM PON, the coarseness factor increases stage by stage (see design convention XII). The primary goal is to calculate the coarseness factor in each stage so that the space demultiplexing function can be guaranteed. The derivation is not an easy task and can be found in (Maier et al., 2000)

$$c_s = \frac{N_{s-1} \cdot M_{s-1}}{I_1} \qquad s \in \left[2, S\right]. \qquad (34)$$

DC WDM PON

In a DC WDM PON, the coarseness decreases stage by stage. The presented method is equivalent to a bottom-up approach which begins at the ONUs and then goes up stage by stage in direction to the OLT. The goal is to calculate the coarseness factor in each stage so that the space demultiplexing function can be guaranteed. The derivation can be found in (Schussmann, 2008)

$$c_s = \frac{M_s \cdot M_{s+1} \cdot N_s}{I_1 \cdot M_1}. \tag{35}$$

Number of Wavelengths Per User

An essential parameter during the design process is the number of wavelengths per user. In a single-stage WDM PON, we are able to calculate the number of wavelengths per user with equations (17) and (18). In a multistage WDM PON, equations (17) and (18) are only valid for one stage and can be used to calculate U_ω stage by stage. It is clear that in the last stage the number of wavelengths per fiber is identical with the number of wavelengths per user.

It is important to know that equations (17) and (18) can reduce the number of wavelengths per fiber but they need not reduce it. Equation (17) and (18) reduce the number of wavelengths in stage s only if $U_\omega < n_\omega$, or seen from another point of view, I/M is less than 1.

It follows that the number of wavelengths per user is the minimum of all U_ω over all S stages:

$$U_\omega = \min\left\{U_{\omega 1}, U_{\omega 2}, ..., U_{\omega S}\right\}. \tag{36}$$

It must be noted that the calculation of U_ω is independent from the coarseness factor. Thus, the presented method can be used in an IC and DC WDM PON.

Capacity Limits

The highest density of number of wavelengths per fiber is at interstage 1 at the root of the tree. The division of the accessible optical spectrum ΔF by the channel spacing Δf leads to the number of wavelengths per fiber n_ω, which can be reused on each port of the WR. Therefore, the capacity limit for a multi-stage WDM PON is then given by

$$B_N = B_0 \cdot N_\omega = B_0 \cdot M_1 \cdot \frac{\Delta F}{\Delta f}. \tag{37}$$

The capacity limit depends on the channel bitrate B_0, the size of the WR in the first stage M_1, the accessible optical spectrum ΔF and the channel spacing Δf. According to the highest density of wavelengths per fiber in the first stage, the capacity limit in (37) is the same as in single-stage WDM PONs and can be seen in Figure 9. Compared with a single-stage WDM PON, the number of users in a multi-stage WDM PON is $M_S \cdot N_S$ and therefore N_S times greater than in single-stage networks.

With an example, the limits are demonstrated. We consider a WDM PON with $\frac{\Delta F}{\Delta f} = 32$, $M_{max} = 64$ and 128 users, each one with a number of wavelengths of 16. Thus, the maximum number of wavelengths in the multi-stage WDM PON is $32 \cdot 64 = 2048$. In other words, the multiplication of the users U with the number of wavelengths per user U_ω does not exceed the number 2048. It must be pointed out that the usage of smaller wavelength spacings allows us to increase the usage of umpteen thousand wavelengths. But smaller numbers of Δf results in higher costs for lasers and WR.

TECHNO-ECONOMIC EVALUATIONS

In most cases of multistage WDM PONs, the technology has worked but the economics has not.

Figure 12. Different levels of modeling the techno- economic evaluation for networks.

Modeling level	Method	Advantages	Disadvantages
Basic	• Average length	• Fast results • Convenient to compare different technologies	• Large margins of errors
Detailed	• Average length and customer densities	• Good assessment for the modeled area • Convenient to compare different technologies	• Medium margins of errors
Highly detailed	• Using geographic positions and existing ducts routes and roads	• Limited margins of errors	• Reliant on the availability of proper data • Highly complex • Manipulation of large amounts of data

Thus, it is important to examine the economical aspects. Thousands of different WDM PON network architectures are technologically possible. The goal is to identify those architectures that are technologically feasible while simultaneously minimizing costs.

In the past, many economic case studies have been published (Bell et *al.*, 1996; Bell & Trigger, 1998; Prat, 2002; Tera, 2008; Tonic, 2008). Each analysis uses a different model to identify the fiber length, ducts, etc. and different cost functions to determine those architectures that minimize the costs. In general, techno-economic evaluations can be performed at different levels as shown in Figure 12.

All listed modeling levels have their own assets and drawbacks. The focus of the techno-economic evaluations is to compare various numbers of different architectures to find those architectures which minimize costs. In this case, the basic modeling level is the best choice despite the drawback of large margins of errors.

Previous works often focused on the investment cost and operational expenses for different first mile solutions. This analysis will focus more on the comparison of network architectural options and influencing factors, which implies setting the focus on a comparison of the investment cost. Costs that have no effect on the comparison were left out (such as central office (CO) building, electricity, lasers, etc.).

Generation of Network Architectures

On the basis of the design conventions, all practicable architectures must be generated. All possible parameter identifications that correspond to an IC or DC WDM PON must be identified. The purpose is to find out which combinations of the number of stages, size of the WR, number of used inputs and the coarseness factor for IC and DC WDM PONs are technologically feasible and minimize the costs.

To determine these combinations, many optimization methods like linear integer programming or other methods can be used. The number of possible network architectures is relatively small. Thus, another approach to solve the problem can be used. All the possible network architectures for a given number of users and for all possible numbers of wavelengths per user are generated. The next step is to compare the costs and identify the optimal parameter choice for a given number of users and wavelengths per user.

A developed design tool is composed of four parts (Figure 13). The first module generates all combinations of $\vec{M}$ and $\vec{N}$ in dependence of S, M_{min}, M_{max} and N_{max}. Another required constraint is the size of the WR. At the moment, due to the technological constraints in the production process, a maximum size of the WR of $M_{max} = 64$ is commercially available which is used to limit the number of architectures. The number of WRs

Figure 13. Flowchart of the functionality of the network design tool for single and multistage IC and DC WDM PONs.

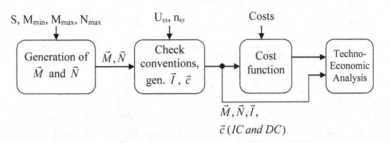

increases stage by stage due to the design convention (V). Thus, the highest number of WRs is in the last stage. The maximum number of WRs in one stage is limited to 64. As a consequence, the maximum number of users can be determined by 64x64=4,096 users.

The second module generates all vectors of $\vec{I}$ and $\vec{c}$ (for both IC- and DC WDM PON architectures) in dependence of U_ω and n_ω and then it moves through all the generated architectures from the first module. Those architectures that do not fulfill the design conventions will be excluded. Within the third module, the costs for all generated architectures are calculated. With the last module, architectures and costs can be analyzed.

In order to verify and to estimate the number of possible architectures for a given parameter configuration, considerations to both questions are explained in the next paragraph.

Starting with the number of theoretically possible architectures, without including the design convention, the number of architectures can be calculated by the well known equation of variations with repetitions

$$V_n^{(k)} = n^k \tag{38}$$

where k is a selection from n different elements.

As a consequence, the number of variations for N is n^{S-1} (S-1 due to the design convention (II)

in which $N_1=1$) and for M it is equal to n_S. The maximum number of N is chosen at 64 which results in an n of 7 ($V_7^{(S-1)}$). The size of the WR is limited between 4 and 64 and will lead to an n of 5 ($V_5^{(S)}$). From this it follows that all theoretical variations can be calculated simply by the multiplication of $V_7^{(S-1)}$ and $V_5^{(S)}$ (Figure 14).

The number of architectures in regard to the design conventions additionally depends on the number of wavelength per user U_ω, the maximum number of wavelengths per fiber n_ω and on the number of stages S. The maximum number of architectures realizable can be calculated with the parameters $U_\omega=1$ and $n_\omega \rightarrow \infty$ (last two columns in Figure 14). The maximum number of architectures in dependence of S generated by the design tool is shown in the last two columns in Figure 14 for IC and DC WDM PONs.

The number of possible architectures depends on S, n_ω, U_ω, M_{min}, M_{max} and N_{max}. As shown in the example in Figure 15, the number of possible architectures rises with the number of users. From this point in dependence of S, the number of possible architectures remains constant or declines. This means that each curve in Figure 15 has its own maximum. This maximum is recommended because it allows a high degree of freedom. This degree of freedom results in a higher number of minimum cost architectures and allows the choice of those architectures that best match the conditions in the real world.

Figure 14. Number of theoretical and practical multistage WDM PON architectures. Theoretically 7,723,805 different architectures are possible. Due to the design conventions, only a fraction of the number of architectures is feasible. For an IC WDM PON, 7,785 possible architectures and 7,660 architectures are possible for a DC WDM PON.

S	$V_7^{(S-1)}$	$V_5^{(S)}$	$V_7^{(S-1)} \cdot V_5^{(S)}$	IC, $n_w \to \infty$, $U_\infty=1$	DC, $n_w \to \infty$, $U_\infty=1$
1	1	5	5	5	5
2	7	25	175	80	80
3	49	125	6,125	782	755
4	343	625	214,375	2,888	2,821
5	2,401	3,125	7,503,125	4,030	3,999
Sum	2,801	3,905	7,723,805	7,785	7,660

Cost Performance

The objective is to identify architectures that minimize the costs. To compare all generated architectures, three distinct cost functions are defined. Only those costs are considered which allow a comparison of different architectures. In the economical evaluations, we have considered the costs for commercially available devices at the end of 2007.

To compare the variety of architectures, the cost function consists of three distinct cost portions. The first one includes the cost of the WR.

The coarseness factor of the WR is determined by the individual design of the star couplers and the integrated waveguides. The reported costs in Figure 16 have been provided by a manufacturer. It is appropriate to expect a drop in costs of WR manufacturing, thus the production volumes should increase. This scenario is likely to happen in the future as these devices will become more and more accepted as devices to set-up optical networks.

The second cost function takes into account the cable cost. The cable cost depends mainly on the number of fibers. The differences between

Figure 15. Example of the maximum number of multistage IC and DC WDM PON architectures in dependence from the number of users U with S between 1 and 5, $n_\omega=64$, $U_\omega=2$, Mmin=4, Mmax=64, Nmax=64. In summary, 1199 different number of architectures are possible.

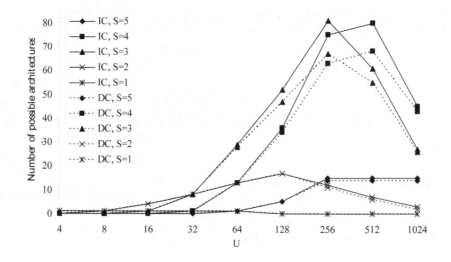

Figure 16. WR cost in dependence on the size at current commercially available devices at the end of 2006

WR size	Cost/port [€/port]	Cost [€]
4	100	400
8	85	680
16	70	1,120
32	55	1,760
64	40	2,560

them are significant. Thus, five different cable costs between 4 and 64 fibers are used in the economic analysis (Figure 17). This implies that each user is linked with four fibers. We suppose for simplicity that all the I_s fibers in stage s are placed in one single cable.

The third cost function accounts for the installation cost (Figure 17) (Tan, 1997). It contains the cost of the installation and of the trenching building operation. Partial cable reuse is not taken into account and therefore it would overestimate the cable and installation costs if networks with more PONs had to be constructed.

For calculating the cable and installation costs, link lengths of all the cable links in a multistage WDM PON must be known to calculate the total cost. This is not an easy task because there are no area-wide commercial multistage WDM PON

Figure 17. Cable and installation cost in € per meter

Cable and installation cost [€/m]	
Cable cost (4 fibers)	0.8
Cable cost (8 fibers)	1.0
Cable cost (16 fibers)	1.4
Cable cost (32 fibers)	3.8
Cable cost (64 fibers)	3.0
Installation cost	15.0

installations today. Therefore, link lengths must be assumed based on more traditional networks. (Mickelsson et *al.*, 2002) shows that with a link length of 1,500m, 50% of all subscribers can be reached. Thus, the sum of the link lengths in a stage is always 1,500m (Figure 18). The length of the interstage links is also difficult to forecast. We assume a distribution as reported in Figure 18.

Performance Analysis

We consider the architectures with U=128, U_ω=2, n_ω=64, M_{min}=4 and M_{max}=64. These numbers derive from a technological perspective and are freely chosen. The data obtained from the cost analysis function were first grouped by the number of stages S, and then each group is ordered by increasing costs per user. To get a clearly arranged view, the resulting costs are numbered in ascending order beginning with zero and act as an index, called architecture index (Figure 19). It should be noted that architectures with an architecture index of zero represents the WDM PON with the best cost performance. The sum of the three cost components is shown in Figure 19 The best cost performance architectures are those with S=5 and then closely followed by those with S=4. This is the result of the relatively higher cost of cable and installation compared with the WR cost.

CONCLUSION

The already enormous and ever increasing bandwidth demand justifies the need for high-speed access networks. In particular, WDM technology is commonly viewed as the preferred choice in building such future high-speed access networks. In WDM networks, multiple wavelength channels operating at electronic transmission rates are supported in parallel to fully utilize the enormous potential bandwidth of a single optical fiber. Furthermore, WDM PONs appear to be the preferred architecture for the development of highly flexible

Figure 18. Link lengths in meters in multistage WDM PONs dependent on the number of WR stages. Links are numbered beginning at the CO and then going in the direction of the ONUs.

Number of WR stages	Link number					
	1	2	3	4	5	>5
1	900	600				
2	900	400	200			
3	900	400	100	100		
4	900	400	100	50	50	
>4	900	400	100	50	30	20

future access networks.

To make sure that multistage WDM PONs can be considered as viable, promising alternatives to current access network developments such as WLAN and xDSL, several requirements have to be fulfilled. First, they have to be realizable and cost-effective. Specifically, commercially available low cost WRs, lasers and receivers are of high interest in this context. Second, they should provide scalability in terms of the number of wavelengths per user and in terms of the number of users. Third, they have to support an independent multi-provider solution and an independent multi-service solution which results in an uncoupling between the network and the service provider.

The paradigm shift of at least one wavelength per user and service is the basis for further considerations. It is a long-term and flexible alternative to the TDM PON currently being deployed by operators to support the extraordinary data traffic growth in access networks.

Further on, techno-economic evaluations of single and multistage WDM PONs are accomplished. Architectures minimizing the costs are identified. The selection depends on crucial requirements such as the number of users and the number of wavelengths per user.

Currently, the researched architectures have been restricted to tree topologies only. However, bus topologies are more fail-safe than tree topolo-

Figure 19. Total costs per user for the architectures received from the design tool with $U=128$, $U_\omega=2$, $n_\omega=64$, $l_{max}=1,500m$, $M_{min}=4$ and $M_{max}=64$. Architectures are ordered by increasing costs.

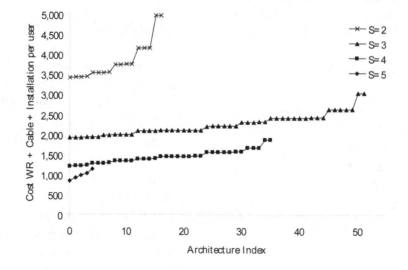

gies (q.v. self-healing in SDH) and thus a topic in optical access networks.

A constraint in the design process is the same number of wavelengths per user for all users. In a real deployment, this assumption limits the employment of multistage WDM PONs. As a result, the design rules must be adjusted. In case of using more than one wavelength per user, standards are needed to define which application is assigned to which wavelength. This means that a protocol must be developed to offer the exchange of the control information needed. Moreover, an intelligent scheduler is needed in the OLT to fairly assign the limited number of wavelengths to users and applications. This is an important topic and has the potential for future research activities.

REFERENCES

Agrawal, P. (2002). *Fiber-Optic Communication Systems*. New York, USA: Wiley-Intersience.

An, F., Kim, K. S., Gutierrez, D., Yam, S., Hu, E., Shrikhande, K., & Kazovsky, L. G. (2004). SUCCESS: A Next-Generation Hybrid WDM/TDM Optical Access Network Architecture. *Journal of Lightwave Technology, 22*(11), 2557–2569. doi:10.1109/JLT.2004.836768

Banerjee, A., Park, Y., Clarke, F., Song, H., Yang, S., & Kramer, G. (2005). Wavelength-division-multiplexed passive optical network (WDM-PON) technologies for broadband access: a review (Invited). *Journal of Optical Networking, 4*(11), 737–758. doi:10.1364/JON.4.000737

Barry, R. A., & Humblet, P. A. (1993). Latin Routers, Design and Implementation. *Journal of Lightwave Technology, 11*(5), 891–899. doi:10.1109/50.233253

Bell, P., & Trigger, P. (1998). Access network life-cycle costs. *BT Technolology, 16*(4), 165–174. doi:10.1023/A:1009612317200

Bell, P., Walling, R., & Peacok, J. (1996). Costing the access network – an overview. *BT Technology, 14*(2), 128–132.

Biraghi, A. M., Cobb, K., Cantone, S. C., Greborio, L., Mercinelli, R., & Walling, R. (2002). *Survey of future broadband services*. Ist for Broadband Europe (pp.1-13).

Bock, C., Prat, J., Segarra, J., Junyent, G., & Amrani, A. (2004). Scalable Two-stage Multi-FSR WDM-PON Access Network Offering Centralized Dynamic Bandwidth Allocation. *ECOC 2004 Proceedings*, (2), 278-279.

Borella, A., Cancellieri, G., & Chiaraluce, F. (1998). *Wavelength Division Multiple Access Optical Networks*. Norwood, Ohio: Artech House.

Coffman, K. G., & Odlyzko, A. M. (2001). *Internet growth: Is there a "Moore's law" for data traffic?* New York, USA: AT&T Labs - Research.

Dixit, S. S. (2003). *IP over WDM, Building the Next Generation Optical Internet*. Hoboken, New Jersey: John Wiley & Sons.

FSAN. (2008). *Full Service Access Network*. Retrieved July 23, 2008, from http://www.fsan-web.org/.

Gillespie, A. (2001). *Broadband Access Technology, Interfaces, and Management*. Norwood, Ohio: Artech House, Inc.

Glance, B., Kamoniw, I. P., & Wilson, R. W. (1994). Applications of the Integrated Waveguide Grating Router. *Journal of Lightwave Technology, 12*(6), 957–962. doi:10.1109/50.296184

Green, P. E. (2005). *Fiber to the Home, The New Empowerment*. Hoboken, New Jersey: John Wiley & Sons.

Gumaste, A., & Antony, T. (2002). *DWDM Network Designs and Engineering Solutions*. Indianapolis, USA: Cisco Press.

Hecht, J. (2005). *Understanding Fiber Optics.* Columbus, Ohio: Prentice Hall.

Iannone, P. P., Reicjmann, K. C., & Frigo, N. J. (1998). High-Speed Point-to-Point and Multiple Broadcast Services Delivered over a WDM Passive Optical Network. *IEEE Photonics Technology Letters, 10*(9), 1328–1330. doi:10.1109/68.705632

ITU-T G.694.1 (2002). *Spectral grids for WDM applications: DWDM frequency grid.* Telecommunication Standardization Sector of ITU.

ITU-T G.694.2 (2002). *Spectral grids for WDM applications: CWDM frequency grid.* Telecommunication Standardization Sector of ITU.

ITU-T G.7041 (2003). *Generic Framing Procedure (GFP), in Series G: Transmission Systems and Media, Digital Systems and Networks.* Telecommunication Standardization Sector of ITU.

ITU-T G.984.1 (2003). *General characteristics for Gigabit-capable Passive Optical Networks (GPON).* Telecommunication Standardization Sector of ITU.

ITU-T G.984.2. (2003). *Gigabit-capable passive optical networks (GPON): Physical media dependent (PMD) layer specification.* Telecommunication Standardization Sector of ITU.

ITU-T G.984.3. (2003). *Transmission Convergence Layer for Gigabit Passive Optical Networks.* Telecommunication Standardization Sector of ITU.

Kanbach, A., & Körber, A. (1999). *ISDN – Die Technik, Schnittstellen – Protokolle – Dienste – Endsysteme.* Heidelberg, Germany: Hüthig Verlag.

Keiser, G. (2003). *Optical Communications Essentials.* New York, USA: McGraw-Hill Professional.

Kettler, D., Kafka, H., & Spears, D. (2000). Driving Fiber to the Home. *IEEE Communications Magazine, 38*(11), 106–110. doi:10.1109/35.883497

Killat, U. (1996). *Access to B-ISDN via PONs, ATM Communication in Practice.* Chichester, England: John Wiley and Sons Ltd.

Kramer, G. (2005). *Ethernet Passive Optical Networks.* New York, USA: McGraw-Hill Professional Publishing.

Kramer, G., Mukherjee, B., & Pesavento, G. (2001). Ethernet PON (ePON): Design and Analysis of an Optical Access Network. *Photonic Network Communications, 3*(3), 307–319. doi:10.1023/A:1011463617631

Kramer, G., & Pesavento, G. (2002). Ethernet Passive Optical Network (EPON): Building a Next-Generation Optical Access Network. *IEEE Communications Magazine, 40*(2), 66–73. doi:10.1109/35.983910

Krishna, M. S., & Subramaniam, S. (2000). *Optical WDM Networks Principles and Practice.* Norwell, Massachusetts: Kluwer Academic Publishers.

Lin, C. (2004). *Optical Components for Communications.* Norwell, Massachusetts: Kluwer Academic.

Lin, C. (2006). *Broadband Optical Access Networks And Fiber-to-the-Home, Systems Technologies and Development Strategies.* Chichester, England: John Wiley & Sons Ltd.

Lund, B. (2001). Fiber-to-the-Home Network Architecture: A Comparison of PON and Point-to-Point Optical Access Networks. *Proceedings of the National Fiber Optic Engineers Conference* (pp. 1550-1557).

Maeda, Y. (2004). Overview of Optical Broadband in Japan. *ECOC 2004 Proceedings, 1*, 4-5.

Mahlke, G., & Gössing, P. (2001). *Fiber Optic Cables.* Munich, Germany: Wiley Verlag.

Maier, G., Martinelli, M., Pattavina, A., & Salvadori, E. (2000). Design and Cost Performance of the Multistage WDM-PON Access Networks. *Journal of Lightwave Technology, 18*, 125–143. doi:10.1109/50.822785

Mayhew, A. J., Page, S. J., Walker, A. M., & Fisher, S. I. (2002). Fibre to the home – infrastructure deployment issues. *BT Technology Journal, 20*(4), 91–103. doi:10.1023/A:1021374532690

McGarry, M. P., Reisslein, M., & Maier, M. (2006). WDM Ethernet Passive Optical Networks. *IEEE Optical Communications, 44*(2), 18–25.

Mestdagh, D. J. G. (1995). *Fundamentals of Multiaccess Optical Fiber Networks*. Norwood, Ohio: Artech House, Inc.

Mickelsson, H., Sundberg, E., Strömgren, P., & Fujimoto, Y. (2002). *IEEE Contribution on loop lengths*. IEEE EFM. Retrieved July 25, 2008, from http://grouper.ieee.org/groups/802/3/efm/public/jan02/mickelsson_1_0102.pdf.

Milorad, C. (2004). *Optical Transmission Systems Engineering*. Norwood, Ohio: Artech House.

Nakanashi, K., & Maeda, Y. (2002). Standardization Activities of FSAN: International Standardization Trends Concerning the Broadband PON (B-PON). *NTT Review, 14*(2), 108–110.

OECD. (2008). *OECD Broadband subscribers per 100 inhabitants, by technology, December 2007*. OECD Broadband Portal. Retrieved July 22, 2008, from http://www.oecd.org/.

Photeon (2008). *Photeon Adds To Extensive Array of AWGs*. Press Release. Retrieved July 24, 2008, from http://www.photeon.com/html/Press/2005 02 28.pdf.

Prat, J. (2002). *Fiber-to-the-Home Technologies*. Dordrecht, Netherlands: Kluwer Academic Publishers.

Ramsey, D. (2005). *World's First International Real-time Streaming of 4K Digital Cinema Over Gigabit IP Optical Fiber Networks*. University of California, San Diego. Retrieved July 22, 2008, from http://ucsdnews.ucsd.edu/newsrel/science/iGrid4K.asp.

Reed, D. (2003). *Copper Evolution. Federal Communications Commission Technological Advisory Council III*. Retrieved July 23, 2008, from http://www.fcc.gov/oet/tac/TAC_III_04_17_03/Copper_Evolution.ppt.

Schussmann, J. (2008). *Design and Cost Performance of WDM PONs for Multi-Wavelength Users*. Unpublished doctoral dissertation, Technical University of Ilmenau, Germany.

Shannon, C. E. (1948). A Mathematical Theory of Communication. *The Bell System Technical Journal, 27*, 623–656.

Simcoe, M. (2002). A Multiservice Voice, Data, and Video Network Enabled by Optical Ethernet. Proceedings of the National Fiber Optic Engineers Conference, (pp.119-129).

Sivalingam, K. M., & Subramaniam, S. (2005). *Emerging Optical Networks Technologies*. New York, USA: Springer.

Smit, M. K., & van Dam, C. (1996). PHASAR-Based WDM-Devices: Principles, Design and Applications. *IEEE Journal on Selected Topics in Quantum Electronics, 2*(2), 236–250. doi:10.1109/2944.577370

Söderlund, M., Kapulainen, M., & Koponen, J. (2004). Techno-economic Comparison of Four Alternative FTTH Approaches in a Greenfield Building Site. *ECOC 2004 Proceedings, 3*, 394-395.

Stern, J. R., Hoppitt, C. E., Payne, D. B., Reeve, M. H., & Oakley, K. A. (1988). TPON - A passive optical network for telephony. *Fourteenth European Conference on Optical Communication, 1,* 203-206.

Takahashi, H., Oda, K., Toba, H., & Inoue, Y. (1995). Transmission Characteristics of Arrayed Waveguide NxN Wavelength Multiplexer. *Journal of Lightwave Technology, 13*(3), 447–455. doi:10.1109/50.372441

Tan, A. H. H. (1997). SUPER PON – A Fiber to the Home Cable Network for CATV and POTS/ISDN/VOD as Economical as a Coaxial Cable Network. *Journal of Lightwave Technology, 15*(2), 213–218. doi:10.1109/50.554326

Tanis, D., & Eichenbaum, B. R. (2002). Cost of Coarse WDM Compared with Dense WDM for Wavelength-Addressable Enhanced PON Access. *IEE Seminar on Photonic Access Technologies.*

Tera (2008). *Techno Economic Results from ACTS Program from the European Commission, AC364.* Retrieved July 24, 2008, from http://www.telenor.no/fou/prosjekter/tera/index.htm.

Tonic (2008). *Techno Economics of IP Optimised Networks and Services, Project within the IST Programme (Information Society Technologies) of the European Union.* Retrieved July 24, 2008, from http://www-nrc.nokia.com/tonic/.

Wan, P. J. (2000). *Multichannel Optical Networks.* Dordrecht, Netherlands: Kluwer Academic Publishers.

Weldon, M. K., & Zane, F. (2003). The Economics of Fiber to the Home Revisited. *Bell Labs Technical Journal, 8*(1), 181–206. doi:10.1002/bltj.10053

Wolf, J., & Zee, N. (2000). *The Last Mile. Broadband and the Next Internet Revolution.* New York, USA: McGraw Hill.

Section 5
The Way Forward

Chapter 12
The New Generation Access Network

Christos Vassilopoulos
Hellenic Telecommunications Organization S.A., Greece

ABSTRACT

This chapter offers a qualitative approach towards the development of the new generation access network, based on FTTx implementations. After a brief description of the current state of traditional access networks and an estimation of the expected data rate per household in terms of services, the chapter examines all the available Network Technologies (FTTx), Access Technologies (xDSL, Ethernet and PON) for both P2P and P2MP development schemes and their relevant implementations. The prospects of NGA are also strategically examined in view of the complicated multi-player environment, involving Telco (ILEC and CLEC), regulators and pressure interest groups, all striving to serve their individual, often conflicting interests. The chapter concludes with an outline of the different deployment strategies for both ILEC and CLEC Telco.

INTRODUCTION - BACKGROUND

NGA stands for New Generation Access, a new concept associated with the future access network architecture that is capable for providing sufficient bandwidth to all present and forthcoming applications, thus removing the bandwidth barrier from the last mile for the next fifty years. It is closely related with NGN, the New Generation Network platform that will transform the existing, traditional

DOI: 10.4018/978-1-60566-707-2.ch012

TDM-based multi-network structure into a universal all-IP network.

Nobody denies that the future access network will eventually be an all-fibre network in a "fibre-to-the-home" (FTTH) architecture, even though the offering of POTS (plain old telephony services) services over optical fibre requires a higher cost and imposes the need for local powering for opto-electronic conversion, which is unnecessary in the case of copper access telephony. Most Telco have already taken the decision to adopt FTTH solutions in "Greenfield" network implementations, and in most countries, new-house cabling regulations are coming in force for assisting the provision for future installation of optical fibres inside the buildings.

What is still under strong debate in the telecommunication industry is not the final outcome of the access network evolution towards fibre, but the evolution stages and duration. If one looks into the history of the conventional copper based access network in Europe and the USA, it is clear that the whole process of development was both costly and highly time consuming. It took more than fifty years to built, the development was gradual and the cost gigantic. To replace the existing copper access network with a new overlay optical fibre network, even in urban areas and over a period of ten years, is a substantial challenge requiring a huge amount of funding that can be hardly justified by present day economics and the heavily competitive Telco environment.

An alternative approach that goes via a "fibre-to-the-cabinet with VDSL2" stage (FTTC+VDSL2) seems a more economic intermediate step, because it takes advantage of the shorter copper lengths (<500m) of the existing distribution networks in order to deliver data rates of above 50 Mb/s to most of the customers at a small fraction[1] of the FTTH cost. If it was only a matter of economics, nobody could honestly argue against the combination of FTTC+VDSL2 as the most viable solution for NGA development, particularly in areas with an existing copper network infrastructure. Nevertheless, in the present day complicated telecommunication environment, the strategy[2] of the individual players in view of their contrasting interests may also play an equally important role that cannot be ignored. It is, therefore, essential to understand the basic NGA development strategies and assess all the different implementation scenarios from the perspective of each player.

The present chapter is divided in six sections examining: the traditional structure of the access network, the present and future services in terms of their bandwidth requirements, the available network and access technologies, different NGA network implementations and deployment strategies. The content of this chapter reflects the personal views of the author, which by no means imply official OTE policy.

DEFINITIONS: THE TRADITIONAL STRUCTURE OF THE ACCESS NETWORK

In accordance with the European Telecommunications Standards Institute (ETSI), the access network is defined as the network part that links the subscriber to its local exchange (LE) including the primary network also known as feeder, the secondary or distribution network and the customer or drop segment as shown in Figure 1. It is also often referred to as the Local Loop, Copper Loop or Last Mile.

Even though, in most cases, the three-segment "modular" design depicted above is followed, in some exceptions and for large customers (i.e. ministries, banks, hospitals, police etc.) the primary network may be directly terminated into the customer's premises in a "fixed-feeder like" network approach[3].

The access network was originally designed and gradually built over the last fifty years to provide ordinary telephony (POTS) services to customers. It employs cables consisting of unshielded twisted copper pairs or quads[4], that start from the Main Distribution Frame (MDF) of the LE and terminate at the Network Termination Equipment (NTE) at the subscriber premises.

The primary network contains a number of high capacity cables of up to 2,400 pairs that start from the LE, split in smaller capacity cables of up to 400 pairs at intermediate junction points in a tree-like topology, covering the LE area and eventually terminate at outdoor cross-connect cabinets (XCC). The cables of the primary network are usually dry, kept under pressure and are

Figure 1. Traditional European Access Network Development

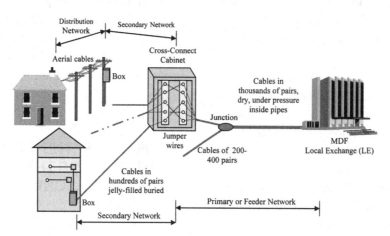

positioned in pipes.

The secondary network starts from the XCC and contains a number of smaller capacity cables (from 10 to 100 pairs) that develop in a star-like topology around the cabinet and terminate either at the termination box of the customer building (inside or outside) or on a pole situated near the customer building. There follows the distribution network with indoor or outdoor cabling using 1 or 2 pairs for each subscriber.

According to the BOBAN[5] project, the access network configuration of different European Public Network Operators (PNO) is outlined in

Table 1.

The access network cables are of different age and technologies[6], they have over the years received numerous maintenance works and their current state is largely uncertified. They use conductors with diameters of 0.4[7] and 0.6mm and more rarely 0.8 and 0.9mm[8]. For uniform spans, the maximum cable length for each conductor diameter is limited to 3.5 − 6.0 − 8.0 and 9km respectively. Even though uniform spans with conductors of 0.4 and 0.6mm diameters are preferred, there are cases where, in order to increase the reach, a mixed use of 0.4 and 0.6mm conductor cables is

Table 1. Results from EURESCOM Project P614 questionnaire to European ILEC

Network Structure	Primary Network 1st Segment	Secondary Network 2nd Segment	Distribution Network 3rd Segment
Average Cable Length (m) *Urban areas* *Rural areas*	500 − 2,000	300 − 1,000 100 − 1,000	50 − 100
Pairs per segment (on average)	400 − 2,400	10 − 200	1 − 2
Conductor diameter (mm)	0.4 − 0.6	0.4 − 0.6	0.4 − 0.8
Percentage of used pairs per section	65%	50%	100%
Average subdivision of copper network			
Urban/Metropolitan: Ducts	80%		
Sub-Urban: Ducts / directly buried	50% / 30%		
Rural: Aerial	80%		

Figure 2. OTE XXC cabinets (new and old types) allowing for up to 1,200 and 750 pair terminations respectively. A primary to secondary cable pair ratio of 2:3 is used.

allowed without the use of impedance matching. Uniform spans with 0.8 or 0.9mm conductors are only used to provide telephony services to isolated settlements distant from the LE.

The XCC is the main passive flexibility point for accessing the copper pairs of the main and distribution networks. It comprises of a metal or plastic outdoor cabinet of the appropriate dimensions[9] situated on a plinth to protect from flooding, and contains cross-connection elements[10] for the termination of the primary and secondary network cables. The total number of customers served by a XCC varies from 150 to 600 for different operators[11], and depends on the population density and overall network design. Denser areas are associated with larger XCCs, while smaller XCC usually indicate shorter secondary network lengths. Figure 2 shows the different XCC used by OTE[12] and their internal organization.

The termination or terminal box (TB) represents the closest to the customer Telco-owned flexibility point. It is situated either within the building, connected to the interior network of

multi-apartment buildings, or in the external plant, wall-mounted or pole-mounted. Both indoor and outdoor TB are metal or plastic and contain copper cable termination and cross-connection blocks or strips. The indoor boxes, also called *"escalit"*, are bigger to allow for the termination of up to 100 copper pairs (for external and internal cabling purposes), while outdoor boxes are smaller and offer a termination capacity of 10 to 20 pairs. The distribution from the outdoor boxes to the individual customer is provided using self-supporting cables[13] of 2 or 4 pairs.

From the above, it is clear that the ability of an operator to provide broadband services through the existing access network is clearly affected by the individual dimensional characteristics of their access network. Some interesting data reported by European Telco[14], related to their access network coverage, are presented in Figure 3 and Figure 4 for the primary and secondary networks respectively.

The corresponding figures related to the average number of household served from XCC and

Figure 3. Reported primary network coverage for ILEC operators [ITU-T FS-VDSL FGTS Part 1: Operator requirements, p. 26].

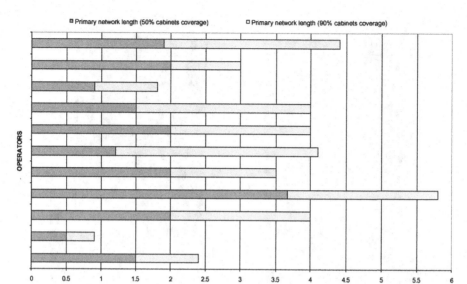

TB are reported in Table 2.

From the above, it is clear that in most countries, about 50% of the total number of customers on average, are situated within 2,000m from the MDF of the LE. The situation appears to be more favorable for the European Operators, having shorter access networks due to denser population, when compared to those in the USA, as shown in Figure 5.

Figure 4. Reported secondary network coverage for ILEC operators [ITU-T FS-VDSL FGTS Part 1: Operator requirements, p. 26].

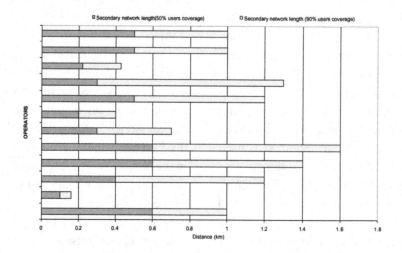

Table 2. Average number of household served at various points of the access network

Operators	1	2	3	4	5	6	7	8	9	10	11	12
XCC	250	600	250	500	250	200	350	500	400	300	350	110
TB	20	25	20	8	4	12	7	8	4	n/a	10	7

NETWORK SERVICES

Services is what adds value to a network, without them, telecom networks are like highways without cars, unused empty lanes without traffic and no significant revenues from tolls to finance maintenance and justify their existence. Some argue beforehand, "Give us the applications that require bandwidth enough to justify the network investment, give us a *killer application*". Others support that by providing the bandwidth by itself at a low price will create the applications and will generate the expected growth in the same way S/W and H/W did for the computer industry.

To the author's knowledge, there is no "killer application" in telecommunications that could, by itself, justify the vast investment required for NGA deployment. There is, however, a considerable number of applications, outlined in Table 3, which, when served by a common IP-based platform[16], can provide sufficient economy of scale for low enough pricing to stimulate growth

and generate revenues.

Table 3 refers to specific services of voice, audio, video and data and their requirements in terms of bandwidth, quality of service (QoS), delay, packet-loss, flow and type. Event though these services are constantly evolving and their transmission requirements over then next five years are likely to change, one can estimate that the future average household bandwidth requirements for 1 HDTV (high-definition television) and 2 SDTV (standard-definition television) channels, 2 gaming channels, 2 voice video calls and a high speed internet connection will exceed 20 Mb/s[17].

To understand the nature and development of future multimedia services one must appreciate the difference between the provision of services over the general Internet, which only supports one "best effort" class of service, and the provision of services from managed networks, that can support a wide range of QoS parameters for individual services and applications.

Figure 5. Cumulative distribution of copper line lengths[15]

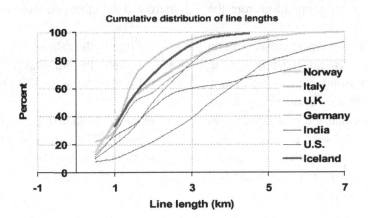

Table 3. Types and QoS requirements of multimedia services [SOURCE: Sigurdsson, H. M. (2007), Ch.2, p. 91]

Category	Service	Download	Upload	QoS	Delay	PLR[19]	Flow	Type
Voice	PSTN	64	64	CBR[20]	150	1%	two-way	unicast
	VoIP	30	30	rt-VBR[21]	400	3%	two-way	unicast
Audio	Audio-conferencing	128		rt-VBR			two-way	multicast
	Audio-on-Demand	256		rt-VBR			one-way	unicast
	Broadcast Audio	128		rt-VBR			one-way	multicast
Video	Broadcast SDTV @ MPEG-2	4.000		CBR			one-way	multicast
	Broadcast HDTV @ MPEG-2	15.000		CBR			one-way	multicast
	Broadcast SDTV @ MPEG-4	2.000		CBR			one-way	multicast
	Broadcast HDTV @ MPEG-4	8.000		CBR			one-way	multicast
	Video-on-Demand SDTV	2.000		CBR			one-way	unicast
	Video-on-Demand HDTV	8.000		rt-VBR			one-way	unicast
	Video-telepony	384	384	rt-VBR			two-way	unicast
	Video-conferencing	384	384	rt-VBR			two-way	multicast
Data	File Transfer	1.000-5.000		UBR[22]			one-way	unicast
	WWW	256-1.000		ABR[23]			one-way	unicast
	E-mail	1.000		UBR			one-way	unicast
	Peer-to-peer	5.000		UBR			two-way	unicast
	Instant Messaging	10		ABR			two-way	unicast
	Gaming-on-Demand	1.500		CBR	100		two-way	unicast

Most European Telco are getting ready to rollout new multimedia services, mainly IPTV (Internet Protocol TV) and VoD (video on demand). Despite great efforts by equipment vendors and high profile industry collaborations there are still several roadblocks, ranging from the lack of an overall standardisation framework, low profit margins and the threat of competition from the Internet.

To evaluate the effect of emerging services on access bandwidth requirements it is important to consider demand forecasting over the next 5 years. Published data from studies using extrapolation techniques[18] point out that the transmission requirements of multimedia services for an average household are likely to increase from 20 to above 50 Mb/s in the next 5 years, subject to individual demand, demographics etc.

NETWORK TECHNOLOGIES

The employment of optical fibre in the access network is served by a group of network technologies referred to as FTTx (Fibre To The x), where x implies the degree of fibre penetration into the network. Even though the term has not been standardized the most common variants include:

- Fibre to the cabinet or curb (FTTC)
- Fibre to the building (FTTB)
- Fibre to the home (FTTH).

Others[24] include FTTN (fibre-to-the-neighbourhood), FTTO (fibre-to-the-office), FTTP (fibre-to-the-premises), FTTU (fibre-to-the-user) and FTTD (fibre-to-the-desk).

Figure 6. FTTC network outline with active cabinet placed on existing XCC location

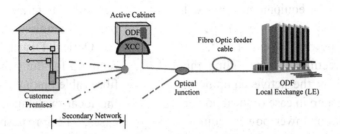

FTTC

FTTC stands for fibre-to-the-cabinet (or Curb) and considers the use of optical fibres up to an active cabinet, placed on the street pavement or curb, serving a group of customers (from 50 to 500) via the existing copper distribution network. Given the present access network topology, the most suitable place for installing the active cabinet seems to be the existing cross-connection cabinet (XCC), with the fibre replacing only the feeder cable, that links the XCC to the local exchange (feeder network), and retaining the secondary network to connect to the customer premises, as shown in Figure 6.

FTTC deployment requires the installation of an outdoor cabinet for housing the active equipment. The cabinet must be such as to house and protect the active equipment from the outside environment[25], ensuring its normal and undisturbed operation. Even though different deployment scenarios may be considered[26], in the general case, the active equipment inside the cabinet must be properly sized to cover the requirements of all the customers, providing services such as POTS, VoIP, ISDN (Integrated Services Digital Network), DSL-based (Fast Internet, IPTV, VoD), leased lines etc.

An outdoor cabinet must contain both active and passive equipment as described below:

- Active equipment for providing both broadband and narrowband services. In the general case two different sub-racks

may be employed, namely a TDM-based flexible multiplexer for providing POTS and ISDN services and a DSLAM (DSL Access Multiplexer) for providing xDSL services (normally VDSL2). In most recent implementations, a single sub-rack in the form of MSAN[27] may be used for providing both narrowband and broadband services, thus saving cabinet space and power requirements, though MSAN technology still lacks maturity. In some cases might be also useful to include the installation of an Ethernet Switch inside the cabinet in order to aggregate flow from the DSLAM and FE services, offered directly to customers from the cabinet over fibre. In all cases, active equipment placed in outdoor cabinets must be temperature hardened, to withstand higher temperatures of up to 60°C during summer without the need for air conditioning, and lower temperatures down to -20°C during winter.

- Transmission equipment for serving the above sub-racks. DSLAM, MSAN Switch or ONU/ONT[28] equipment requires an Ethernet connection (FE or GE) for all Ethernet based services, usually served by a dark fibre connected to an access switch at the local exchange. TDM based flexible multiplexers require V5.2 NxE1 connections to an OLT[29] equipment at the local exchange, served by optical "widelink modems"[30] or SDH (Synchronous Digital Hierarchy) transmission equipment.

- A powering unit, local or remotely fed, for powering the active equipment. For local powering a meter is usually necessary, for measuring the power consumption[31], together with a rectifier, for providing the -48 V required by the active equipment, and batteries backup in case of local power failure[32]. In case of lower power requirements (<500W), remote power from the local exchange may be used[33] with sufficient copper pairs[34] from the feeder cable. Both dc and ac remote powering have been considered[35], either from the exchange or from an outdoor power cabinet using clustering techniques[36]. Remote power feeding requires the installation of a remote power unit at the cabinet, which is fed by the central office unit using telecom copper pairs. No batteries are required with significant space and maintenance cost savings, since, in case of power cuts, back-up powering is provided by the local exchange. The choice between local and remote powering depends on power requirements[37], cost, availability and national safety limitations. In both cases, a PDU[38] is useful for providing powering connections to the active equipment with the appropriate safety fuses.

- A Local Distribution Frame (LDF) is required for the provision of both narrowband and broadband services to the customers. In most cases, it is highly desirable to maintain the LDF of the XCC by integrating it into the cabinet structure, thus avoiding the construction work necessary for connecting the existing distribution network to the new cabinet. The cabinet space required for the LDF depends on the dimensioning of the active equipment. Sufficient space must be available for the termination of the narrowband and broadband sub-racks, the housing of the MDF splitters and the termination of the distribution network. For a cabinet of up to 300 customers and a 2/3 ratio between feeder and distribution networks, a termination space for 1,050 pairs[39] is required at the LDF.

- An Optical Distribution Frame (ODF) is required for the termination of the optical fibre cable linking the outdoor cabinet with the local exchange. Even though in most cases, a 12 fibre termination may seem sufficient, higher termination capacities may be useful for future applications (i.e. for providing from the cabinet FE services to customers over fibre etc.).

- A temperature controlling unit for the active equipment inside the cabinet; its type and dimensioning depends on the power dissipation requirements of the cabinet. Cooling is more important to southern countries, where during the summer the external temperature may rise as high as 50°C. Different cooling schemes can be used, varying from free-cooling with fans and dust filters, to heat-exchangers of various sizes and air conditioning units. The latter are usually avoided, since their power requirements quickly exhaust the battery back-up power in case of power cut. For smaller cabinets free cooling with dust-filter offers a good low cost solution, with additional maintenance cost required for the filters and the fans. The heat exchanger seems a more attractive solution for medium and large size cabinets, with improvements in efficiency and fan noise reduction. A recent 3M product, with the commercial name "Thermosyphon™"[40], is also expected to provide an efficient attractive alternative for cooling. Heating is also important to northern countries, where during winter the outdoor temperature can go as low as -30°C. Since in normal operation, the equipment generates enough heat to be within its normal temperature operating levels, it is often only necessary to provide a heating element with a fan for

Figure 7. FTTC network outline with one active cabinet serving more XCC cabinets

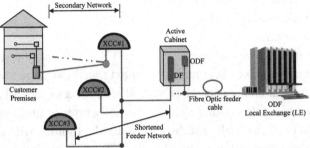

cold start up[41].

- Finally, a unified management system that will control the operating conditions inside the outdoor cabinet is required. A group of sensor elements will be used to detect un-authorized entrance, temperature, humidity and smoke levels, while other detectors will manage the cooling fans, the power supply and the batteries. A remote management unit will multiplex the individual sensor signals into a single management signal (usually with FE interface), which will be transmitted as in-band or out-band information to the local exchange central management unit. Note that the active equipment (DSLAM, MSAN, switch and transmission system) also has its own management system, which is usually transmitted to a dedicated NMS.

Even though significant work has been carried out in attempts to standardize a broadband cabinet for outdoor use[42], it usually acts as basic guidelines and each Telco prefers to design cabinets best suited to its individual requirements.

Some Telco have even considered the possibility of grouping customers from more XCCs into a bigger group of 500-1,000 customers, to be served by one active cabinet (see Figure 7), thus reducing the overall number of active cabinets required, with the following implications:

- A bigger customer group requires bigger space for the active equipment, increasing the cabinet size and the space requirements on the pavement, which may also be subject to additional "space coverage fees" to the local authorities or municipalities.
- By grouping customers from more XCCs it is necessary to connect the distribution network of each XCC to the active cabinet, requiring additional construction work that increases the overall cost.
- Higher customer groupings need active equipment of bigger capacity, that requires higher power consumption, making local powering with power meter and battery buck-up a unique solution (remote-powering is not an option for power requirements above 500 W), which will increase both space and cost.

It seems, therefore, more attractive to consider a one to one correspondence between active equipment cabinets and XCCs, with the possibility of integrating the existing LDF of the XCC into the new cabinet, even at the expense for more active cabinets[43]. This approach has the following advantages:

- Small DSLAM or MSAN equipment for serving from 100 to 200 customers is usually sufficient, requiring limited space and low power dissipation. This has a marked effect on the overall cabinet dimensions. Lower power requirements make remote powering a viable alternative, further reducing the space requirements by removing the need for meter and back-up batteries.

- Lower power dissipation makes heat management inside the cabinet easier. For small DSLAM/MSAN a free cooling system with a filter and two fans is often sufficient, though some vendors[44] allow the use of closed cabinet systems, eliminating the maintenance cost for filters and fans.

- A small active cabinet can be more easily integrated with an existing XCC into a single unit. The new cabinet can be usually situated on the same footprint of the existing cabinet, with only an increase in height required. Some small increase in depth and width might also be necessary, though the changes in footprint are not severe enough to require additional licensing from the local authorities.

- By utilizing the LDF of the existing XCC there is no need for additional network works, in order to gain access to the distribution network, minimizing both construction cost and service interruption during implementation.

The use of a large number of outdoor cabinets has also a marked effect on the OPEX, both for service provisioning and equipment maintenance. The provisioning of a DSL service normally requires some patching at the local exchange, which now has to be done at the cabinet site, increasing the operational cost. This can be partly or fully eliminated at the expense of CAPEX by employing AMDF equipment or using DSL-ready techniques[45]. Equipment maintenance, that includes both proactive (filter cleaning, fan replacement, battery servicing etc.) and reactive maintenance is a substantial OPEX contribution which characterizes FTTC deployment. This can be reduced by employing environmentally hardened equipment

for external use.

FTTB

FTTB stands for fibre-to-the-building and is the next transition stage towards the evolution of an all-fibre access network. By the term building, we usually refer to multi-apartment houses and not individual homes, where FTTB becomes identical to FTTH. Though the number of apartments per building may vary at different countries[46], for most European countries a typical size of 10 apartments per building may be considered satisfactory.

As shown in Figure 8, for a FTTB implementation an optical fibre cable must link every building to its local exchange. Point-to-point (P2P) or point-to-multipoint (P2MP) topologies may be employed, with 2 to 3 fibre pairs dedicated per building for redundancy purposes. The optical fibres must enter the building and terminate inside a small ODF at the building entrance, close to the building internal cabling box. An agreement of the building owner or owners is necessary for gaining access to the building.

The active equipment to be placed inside the building may consist of a small DSLAM, MSAN, switch or other active equipment with a capacity to offer the building customers with all the existing services, including POTS, ISDN, DSL-based (Fast Internet, IPTV, VoD), leased lines etc. This equipment has to be properly installed inside a small indoor cabinet, positioned at an appropriate place inside the building, close to the ODF and the internal cabling distribution frame (ICDF) that gives access to the individual apartments. The power for the active equipment must be provided locally (using a common or separate power meter). For placing the small indoor cabinet inside the building, the Telco need to be granted access from the building owner or owners, which may not be easy for multi-ownership buildings.

The basic structure of an indoor cabinet, to be placed inside the building, must contain both active and passive equipment, as described below:

Figure 8. FTTB network outline for P2P and P2MP implementations

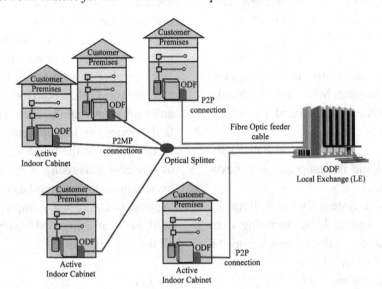

- Active equipment in the form of a mini DSLAM/MSAN or switch or other active equipment for providing services to the building customers. Even though different deployment FTTB scenarios have been examined[47], the active equipment will have eventually to provide all the offered services (legacy and DSL-based), cutting-off the copper link to the building from the local exchange. For space economy, it is desirable to use a single sub-rack (namely MSAN) for all services. A capacity of 12 to 24 ports should be enough to cover all the requirements of a medium sized building.
- The active equipment must have a FE/GE uplink for connection to the local exchange via a dark fibre pair, so no transmission equipment will be required.
- The active equipment will be directly connected to the building ODF using a pair of patch-cords, so there will be no need for an additional small ODF inside the cabinet. The patch-cord path between the cabinet and the building ODF must be relative short and carefully protected.
- Power for the active equipment must be provided locally by the building owner, using a common or separate meter. Remote powering seems an unlikely option. The use of a separate meter seems compulsory in a multi-Telco building environment, in which case sufficient space must be reserved inside the cabinet. The active equipment must be preferably ac powered, or alternatively a small rectifier must be used to provide the -48V, required for operation. In this case a small PDU may also be useful for providing powering connections to the active equipment with the appropriate safety fuses. The use of batteries for backup powering must be also considered, particularly for the provision of POTS services (life line) in case of power cuts.
- A Local Distribution Frame (LDF) is required for terminating the active equipment ports and connecting to the ICDF building box. The size of the LDF must be such as to allow:
  ◦ A 50-pair IDC termination block for the active equipment (12 to 24 ports).
  ◦ A 50-pair IDC termination block for

bridging to the ICDF building box. A 50-pair cable must be used for the LDF- ICDF bridge.

○ Sufficient space for the splitters. Some space savings can be achieved by employing MDF splitter blocks that combine narrowband and broadband services.

• Since the active equipment is housed inside the building no temperature controlling unit is necessary.

• A management system for controlling the active equipment and the operating conditions inside the indoor cabinet may be considered. Sensor alarms for unauthorized cabinet entrance, high temperature, humidity and smoke could be monitored particularly in a multiple-Telco building environment. The active equipment must also be individually monitored by its NMS system.

For new buildings, it might be easier to standardize[48] the building entrance, the type and position of the ODF and the indoor cabinet with the active equipment. This can be further complicated if multi-Telco access to different building apartments is required. The use of a Telco room, for allowing different Telco to position their active equipment does not seem a likely option, since it increases the space requirements inside the building.

The situation is even more complicated in existing buildings, where some construction work inside the building is necessary, the position of the ODF and the indoor cabinet placement is not uniquely defined, and an agreement of all the apartment owners is required before entering the building. Experience shows that entering an existing multi-owner building will be a rather complicated and time consuming issue.

In economic terms FTTB deployment is characterized by increased CAPEX and OPEX, compared to FTTC. The smaller capacity[49] of the active equipment placed inside the building increases the cost per port and the small number of potential broadband customers allows for poor port utilization. Additionally, the presence of a very large number of active nodes (one per building) increases the OPEX, both for service provisioning and equipment maintenance. The provisioning of DSL services and equipment maintenance has to be performed at the building site, increasing the overall cost and raising questions on easy access. Service provisioning may be greatly assisted, at the expense of CAPEX, by employing small AMDF equipment or using DSL-ready techniques[50] for all the potential customers.

FTTH

FTTH stands for fibre-to-the-home and is target of the evolution towards an all fibre network. By the term home we usually refer to an apartment inside a multi-apartment building or an individual house.

As shown in Figure 9, for a FTTH implementation an optical fibre pair (normally two pairs for reasons of redundancy) must link every house to its local exchange. In the external network either P2P or P2MP topologies may be employed. An optical fibre cable of sufficient capacity must enter the building and terminate inside the "main building ODF", which acts as the interface between the external and internal optical networks.

For multi-apartment buildings an internal fibre optic network must be developed inside the building, linking each apartment to the "main building ODF". One or more fibre pairs must be reserved for each apartment.

For new buildings the network can be pre-installed using single mode fibres or provisions can be made with the installation of multi-tube ducts and sub-ducts, which allow fibre bundles to be installed by blowing at a later stage. In most European countries standardization for new buildings provides all the necessary utilities for the future installation of optical fibres to all the

Figure 9. FTTH network outline P2P and P2MP implementations with internal optical network

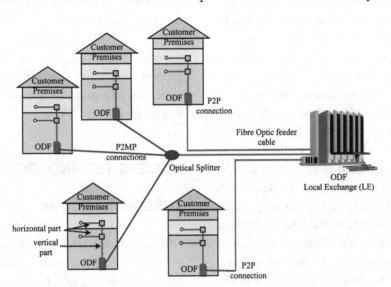

apartments with the minimum of effort.

The situation is significantly more difficult with old buildings, most of which do not have the available path-ways for the installation of fibre cables to the apartments, and require the construction of internal wall-mounted cabling. The outline of such a network consists of:

- A vertical part, that rises from the main building ODF to the intermediate "floor ODF-boxes" (one per building floor), for connecting to the floor apartments. The vertical part is wall-mounted, usually following the path of the building stair-case or elevator. Ideas for wall-mounting outside the building have also been considered, though their installation is more difficult and may cause reaction from the owners, since it may alter the building appearance.
- A horizontal part that extends from each "floor ODF-box" to the "apartment ODF-socket" in a star-like configuration. The horizontal part is also wall mounted on path-ways, carefully constructed to follow wall crossings, while retaining the minimum curvature required for optical

fibres[51], and enters through wall drillings to reach the "apartment ODF-socket" with the minimum of disturbance.

- The in-house distribution part, from the ODF-socket of the apartment to the terminations of the different apartment rooms, namely the office, sitting-room, bedrooms etc. Once the fibre pair is terminated at the active equipment or ONT[52], different technologies can be used to distribute the different services (POTS, Fast Internet, IPTV/VoD etc) to the relevant home appliances, including UTP (Unshielded Twisted Pair) cabling, WiFi or PLT[53].

A more popular approach utilizes multi-tube sub-ducts to construct both the horizontal and vertical parts of the inter-building network, allowing spare tubes at each "floor ODF/Box" for future apartment connection. To connect an apartment all that is required is to establish tube continuity between the relevant horizontal and vertical parts (at the "floor ODF-box") of the internal network and blow a fibre bundle between the main building ODF and the apartment ODF-socket, thus eliminating the need for an intermediate fibre splice per connection. Note that all ducting and fibre cable components must meet the safety regulation for internal use regarding flammability[54], having adequate fire resistance and smoke producing characteristics.

The construction of an internal fibre network inside an existing multi-apartment building is expensive[55] and may also prove to be difficult, since major works (wall drillings, wall mountings etc.) have to be performed in areas of common ownership, and the agreement of all the owners is usually required before access can be granted. It is still debatable to whom the internal cabling cost must be charged; the Telco, the building owners[56] or the government in a form of subsidy. Even though the Telco might be the more likely candidate, the situation becomes more complicated in the case of multiple-Telco access in the same building.

The active equipment (ONT) in a FTTH implementation is in a form of an optical modem or switch, with all the relevant interfaces for the applications provided, namely POTS, possibly ISDN, VoIP, Fast Internet, IPTV/VoD. For some applications multiple interfaces may be required, including WiFi. The equipment is ac powered, and some form of built-in back-up powering[57] must be provided to ensure that at least a minimum POTS life-line service is maintained in case of electric power outages.

Some form of centralized ONT management by the Telco is also desirable.

The ONT is positioned close to the "apartment ODF/socket" and is connected to it using a pair of short patch-cords. The connection of the ONT interfaces with the relevant home appliances is made via the in-house distribution network.

In terms of Telco CAPEX, the major cost concerns the construction of the outer and inner networks, while the cost of the active equipment may be transferred to the customer[58].

In terms of OPEX, FTTH promoters claim to be the minimum possible[59], since the ONT is not maintained by the Telco. This, however, is only true, provided the active equipment reaches a form of standardization and maturity comparable to that of a DSL modem, which is not currently the case. Since then, a form of 24-hour technical support must be provided to the customer services, which for an extended customer base may result in a much higher OPEX.

ACCESS TECHNOLOGIES

xDSL Technologies

DSL stands for Digital Subscriber Line and is today's dominant technology for providing broadband services over telephony to a large number of customers using the traditional access network. Its major success may be attributed to the fact that the service can be directly provisioned from the local exchange with minor network operations. The outline of a DSL broadband network is described in Figure 10.

A DSL broadband connection is set between the subscriber modem and the DSLAM[60] at the local exchange. The DSLAM acts as multiplexer, combining data streams from a number of different subscribers into a single data stream signal, and, using ATM or Ethernet transmission technology[61], forwards it towards a BRAS[62] server, which provides access both to the ISPs[63] and the IP core network.

ADSL2+ is the most recent standardised[64] version of the ADSL variant. ADSL stands for Asymmetric DSL and offers asymmetric data transmission over telephony[65] over a conventional telephony line, with a significantly higher data rate in the downstream[66] direction. ADSL2+ offers a downstream of up to 24 Mb/s, with a maximum upstream[67] of 2 Mb/s. This can be achieved over 2.2 MHz of the copper wire spectrum, by employing an advanced modulation technique known as DMT[68].

The data rates achieved by ADSL2+ connections are seriously affected by a number of factors,

Figure 10. Basic ADSL architecture

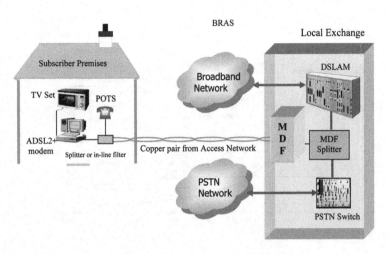

the more serious being the cable distance between the subscriber and the DSLAM. Even though a nominal data rate of up to 24 Mb/s may be offered to a subscriber, the modem can hardly deliver more than 2 Mb/s if the cable distance exceeds 3 km, for a copper pair diameter of 0.4 mm. The performance becomes even worse if the multi-pair feeder cable is filled up with a large number of DSL services, which interfere with each other, reducing the S/N (service to noise ratio) with a marked effect on the effective bit rate of all the connections[69]. Thus in practice, the high bit rates of ADSL2+ are only available to a small percent-

age of the customers that are situated close to the local exchange.

VDSL2 is the latest development[70] of the xDSL family and has the potential of replacing ADSL2+, becoming the major DSL variant. VDSL2 stands for the 2nd version of Very high bit rate DSL and can be considered as an extension of ADSL2+. It can operate at different spectral profiles, as shown in Figure 11, providing both symmetric and asymmetric services.

The asymmetric operation offers downstream data rates in excess of 100 Mb/s, effective over cable distances of up to 500m from the DSLAM.

Figure 11. VDSL2 band-plan – profile 30a (30 MHz bandwidth – 8.625 kHz tone spacing)

Profile	8a	8b	8c	8d	12a	12b	17a	30a
Bandwidth (MHz)	8.832	8.832	8.5	8.832	12.	12.	17.664	30.
Tones	2048	2048	1972	2048	2783	2783	4096	3479
Tone Spacing (kHz)	4.3125	4.3125	4.3125	4.3125	4.3125	4.3125	4.3125	8.625
Line Power (dBm)	+17.5	+ 20.5	+ 11.5	+14.5	+14.5	+14.5	+14.5	+14.5

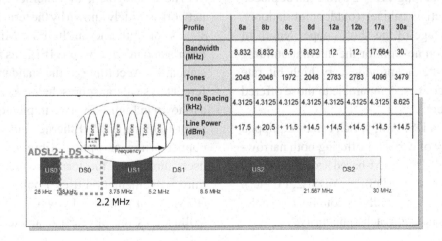

Figure 12. Comparative performance of ADSL2+, VDSL and VDSL2 in terms of span[73]

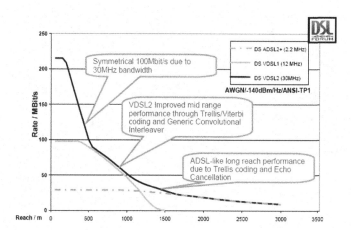

Figure 12 shows a comparison between the performances of ADSL2+, VDSL[71] and VDSL2 in terms of downstream data rates as a function of the cable distance between subscriber and DSLAM[72]. Note that beyond a certain distance of 1.500m the VDSL2 operation is fully compatible with that of ADSL2+.

The symmetric operation offers data rates of above 50 Mb/s over cable distances of up to 500m.

It is clear that in order to take full advantage of the very high bit rates available by the VDSL2 technology, the maximum cable distance between the DSLAM and the subscriber must be limited below 500m. This can be achieved by including VDSL2 in FTTC/B implementations. A small DSLAM, containing VDSL2 cards, can be placed inside the active cabinet (outdoor or indoor), taking advantage of the short copper length of the distribution network or the internal building cabling in order to provide very high bit rate broadband services, comparable to those offered by optical fibre.

MSAN is a DSLAM, which can be equipped with a variety of cards for offering both narrowband (POTS, ISDN, TDM-based leased lines) and broadband services (ADSL2+, VDSL2, FE etc.). The technology is rapidly becoming available, though it has not yet reached maturity[74].

Ethernet Technologies

Ethernet has over the recent years become the most widely used technology. It started as a protocol for local area networks and has been extended into both metro and wide area networks. There is an installed base of over 500 million Ethernet ports. As a result, Ethernet components are highly reliable and low cost, available worldwide. Using Ethernet as an access technology allows IP and Ethernet protocols to be directly applied in the access network[75], avoiding the cost and complexity of protocol conversion necessary with other implementations. In addition network managers can take advantage of available management tools designed for the Ethernet network.

The employment of Ethernet in the access network is widely known by the term EFM, which stands for "Ethernet in the First Mile", and has been standardized by the IEEE as the 802.3ah standard[76]. According to the standard, three different physical transport schemes are considered, as shown in Figure 13. Two implementations use point-to-point (P2P) Ethernet links over either copper wires or optical fibre for connecting users directly to the local exchange, while the third implementation uses a point-to-multipoint (P2MP) configuration known as EPON[77]. Table 4 illustrates the physical characteristics of the

Figure 13. Three different EFM physical implementations

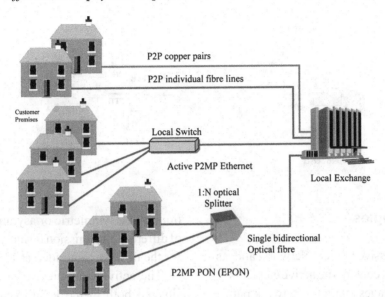

three implementation schemes according to IEEE standardization.

In addition to IEEE, ITU-T recommendation G.985[79] gives the specifications for a 100 Mb/s P2P Ethernet-based optical access system. The main P2P scenario considers dedicated fibre running between the local exchange and the individual subscribers. Such a scenario, favoured mainly by Ethernet switch manufacturers like Cisco and Telecom Regulators (for obvious reasons), requires a very large number of fibre optic lines, with each line having its own optical transceivers in both ends. The implementation is, therefore, expensive and is useful only if the subscribers require bandwidth close to the full capacity offered by the fibre.

Another option, known as active P2MP (not favoured by Telecom Regulators; Ethernet switch manufacturers still satisfied), is to run a pair of fibre from the local exchange to an Ethernet switch located inside an outdoor cabinet (FTTC configuration), with individual fibre pairs running from the switch to the customer premises. This layout reduces the number of fibre pairs terminated at the local exchange, but now all the subscribers share the bandwidth of the same optical fibre from the local switch to the local exchange.

A third option is EPON, which is a part of PON technologies, is considered next.

Table 4. Main Physical Layer Characteristics of EFM options

EFM Option	Physical Layer Options
EPON	- 10-km distance; 1Gb/s; 1x32 splitter; one bidirectional single mode fibre. - 20-km distance; 1Gb/s; 1x16 splitter; one bidirectional single mode fibre.
P2P over fibre	- 1000Base-X (10-km) or LX (40-km); 100 Mb/s or 1 Gb/s; one pair of unidirectional single mode fibres. - 1000Base-X (10-km) or LX (40-km); 100 Mb/s or 1 Gb/s; one bidirectional single mode fibre.
P2P over copper	- 750-m distance; 10 Mb/s full-duplex transmission over single voice-grade copper pair[78].

Figure 14. FTTx implementations based on PON technology

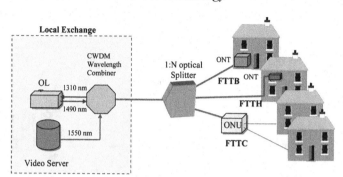

PON Technologies

PON stands for Passive Optical Network, and, as the name implies, it contains no active elements in the intermediate stages along the network paths, following a P2MP topology. The basic architecture of a typical PON is illustrated in Figure 14, in which a fibre optic network is used to connect telecommunication equipment[80] in the local exchange with a number of customers.

Starting from the local exchange, a single mode optical fibre runs to a passive optical power splitter that simply divides the incoming power into N separate paths to the customers. The power splitting ratio of the splitter can vary from 2 to 64, though typical values of 8, 16 and 32 are more often used. The splitter outputs connect to individual single-mode fibres that run to the customer equipment. The maximum fibre span from the local exchange to the user can be up to 20 km.

PON operation is based on simultaneous transmission of separate service types on the same fibre, which is enabled by using different wavelengths for each direction (as shown in Figure 14). For downstream transmission PON uses a 1490 nm wavelength for combined voice and data traffic and a 1550 nm wavelength for video distribution. Upstream data and voice transmission is served by a wavelength at 1310 nm. The combination and separation of the different wavelengths is performed by low-cost WDM[81] couplers. Depending on the particular implementation, the PON opera-

tion can be symmetric or asymmetric. A number of different transmission formats can be employed for the downstream video at 1550 nm.

The active equipment of a PON network is situated both in the local exchange, known as Optical Line Terminal (OLT), and the customer premises, known as Optical Network Unit (ONU) or Optical Network Terminal (ONT). The term ONT is mainly used when the fibre serves an individual customer in a FTTH implementation, whereas the term ONU is used when the fibre terminates in an outdoor or indoor cabinet for FTTC/B implementations. Connections from the ONU to the premises are made using the distribution copper network.

The OLT is normally situated in the local exchange and controls the bidirectional flow of information across the PON. In the downstream direction the OLT multiplexes traffic, including POTS, data and video, from different long-haul or metro networks and broadcasts it to all the ONT and ONU modules connected to the PON. In the upstream direction, the OLT accepts multiple types of traffic from the network users using TDMA[82] techniques, and directs each type to the appropriate network interface. A typical OLT is designed to control more than one PON. PON vendors offer products that can support more than 56 PONS per sub-rack[83]. OLT equipment must meet specific PON standards[84] in order to allow connection with ONT and ONU modules of different manufacturers.

Table 5. Major PON Technologies and Characteristics

Characteristic	Passive Optical Network Type		
	BPON	EPON	GPON
Standard	ITU-T G.983	IEEE 802.3ah	ITU-T G.984
Protocol	ATM	Ethernet	ATM and Ethernet
Transmission speeds (Mb/s)	622/1.244 down 155/622 up	1.244 down 1.244 up	1.244/2.488 down 155 to 2.488 up
Span (km)	20	10	20
Splitter ratio	1/32	1/16 nominal 1/32 allowed	1/64

The ONT is an optical modem located at the customer premises. It provides an optical connection to the OLT via the PON network on the upstream side, offering a variety of electrical interfaces for connection to the electronic equipment of the customer. Depending on the communication requirements of the customer the ONT typically supports a mix of telecom services including POTS, ISDN, Fast Internet connections, digital and analogue video formats. A wide variety of ONT terminals are available for different levels of demand and customer requirements. Most of them are small and sophisticated, ac powered for in-house use. The ONU is basically a higher capacity rack-mounted ONT, in the form of a PON-driven DSLAM/MSAN[85], installed in an outdoor or indoor cabinet and used to serve a larger number of customers in FTTC/B PON implementations. It is usually dc powered and can be equipped with a combination of different service cards for different applications.

Over the last fifteen years there have been several PON implementation schemes, the three main ones being: Broadband PON or BPON, Ethernet PON or EPON, and Gigabit PON or GPON. Even though all follow the basic PON architecture of Figure 14, the individual characteristics of each implementation are outlined in table 5.

GPON appears to be the dominant PON standard, since it offers higher transmission speeds, improved security and customer-driven operation, while retaining many of the functionalities of the BPON and EPON schemes, such as dynamic bandwidth assignment (DBA) and the use of operations, administration and maintenance (OAM) messages.

GPON appears to be the favoured access technology for incumbent Telco[86] (ILEC) and major telecom equipment vendors[87] for use in all FTTC/B/H implementations. It appears to be significantly more economical[88] than P2P Ethernet schemes, both in terms of fibre utilization and transceiver requirements. It is, however, disliked by regulators[89] and Ethernet switch manufacturers[90], who claim that it leads to a "closed access" network, against the rules of open competition in the telecom market.

The only obvious disadvantage of a GPON scheme has to do with the fact that all customers share the same fibre line[91] between the splitter and the OLT, which may result in future bandwidth limitations, particularly when a GPON is used to serve a large customer base with ONUs of GPON-driven DSLAM/MSAN. Thus, future developments on PON schemes are expected to increase the uplink bandwidth. Even though a 10GPON standardization is under way[92], a vast increase in bandwidth is expected by the application of DWDM technology on a PON network, which will effectively allow each customer (at the ONT and ONU level) to operate on a dedicated wavelength pair[93], taking full advantage of the available bandwidth.

The effect of DWDM technology on the PON

Figure 15. Overbuilt cabinet FTTC implementation[103]

scheme together with alternative PON architectures[94] have been also extensively examined in a large number of mainly EC funded European projects and field trials[95] over the last ten years.

NGA NETWORK IMPLEMENTATIONS

FTTC Implementations

In urban areas, FTTC implementations seem to be the most straight forward approach towards NGA. By placing the active equipment in cabinets in the location of the XCCs, the main task is to replace the copper feeder network by a fibre optic network. This can be further assisted by the existence of ducts in a good percentage of the feeder network[96], allowing for quick and low-cost installation of fibre optic cables by blowing. For the rest, micro-trenching techniques can be used to install fibre optic cables or multi-tube ducts in a relative short time.

Even though either P2P or P2MP GPON topologies may be used for connecting the active equipment to the local exchange, P2P with medium sized DSLAM/MSAN has a clear advantage in terms of bandwidth availability[97] for future applications.

Two[98] different approaches may be used to-

wards FTTC NGA implementation. Both use the present location of the XCC to install the active equipment, taking advantage of the existing distribution network. They vary, however, in terms of complexity, time-of-life, flexibility, deployment time and cost.

The first is known as "overbuilt cabinet" approach and positions of a small-sized broadband, double-wall aluminium cabinet on the top of the existing XCC, as shown in Figure 15. The width and depth of the new cabinet match exactly the foot-print of the existing XCC[99], allowing for direct and secure bolting of the two cabinets with minimum effort. The height of the new cabinet, though limited by practical considerations, must be sufficient[100] to contain the active equipment, namely a temperature hardened mini-DSLAM with up to 2x48=96 VDSL2 ports, remote-powering equipment, an ODF and MDF splitter boxes. The two cabinets communicate through a set of holes for the passing of cables (optical, power and VDSL2 subscriber cables). Even though remote-powering from the local exchange, using available copper pairs of the feeder cable, seems better suited in terms of simplicity and space-saving requirements, local powering can also be considered, with additional space requirements for a power meter[101] and back-up batteries[102].

The "overbuilt cabinet" approach seems

Figure 16. Replacement cabinet FTTC implementation

straight forward enough to ensure massive deployment with little annoyance[104] in a relative short time, provided the existing XCC casing is metal[105] and in good condition to securely lift the weight[106] of the new cabinet. Due to space limitations of the overbuilt cabinet and remote-powering limitations to below 300 W[107], the capacity of the active equipment has to be limited to below 100 ports. This means that the active equipment can only partly satisfy the complete telecommunication requirements of the under-laying XCC (typically 250 to 300 subscribers) and the "overbuilt cabinet" approach can only work in an overlay scenario, where legacy POTS services are provided from the local exchange and FTTC is only used to provide broadband VDSL2 services, with a maximum potential penetration of 34%[108].

The second approach, known as "replacement cabinet", replaces the casing of the existing XCC with a new bigger cabinet, which is higher, slightly deeper and somewhat wider. For practical reasons two different cabinet designs may be considered, one extended mainly upwards, the other sideways, as shown in Figure 16. The new cabinet must be spacious enough to accommodate all the necessary equipment, namely a midi-DSLAM/MSAN of potential capacity of above 300 ports, local powering equipment (including rectifier, PDF (power distribution frame, power meter and back-up batteries), ODF and MDF splitter boxes, retaining the distribution frame of the XCC.

The "replacement cabinet" approach requires substantial more construction work, so that massive deployment in a short time becomes rather unlikely. In some cases the concrete base of the XCC need to be extended sideways by up to 0.60 m, requiring permission from the municipality or the local authority. Even though the overall cost[109] of the "replacement cabinet" FTTC approach is substantially higher, it constitutes a much more secure implementation, since the active equipment can potentially fully support the telecommunication requirements of the XCC customers[110]. In this respect, the copper feeder cable that ties the cabinet to its local exchange becomes obsolete and the cabinet is "virtually" independent[111].

FTTC implementations suffer from higher OPEX in terms of DSL provisioning, as discussed in section 4.1. This can be substantially reduced by using two possible solutions: using an AMDF inside the cabinet, or by applying the "DSL-ready" concept, both at the expense of CAPEX. Note that both solutions apply only to the "replacement cabinet" FTTC approach.

AMDF stands for Automated Main Distribution Frame and is a remotely managed electromechanical switching matrix of N inputs and M outputs (M>N), its principle of operation outlined in Figure 17. It is a bulky device that requires additional cabinet space that greatly influences the overall cabinet dimensions. It is also expensive, at a cost per port equal to that of a DSLAM port, and

Figure 17. Block diagram explaining the AMDF principle of operation. AMDF installed in DT FTTC XCC replacement cabinet

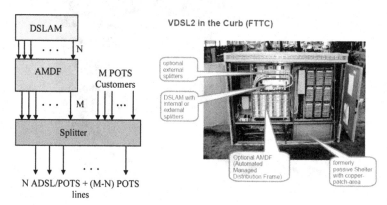

there are recent reports on operational degradation of VDSL2 services due to impedance mismatch with the distribution copper line[112].

Comparatively the "DSL-ready" concept is simple and easily applicable. The basic idea, outlined in Figure 18, is to reserve a VDSL2 port for each POTS customer served from the active cabinet. Since the capacity of the DSLAM/MSAN is such as to accommodate the full needs of the XCC, it is easy to pre-wire every POTS connection to a VDSL2/POTS port through a MDF splitter box. The VDSL2 port is at an "inactive" default state and can be activated from the NMS, as soon as an order for a new connection is made, thus

significantly reducing both provisioning time and cost.

FTTB and FTTH Outside Plant Implementations

FTTB and FTTH implementations are similar in terms of the outside fibre optic network development, since they both ignore the existing copper cable access network and develop an overlay network, with fibre cables coming from the local exchange and terminating at the "main building ODF", as described in previous sections 4.2 and 4.3. FTTH has the additional requirement[113] for

Figure 18. Block diagram explaining the "DSL-ready" principle of operation

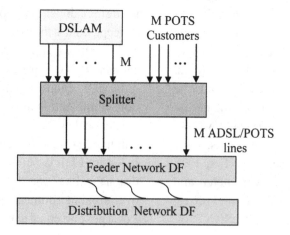

constructing the internal fibre optic building network that links the individual apartments with the main building ODF, as clearly discussed in section 4.3.

If one could design and build a new access network from scratch today[114], it is highly likely that it would be a fibre optic network identical in architecture to the existing copper network, with a pair of optical fibres used to connect each subscriber to a central point of network convergence, in a similar fashion to the copper pair that is conventionally used to connect the subscriber to its local exchange. The similarities are even stronger if the new fibre optic network adopts a P2P architecture, in which case a fibre optic cable directly replaces the multi-pair copper cable and an optical distribution frame (ODF) the copper cable patch panel (MDF).

A direct implication of the above implementation is the need to develop higher capacity fibre optic cables equivalent to the copper cables, used in the main-part of the access network. Given that copper cables of up to 3,000 pairs may be used in the access network, it is difficult to envisage a fibre optic cable of 6,000 fibres[115], even with fibre-ribbon technology which provides the highest fibre optic cable capacity[116]. Consequently, a larger number of individual fibre optic cables will have to be used on multi-tube ducts to provide the required fibre capacity.

Another implication of the point-to-point fibre optic implementation is related to the size of the central point of network convergence. Taking into account that a typical local exchange serves more than 40,000 subscribers, a point-to-point FTTH implementation would require the termination of 80,000 fibres at the Optical Distribution Frame (ODF), which will replace the existing Main Distribution Frame (MDF) at the local exchange. Note that there is a significant difference in size between fibre optic and copper pair termination modules. A typical ODF with a termination capacity or 1,000 fibres (500 subscribers), including cable and patch-cord management facilities, has dimensions of: 90cm x 30cm x 220cm (WxDxH), requiring a central office footprint of 0.27m². For comparison purposes consider that the termination of 500 copper pairs in IDC[117] strips requires a significantly smaller space of dimensions: 18cm x 15cm x 150cm. Even though the new SFF[118] fibre optic connectors offer a marked reduction in size and higher termination densities, compared with older FC (fibre connector) and SC (subscriber connector) products, it is very difficult to envisage a functional ODF[119] with comparable termination capacity. Note also that optical terminations and patch-cords are more sensitive than copper wire used for patching. They require a much cleaner environment (very sensitive to dust), they cannot be fabricated on-site and have to be ordered at specific lengths, often requiring space for extra length storage and management purposes. The above complications lead to smaller sizes of network convergence points. Relevant studies[120] indicate typical fibre optic convergence points of up to 15,000 customer lines. This, however, has a direct implication in increasing the overall telecommunication points of presence, i.e. local exchanges, by about 50%, which is contrary to the tendency for reducing them by creating bigger nodes.

Both "fixed" and "modular" network designs can be implemented, using optical fibre cables and or multi-tube ducts, where in a "fixed" approach the fibre pair terminates directly at the "main building ODF" from the local exchange, while in a "modular" approach an intermediate termination stage (ODF) at an outdoor cabinet is used to provide additional cross-connection flexibility.

When developing an overlay fibre optic network in a fully developed, saturated[121] urban area, the "fixed" approach is preferred, since there will be no need for additional future connections.

Compared to P2P, the P2MP GPON FTTB/H implementations have fewer implications on the existing network dimensioning, reducing the overall implementation cost. An important aspect that

Figure 19. GPON FTTH implementation with two optical splitter stages

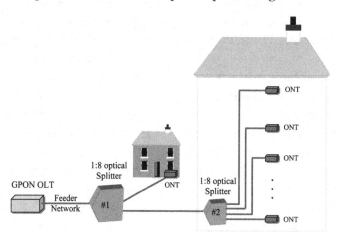

influences the network development has to do with the positioning of the 1/64 optical splitter, used in a GPON scheme to divide the power from the feeder fibre to the customer lines (ONTs). Since FTTB/H implementations may ignore the existing copper network infrastructure, the optical splitter can be placed at a variety of different convenient positions, namely a fibre optic splice closer, an outdoor ODF cabinet or at the "main building ODF", depending on the required density of the overall fibre network. For FTTH implementations it is also possible to employ more than one splitter stages (with overall splitter ratio 1/64), as outlined in Figure 19.

FTTB/H NGA implementations are characterized by a substantially higher cost than "equivalent" FTTC+VDSL2 implementations, and can be viewed either as long term strategic decisions or as the final evolution stage of a long term FTTx implementation strategy.

NGA DEPLOYMENT STRATEGIES

Significant Players

The selection of the appropriate FTTx implementation is clearly a matter of Telco strategy, available funding, present access network state, competition,

regulation and commitment. Even though FTTH seems to be the "final stage" towards an all-fibre access network transformation, the reluctance of Telco to invest the vast amounts of funding required in the present time of increased competition and regulation uncertainty, has prompted some local authorities[122] and alternative carriers to seek public funding in order to develop their own FTTH network infrastructures in order to stimulate growth by creating a "high tech" environment.

On the other hand, European governments and the EC (European Commission) have finally come to terms with the understanding that large infrastructure investment with long payback periods can only come from "public funding" and have expressed their willingness to subsidize FTTH deployment plans[123], based on joint ventures between public and private sectors, provided they are supported by adequate business plans.

This has, however, contributed to the overall confusion disrupting Telco's original business plans and creating a complicated but interesting multiple-player puzzle, where both incumbent (ILEC[124]) and new comer (CLEC[125]) Telco, alternative carriers, local authorities and pressure interest groups, governments and regulators, equipment manufacturers and civil construction companies all strive to influence the FTTx choices in accordance with their individual interests.

ILECs retain the most conservative attitude towards FTTH. Recovering from the initial shock of privatization and the intense competition of CLEC, during which some lost nearly one third of their customer base to their competitors, now comfortably established inside their local exchanges thanks to the LLU[126], they appreciate ADSL-based broadband, as a new, though eroding, source of revenue, but still bear the wounds from earlier abandoned, massive FTTx rollout attempts in the early 1990s[127]. Nevertheless, they clearly recognize the need to push fibre deeper into the access network and possess the funds to support it, but most of them prefer to do in steps going through the FTTC+VDSL2 stage with P2MP PON-like fibre deployment at a significantly lower cost[128], taking advantage of their existing infrastructure in ducts and the relatively short length of their copper sub-loop.

On the other hand, the CLECs appear to have more serious reasons to worry about future developments in the access network. Having invested heavily in LLU, they have gained access to the ILEC customer base at the Local Exchanges, but most of them operate under loss in a heavily competitive telecom market and, furthermore, they lack the additional huge funds needed to invest into FTTx technologies. They clearly prefer FTTH P2P deployments from the ILEC exchanges to the final customers, but they would like somebody else, namely the ILEC or public funding, to pay the excessive bill for the necessary infrastructure so that they can access it through a regulated LLU-like arrangement. They clearly worry about ILEC FTTC/VDSL2 PON deployment plans, which they will eventually limit their role to that of a bit-stream wholesale service provider, and argue for strict sub-loop unbundling enforcement policy, pipe sharing, fibre unbundling and even functional separation of the ILECs.

The regulators have, in the past years, put considerable effort in stimulating competition by opening the telecom market in accordance with EC directives and were, in a good degree,

successful, judging from a marked cost reduction in telecom services. Their most effective tool was LLU, which was imposed on the ILEC as a temporary measure, but has ever since become permanent, based on the argument that the copper access plant was built by public money and must be, therefore, open equally to all. By doing so, however, they triggered the defensive instincts of the ILEC, who consider them fully biased in favour of CLEC and claim that regulation has become another form of protectionism, prohibiting future investments in the access network. Regulators clearly prefer the new generation fibre based network to follow a P2P implementation topology from the ILEC local exchanges (where CLECs have also established presence) so that they can regulate it in a way similar to LLU. They are, therefore, very sceptical towards ILEC P2MP FTTC+VDSL2 NGA rollout plans, since they find them almost impossible to regulate and wave the threat of structural separation towards any ILEC plans for vacating local exchanges buildings and for network restructuring in a way that will close the access network to the CLECs.

Alternative carriers are usually utilities (i.e. cable, electricity, gas, water and sewage) with established customer bases that operate on national or local level. Nearly all of them have their own duct infrastructure in the access network, while some of them have direct access to the customer premises. Over the last ten years, encouraged by access network equipment vendors, alternative carriers have attempted to enter the telecom market as network providers seeking a new source of income. Even so, with the exception of cable companies, who have direct access to the customer and an infrastructure compatible to that of Telco, the rest have not been very successful. Power Line Technology (PLT), as a distribution technology, has not progressed far beyond pilot projects, while gas and sewage utilities have found it very difficult to reach the customer, since they have no access to a compatible inter-building distribution network. Thus, from the alternative carriers only cable companies can play a significant role in FTTx employment plans. After all HFC[129], the basic technology employed in cable networks, is clearly a FTTC/B variant.

Access network equipment vendors have a strong interest in FTTx technologies and possess significant influence and lobbying power at the EC and national government levels. They have, however, conflicting interests towards network deployment (fibre manufacturers and Ethernet switches equipment vendors clearly prefer a P2P topology, while telecommunication equipment vendors prefer MSAN, VDSL2 and GPON P2MP topology), but a common interest and thirst for investment[130] into the access network that will create a large market for their products.

Another pressure group with substantial influence at the national and local government levels are the civil construction companies, who during the recent years have greatly prospered by building roads and highways through joint ventures with the public sector. Having gained expertise in laying fibre optic cables along highways and renting infrastructure to Telco they feel that they can mostly benefit from the vast construction program of a massive rollout in the access network.

Incumbent Telco Strategy

The European ILEC are the dominant players in the access network, since they owe most of the existing infrastructure and are expected to play a key role in NGA development. Even though their strategy varies, subject to local variations, they all have to select between a set of deployment strategies[131], which are characterised by how much of the copper loop they will eventually use and consequently how far towards the customer they will deploy the new fibre network.

The basic ILEC strategy towards NGA, the implementation scenarios and the evolution paths are outlined in Figure 20.

Even though most European operators deploy ADSL2+ from Ethernet DSLAM[132], more than 80% of their installed DSL port-base comprises of ADSL lines, offered from ATM-based DSLAM situated at the local exchanges, with severe limitations in terms of bandwidth and multicasting capability. For this reason ILEC eventually move towards scenario 1, replacing legacy ATM-DSLAM with Ethernet-DSLAM equipped with ADSL2+ ports[133] in order to offer broadband services such as IPTV and VoD, with the minimum infrastructure changes. Their experience with the new multimedia services will provide technical and operational experience and will test their business models based on customer satisfaction, take-up rates, willingness to pay etc.

With good knowledge of the required investment and the expected revenues the ILEC are pushed by the need to differentiate their services from those of their competitors in terms of higher bit rate, reverse the stream of customer losses and probably win some back. Their evolution path towards scenario 2 is an aggressive move, which has to be taken soon (in the next two years) and on a massive scale, to allow the ILEC to retain a tight control over access network developments. A key decision on the migration towards FTTC is how close to the customer the fibre will be deployed. It is generally accepted that the deeper the fibre penetration, the higher the bit-rate offered, but at an increased expense from civil construction work, cables and cabinets. From strategic reasons the ILEC FTTC implementations are most likely going to be based on the existing network topology, replacing the XCC with a fibre fed active cabinet and reducing the cost by taking advantage of the existing infrastructure of pipes and copper distribution network.

Another important aspect is whether the ILEC will transfer legacy telecom services to the new fibre network or retain both networks in an overlay scheme. By transferring all the telecom services to the fibre network, the active cabinet becomes "independent" from its local exchange. In this way the ILEC have the opportunity to cluster more active cabinets to fewer larger nodes of concentration,

Figure 20. NGA strategy implementation scenarios for ILEC, based on Sigurdsson, H. M. (2007)

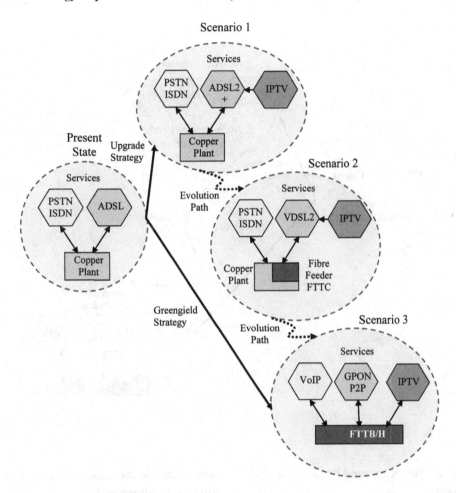

thus eventually "closing down" local exchanges and economizing in operational cost and from selling their real estate property[134]. They have, however, to spend more on bigger active cabinets with MSAN equipment, local powering and power back up facilities, whereas the existence of LLU will most likely delay the "closing-down" of local exchanges for a long period. On the other hand, the decision to go for a fibre overlay network only for broadband (DSLAM with VDSL2)[135] based on XCC overbuilt schemes has a marked advantage in terms of implementation time and cost. If one also considers that legacy services are going through an evolution phase towards IP, and are most likely to be eventually offered from VDSL2,

a lower cost FTTC VDSL2 DSLAM implementation ensures a faster return of investment and seems likely to be a better choice, particularly for a relative short transition phase towards scenario 3 of 5 to 10 years.

Scenario 3 is the final step towards an all-fibre network, replacing the existing copper plant with fibre in FTTB/H implementations. Even though FTTH enthusiasts argue from a direct transition from scenario 1 to scenario 3, the implementation cost associated with a massive FTTB/H rollout is so high that the ILEC are not prepared to pay in today's environment. In this respect ILEC FTTH NGA implementations can be only viewed as a long term prospect. A direct transition to scenario

Figure 21. NGA strategy implementation scenarios for CLEC and new entrants based on Sigurdsson, H. M. (2007)

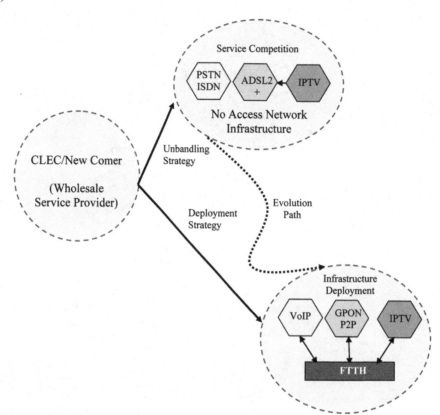

3 may, however, be encouraged by public or EC financing plans.

Even though FTTB/H implementations can be based on P2P Ethernet, P2MP Active Ethernet or PON schemes, most ILEC operators tend to select PON mainly for reasons of economy.

The above apply mainly on FTTx implementations in urban environments. For Greenfield environments the ILEC will probably go directly to scenario 3 skipping the intermediate steps.

Alternative Carrier Telco Strategy

CLEC and new entrants in telecom communications that do not possess an access network infrastructure have three options: rent infrastructure from ILEC and offer wholesale services, install their own active equipment at the ILEC exchanges and access the customers using LLU, deploy their own new infrastructure.

The overall CLEC strategy[136] is outlined in Figure 21.

The role of a wholesale service provider is rather weak, since it resembles that of a bandwidth broker rather than that of a telecom operator. Relying only on low prices, with a small profit margin and with no possibility for service differentiation to the customer, the future of a wholesale provider is dim and uncertain.

This is why CLEC tend to prefer LLU as their main strategy for establishing a substantial customer base. By renting space and installing Ethernet DSLAM/MSAN equipment with ADSL2+ lines inside selected local exchanges, the CLEC can provide 2-play and 3-play services at very low prices[137] to attract customers. Once establishing a customer base through service competition and differentiation they may embark on expensive infrastructure deployment.

In the present state of affairs the CLEC have no intension for investing in access network infrastructure. They have recently invested in LLU and what they want is to continue growing under the protection of the Regulator, taking customers from the ILEC mainly by offering lower prices. This is why they are greatly disturbed by any ILEC plans for FTTC with VDSL2 and "closing down" of local exchanges, which will put them back to the role of wholesale provider.

Theoretically, the CLEC could build their private access networks if they possessed the required funds[138]. If they did so, it is more likely to assume that they would embark on FTTB/H implementations, independently from the existing infrastructure of the access network. In fact, they have done so with "key-account" customers, but for them, a massive FTTB/H rollout for common subscribers is out of the question, due to the very high cost of the required investment. Some propose to use lower cost FTTC implementations with sub-loop unbundling, but their case is rather weak, since in order to make sense, each active cabinet must serve customers from a number of XCC, effectively increasing the distribution network length. If one also takes into account the substantially higher required CAPEX (i.e. outdoor cabinet, active equipment, local powering, network construction and sub-loop unbundling fee) and OPEX (service provisioning and maintenance) it becomes clear that massive implementations of this type are unacceptable for CLEC.

Once CLEC emerge on a FTTx deployment strategy, more often in a Greenfield area, they prefer to deploy FTTH, mainly of the active Ethernet type for reasons of simplicity and higher flexibility. In some countries[139] CLEC announce that they will develop their own parallel access infrastructure, based on pipe sharing or in collaboration with a utility provider. Ideally, a CLEC would like to rend fibre from the ILEC in a sort of LLU-like agreement.

CONCLUSION

NGA is a qualitative term indicating the successive transformation of the existing access network into a broadband network by massive implementation of FTTx technologies, where x, the factor indicating the amount of fibre penetration, becomes "a function of Telco strategy, regulation, available CAPEX, OPEX, Mbps/€, Competition and Expected services"[140].

Even though FTTB/H is expected to be the final stage of this transformation, the excessive investment, required mainly in construction works for rebuilding the access network, indicates that this evolution is going to happen gradually over a long period of time[141] and will most likely go through the intermediate step of FTTC with VDSL2. After all, the 50 Mb/s bandwidth per household anticipated in 5 years time can be easily supported by VDSL2 technology. This is why FTTC+VDSL2 must not be seen as an obstacle to FTTB/H development, but as an intermediate step that will allow for a gradual network transformation over a longer period, based on a "pay as you grow" approach.

FTTH enthusiasts and regulators argue against the need for this intermediate step, which may also create regulatory implications, and try to speed up the transformation process by attracting public and EC funding. If they are successful, this will balance the scale in favour of FTTH and lead to a rollout of massive scale for a period of time of 5 to 10 years, introducing a new "single" infrastructure provider model with many competing service providers. This will also have a marked effect on the overall telecommunication environment, with the competition of two infrastructure monopolies, the old copper infrastructure owned by the ILEC and rented to the CLEC by LLU, and the new fibre infrastructure owned by the new provider and rented equally to all Telco.

In any case, what we are likely to see over the next few years are ambitious plans for massive investments into the access network. History shows that ambitious, highly expensive and long term plans in telecommunications usually fail, if not well defined and supported by all players (governments, regulators, operators, competition authorities, investors etc). We hope that this time they will be successful, since the copper access plant is near its end of lifetime and a new fibre optic access network will be a major tool for development for the next fifty years.

REFERENCES

AT KEEARNEY & PLANNING SA, (2008, May 16). *Developing the Hellenic Ministry of Transport and Communications 5-year broadband strategy for Greece*. Workshop for public discussion, Athens.

Blondel, E., & Rubin, R. (2001). Evolution of the telecom powering strategy: moving to a cost effective and reliable solution using remote powering. INTELEC: *Telecommunications Energy Conference, 23,* 532- 539.

Chanlou, P. (2008, April 11). La Fibre: FTTH deployments and plans of France Telecom, *International Workshop on FTTx architectures, technologies, business and economic aspects,* Athens.

Clarke, R. N. (2008, April 10). FTTN/VDSL2 Broadband Networks Capabilities and Economics. *Fibre Investment and Policy Challenges OECD Workshop*, Stavanger, Norway.

Dal Bono, P. (2008, February 28). Telecom Italia domestic NGN2: the first important steps toward FTTH. *FTTx Council Europe Conference,* Paris.

Diab, W. W., & Frazier, H. M. (2006). *Ethernet in the First Mile – Access for Everyone.* USA: IEEE Press

EURESCOM. Project P917, Building and Operating Broadband Access Networks (BOBAN) (2000). *Deliverable 8,* Volume 1 of 2: Main Report, *Broadband cabinet survey, specification and demonstration.* (p. 61)

Felten, B. (2008, April 10). Municipal fibre – an overview. *Fibre Investment and Policy Challenges OECD Workshop*, Stavanger, Norway. Yankee Group

ITU-T SG-16. (2002). Full-Service VDSL Focus Group Technical Specification 1 *Part 1: Operator requirements* (p.33)

ITU-T SG-16. (2002). Full-Service VDSL Focus Group Technical Specification 1 *Part 3: Customer Premises Equipment specification.* (p. 49)

Keiser, G. (2006). *FTTX Concepts and Applications*. USA: IEEE Press, J Wiley Interscience.

Prat, J., Balaguer, P. E., Gene, J. M., Diaz, O., & Figuerola, S. (2002). *Fiber-to-the-Home Technologies*. The Netherlands: Kluwer Academic Publishers.

Sigurdsson, H. M. (2007). *Techno-Economics of Residential Broadband Deployment*. PhD Thesis, Technical University of Denmark, Denmark

Weller, D. (2008, April 10). A Fiber Future Challenges for markets and policy, *Fibre Investment and Policy Challenges OECD Workshop*, Stavanger, Norway. Verizon.

ENDNOTES

[1]   Ranging from 1/5 to 1/10 of the FTTH cost

[2]   "STRATEGY is what call in order to justify something that we cannot otherwise jus-

tify using technical and/or financial terms" FITCE (Federation of European Telecommunications Engineers) 2002.

[3] In Europe 15% from a total of 287,209,000 access network lines are directly connected to the MDF. SOURCE: EURESCOM (European Institute for Research and Strategic Studies in Telecommunications).

[4] Even though in USA, UK and most countries copper pair cables are used, central European countries like Germany, Austria etc. quad copper formations are preferred. OTE having a history of strong German influence employs quads.

[5] EURESCOM P.917 (1999)

[6] Even though copper conductors are mostly used, copper clad aluminium was also common in the 1960's and 1970's. Until the 1980's paper was the main insulation material until finally replaced by polyethylene.

[7] In USA and UK a 0.5mm conductor diameter is used.

[8] In USA the AWG (American Wire Gauge) unit is used. Thus 0.4, 0.5 and 0.6mm are 26, 24 and 22 AWG.

[9] OTE uses XCC with dimensions (HxWxD) of 1,400x730x270 mm

[10] Older cabinets used pre-terminated wire wrap blocks while newer cabinets employ IDC (Insulation Displacement Contact) blocks.

[11] ITU-T FS-VDSL FGTS Part 1: Operator requirements, Table 4 p. 27

[12] OTE stands for Greek Telecom Operator and is the Greek incumbent

[13] In some countries drop-wires are used instead.

[14] ITU-T FS-VDSL FGTS Part 1: Operator requirements, Table 4 p. 26

[15] SOURCE: Sigurdsson, H. M. (2007)

[16] A transformation from a "data-centric" to a "multimedia" platform

[17] P. Dal Bono, "Telecom Italia domestic NGN2: the first important steps toward FTTH", 2008, FTTx Council Europe Conference"

[18] H. M. Sigurdsson, Ph.D. Thesis, Kongens Lyngby 2007, Ch.2

[19] Packet Loss Rate - PLR

[20] Constant Bir Rate - CBR

[21] Real time Variable Bit Rate – rt-VBR

[22] Unspecified Bit Rate - UBR

[23] Available Bit Rate - ABR

[24] Keiser, G. (2006), p.15

[25] Both climatic conditions (heat, cold, moisture, dust etc) and vandalism must be taken into account when designing the cabinet.

[26] An alternative fibre overlay deployment scenario considers the provision of only broadband services from the outdoor cabinet, retaining the POTS services over copper from the local exchange.

[27] Multiple Service Access Node – MSAN

[28] Optical Network Unit – ONU; Optical Network Termination – ONT; Refers to ONU or ONT terminal units for GPON schemes

[29] Optical Line Termination – OLT

[30] "Widelink" is equipment used for the transmission of 2xE1 per optical fibre

[31] Usually housed in a different compartment for independent accessing by the powering utility

[32] A 3 to 6 h back-up time may be sufficient, though this may depend on the SLA with the power utility.

[33] Blondel, E. & Rubin R. (2001)

[34] 25 copper pairs are sufficient for providing about 300W over a copper pair cable distance of 3km from the local exchange.

[35] dc remote powering with up conversion to 300V at the local exchange and down conversion to -48V at the remote side can be used to transmit up to 15W per copper pair.

[36] A central powering unit is installed in a powering outdoor cabinet, used to remote-power a cluster of telecom outdoor cabinets (usually up to 10 per cluster).

[37] Today about 3W power consumption per customer may be estimated for VDSL2 broadband services. In the future, a reduction of this figure to about 2 W is anticipated.

[38] Power Distribution Frame – PDU

[39] It requires 300 pairs for narrowband sub-rack, 300 pairs for broadband sub-rack and 450 pairs for the distribution network. Significant space savings can be achieved by using MDF splitter blocks combining broadband and narrowband termination requirements.

[40] Warm air from the DSLAM heats up the fluid in the evaporator. By natural convection the fluid flows to the condenser in the roof here dissipating the heat to the environment. The fluid being cooled down flows back to the evaporator thus forming a hermetically sealed loop.

[41] Note that a start procedure at low temperatures greatly influences the laser operation, reducing the expected lifetime of the active equipment.

[42] BOBAN P917GI Eurescom project

[43] This will result on additional CAPEX due to the cost of more cabinets and higher cost per port due to smaller DSLAM capacity.

[44] Ericsson offers active equipment that can operate at temperatures up to 70°C, which can be placed on a closed cabinet without ventilation.

[45] Both techniques will be further analyzed at later sections of this chapter.

[46] In South Asia multi-apartment buildings of more than 100 apartments are commonly met.

[47] In the overlay scenario the existing copper network is retained for POTS and ISDN services and a DSLAM is installed inside the building for broadband services and VoIP.

[48] In most European countries the entrance of FO cables, the internal building cabling and the positioning of active equipment have been standardized or are under standardization.

[49] Mini DSLAM/MSAN using cards of 12 to 24 ports per card capacity.

[50] Both techniques will be further analyzed at later sections of this chapter.

[51] Special types of fibre for internal cabling are available by vendors like Corning, which can sustain a much tighter curvature.

[52] Optical Network Termination – ONT

[53] Power Line Technology (PLT) uses the installed power lines inside the house to distribute FE to all the electrical sockets. A special plug is used to access the FE interface for connecting to a PC, Set Top Box etc.

[54] They must be constructed by low-smoke, halogen free, fire retardant materials (LSHF) in accordance with existing standards.

[55] A net cost of about 300€ per connected apartment has been reported.

[56] This seems highly unlikely for ordinary subscribers.

[57] Long-life rechargeable batteries are often used.

[58] After all, the customer has ownership of his modem.

[59] Chanlou, P. (2008).

[60] DSL Access Multiplexer – DSLAM

[61] More recent systems use Ethernet technology DSLAMs with GE uplink interfaces, which allow for multicasting services.

[62] Broadband Access Server - BRAS

[63] Internet Service Provider – ISP

[64] ITU G992.5 published on 01/2005

[65] POTS or ISDN

[66] Towards the subscriber modem

[67] Towards the DSLAM at the local exchange

[68] Discrete Multiple Tone - DMT

[69] The further connection may suffer more by the interference of a stronger on a weaker signal.

[70] Standardized by ITU as G.993.2 on 02/2006

[71] An earlier version of VDSL, not backward compatible with ADSL2+, whose perfor-

mance would degrade considerably beyond 1,000m

[72] The same DSLAM sub-rack can house different xDSL cards including ADSL2+, VDSL2, SHDSL etc.

[73] SOURCE: DSL Forum (2007)

[74] POTS services are provided by either the H.248 or the SIP protocols, ISDN and TDM-based leased lines not yet offered by most vendors.

[75] Diab, W.W. & Frazier, H. M. (2006), Ch.2, p.15

[76] Standard approved in June 2004, Diab, W.W. & Frazier, H. M. (2006), Ch.3, p.37

[77] Ethernet over Passive Optical Network – EPON

[78] With copper pair bonding of up to 8 pairs it is possible to increase the effective data rate up to 80 Mb/s.

[79] Standard approved in March 2003.

[80] Including PSTN switches, ATM switches, IP routers, Ethernet switches, Video servers etc.

[81] Wavelength Division Multiplexing (WDM) is a technique for multiplexing different wavelengths on the same fibre, increasing the information carrying capacity of optical fibres.

[82] Time Division Multiple Access - TDMA

[83] Products available by major vendors like Alcatel, Ericsson, NSN, Huawei and ZTE

[84] BPON, EPON, GPON discussed later in this section.

[85] Most DSLAM/MSAN vendors offer a GPON uplink interface card.

[86] Also because it makes fibre LLU almost impossible

[87] Since despite standardization ONT and ONU terminals have to be of the same technology as the OLT

[88] A 40% reduction in cost in favour of P2MP topology is the result of an economic case study reported in Prat, J., Balaguer, P.E., Gene, J.M., Diaz, O.& Figuerola, S. (2002),

Ch.7.

[89] Find it almost impossible to regulate a shared access other than bit-stream.

[90] Not so many Ethernet switches required.

[91] 2.5 Gb/s upstream.

[92] Work in progress by ITU/FSAN group 2008-2011

[93] The scheme has been already implemented by vendors leading to commercial products.

[94] SuperPON, HyperPON, SMP-PON etc.

[95] A typical but not conclusive list may incude: PLANET (ACTS-2000), TOBASCO (ACTS-1998), HARMONICS (IST-2001), SONATA (2001), BONAPARTE (ACTS-1998), FIBERVISTA (1999), RINGO (2001) etc.

[96] Though it varies between countries, a 30 to 40% might be a good assumption.

[97] By considering GPON with a splitting ratio of 1/64 and ONUs, each with a capacity for up to 300 subscribers the, 2.5 Gb/s fibre uplink to the OLT will potentially serve up to 19.200 customers.

[98] More may be applied, though the author thinks those two to be the most effective in terms of cost and implementation time.

[99] This influences the choice and positioning of the active equipment inside the cabinet.

[100] An extra height of up to 0.50 m seems reasonable, for an existing XCC height of 1.20m.

[101] Facilities for accessing the meter by the power utility may also be provided, i.e. separate door.

[102] It will not be right to place back-up batteries on the top cabinet, because their weight will lift the centre of gravity of the joint cabinets, causing instability.

[103] SOURCE: OTE 2008 VDSL2 pilot projects

[104] No permission of the municipality or the local authority is required.

105 In countries where plastic casings for XCC are used tests have to be carried out.

106 A total weight of 40 Kg is anticipated for both the cabinet and the active equipment.

107 For 25 feeder pairs with about 12 W/pair

108 A second VDSL2 mini DSLAM of 96 ports can also fit into the cabinet, and provided the powering issue for 600 W can be met, the maximum broadband penetration can increase to about 70%.

109 Includes both equipment and construction cost.

110 The MSAN can be equipped with different cards for both narrowband and broadband services including POTS.

111 Plans of reducing the number of local exchanges can be easily implemented by connecting the cabinet to another exchange.

112 DT has recently stopped the employment of AMDF in its FTTC T-Com network.

113 And the additional cost

114 Greenfield situation

115 Two fibres are usually used per customer. There is a possibility of using one fibre per customer with different wavelengths for uplink and downlink at a much higher cost.

116 Higher ribbon fibre optic cable reported capacity of up to 2,000 fibres.

117 Insulation Displacement Contacts

118 Small Form Factor fibre optic connector families include LC, MT-RJ, etc. connectors

119 Which will allow for easy and safe patching

120 Typical study by AT KEEARNEY and PLANNING SA on "Developing the Hellenic Ministry of Transport and Communications 5-year broadband strategy for Greece" – Workshop for public discussion, 16/05/2008.

121 No more prospects for additional construction.

122 40.000 homes in the city of Amsterdam, 200,000 in Cologne, 450,000 in Munchen,

Hauts-de-Seine etc in Felten, B., Yankee Group (2008)

123 AT KEEARNEY & PLANNING SA, (2008). Developing the Hellenic Ministry of Transport and Communications 5-year broadband strategy for Greece

124 ILEC – Incumbent Local Exchange Carrier

125 CLEC – Competitive Local Exchange Carrier

126 LLU - Local Loop Unbundling

127 Early attempts for massive FTTx rollouts in the 1990's include DT OPAL project, FT DORA project, TI SOCRATE project and BT OTIAN project.

128 Clarke, R. N., AT&T (2008).

129 HFC - Hybrid Fibre Coax

130 Somebody else's – public or private they do not care

131 Sigurdsson, H. M. (2007), Section 3.8

132 Some are also experimenting with VDSL2 in FTTC implementations.

133 That offer speeds up to 24 Mb/s to customers in the vicinity the local exchange

134 On early September 2008 Wirtschaftswoche reported details on a plan by DT to replace its current 7,900 local exchange buildings (MDFs) with 800-900 high level switches. According to the paper, this will make half of DT's 17,000 network technicians redundant, and allow the company to raise €3.5bn from the sale of its exchange buildings. This would of course jeopardise competitor unbundling facilities. The plan follows Dutch and EU approval for KPN's plan to shut down MDFs as part of its VDSL/FTTH build. The EU's supportive position reflected a view at the time that incumbents should not be forced to maintain legacy infrastructure simply to support competitors. Report by Credit Suisse Equity Research.

135 Retaining the legacy services to the copper plant

136 Sigurdsson, H. M. (2007), Section 3.9

137   Even at loss, effectively "buying" customers

138   The author's position is that even if they had they would never invest in passive infrastructure because of the very long period in the return of investment. According to AT KEARNEY in the business plan for the 2008 "5-year broadband strategy of Greece" for a FTTH access infrastructure provider a 17 year return of investment period is anticipated.

139   Like France, where Iliad/Free, France Telecom, Neuf Cegetel and Numericable all go for independent implementations.

140   Noted by Dr. T. Doukoglou in his presentation "FTTx the Choices, the Roles & the Players (Incumbent Operator's View)" at the Athens Workshop on New Generation Access, 11/4/2008

141   Over ten years from now

Chapter 13
Next Generation Home Network and Home Gateway Associated with Optical Access

Tetsuya Yokotani
Mitsubishi Electric Corporation, Japan

ABSTRACT

As optical broadband access networks have been popularized, triple play services using IP technologies, such as Internet access, IP telephony, and IP video distributing services, have been also popularized. However, consumers expect new services for a more comfortable life. Especially, when QoS guarantee and high reliable services are provided in NGN (Next Generation Network) era, various home network services over NGN are deployed. For this purpose, the home gateway has been installed in consumer houses for the connection between access and home networks, and providing various services to consumers. Even though, the broadband router currently plays a role similar to the home gateway, this home gateway should comprehend functionalities of the broadband router, and should have additional features. The functional requirements of such home gateway have been discussed in standard bodies. That is, the next generation home gateway in NGN era generally should have four features as follows; High performance for IP processing, Compliance with the interface of carrier grade infrastructure including NGN, Flexible platform for various services, and Easy management and maintenance. This chapter describes the standardization of the home gateway and, proposes its evolution scenario the present to the future. Then, it also proposes these four requirements, and technologies to comply with features described above.

INTRODUCTION

Internet services have been popularized widely by the deployment of broadband communication

DOI: 10.4018/978-1-60566-707-2.ch013

infrastructure. In particular, the performance of broadband communication has been improved by optical fiber transmission technologies. Although transmission rate by metal transmission technologies including ADSL will approach the upper bound, e.g., less than 100Mb/s, rate by optical fiber transmission rate has been still improved. 1Gb/s transmission in optical access networks is currently feasible. The deployment of 10Gb/s transmission will be started within the next several years. Moreover, when NGN discussed in ITU-T is deployed, various functionalities will be supported in addition to an improvement of the performance. For example, QoS control and high reliable transmission control will be available. In this situation, consumers expect an enhancement of the existing services and a deployment of new services for a more comfortable life. For this purpose, the "Next Generation" home gateway will be installed for consumers. The home gateway currently provides IP processing, mainly. It looks like the broadband router. However, in NGN era, the home gateway comprehends functionalities provided by the conventional one, and provides an additional functionality and the high performance.

In short, this home gateway provides a high performance transmission without the degradation of a transmission performance of an optical access network, and various services over broadband networks for consumers. Therefore, as this home gateway is one of the key components for the high performance communication and new services across telecom networks, most telecom operators are interested in requirements of such home gateway.

ITU-T and other standard bodies activated by telecom operators currently have discussed these requirements actively. This Chapter summarizes a discussion about the next generation home getaway focusing on the standardization, and presents its evolution scenarios. Then, it describes key features of the next generation home gateway.

POPULARITIES OF INTERNET AND BROADBAND SERVICES

This section describes the present and the future broadband services which motivate an installation of home networks and the home gateway.

Worldwide Trends of Internet and Broadband Services

The growth of Internet has been continued worldwide. Figure 1 shows report of ITU-D (See, ITU-D (2008)) about the popularity of Internet worldwide. The number of Internet users currently occupies only 17% of worldwide residents. However, its growing rate is rapid. In particular, broadband services by DSL (Digital Subscriber Line), FTTH (Fiber To The Home), and cable modem has been installed. As shown in Figures 2 and 3 (See, ITU-D (2008)), the number of users of broadband services has been increased. Especially, in US, Asia, and European countries, the number of users is grown up rapidly. As broadband services provides high speed IP based communication by economic and fixed rate for consumers, consumers enjoy IP centric triple play services, such as high speed Internet access, IP telephony, IP video distribution services.

Evolution of Broadband Services

Broadband services have been enhanced by the evolution of optical fiber communication technologies. Figure 4 is the Japanese case in broadband service transition (Yokotani, Nakanishi, Ogasawara, & Maeda, 2006) as one of the examples. Since Japan is one of the countries which initiatives the deployment of broadband services, it can looks like a good guideline for a future prediction. In Figure 4, although transmission rate by metal transmission, mainly ADSL is saturated to less than 100Mb/s, a fiber transmission technology for FTTH still contributes increasing of transmission rate. Especially, PON (Passive Optical Network) (Maeda, Okada, & Faulkner, 2001) architecture contributes a promotion of FTTH. As PON architecture provides economical FTTH services by many advantages described in the later. Although STM-PON was proprietary, B-PON (Broadband PON) (ITU-T G.983, 2001) and G-PON (Gigabit PON) (ITU-T G.984, 2002) have

Figure 1. Transition of the number of worldwide Internet users

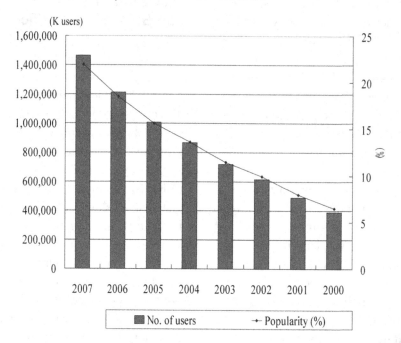

been specified in ITU-T SG15. E-PON (Ethernet PON) has specified in IEEE 802.3 (IEEE802.3, 2005). These standardized PON systems promote the growth of FTTH. In the near future, NG-PON (Nakagawa, 2006) will be specified by ITU-T SG15 and/or IEEE 802.3. This system also will

contribute the promotion of FTTH (Nakamura, Ueda, Makino, Yokotani, & Oshima, 2004).

The advantages of PON architecture are described as follows using Figure 5. PON architecture invokes the Point to Multi-point topology as shown in Figure 5. This topology has many

Figure 2. Transition of the number of worldwide users of broadband services

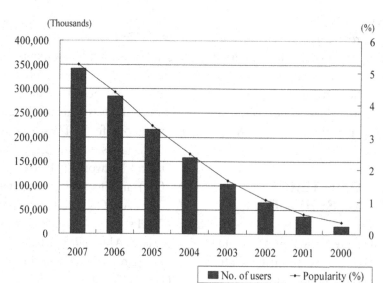

Figure 3. Transition of the number of users of broadband services (Top 10 countries)

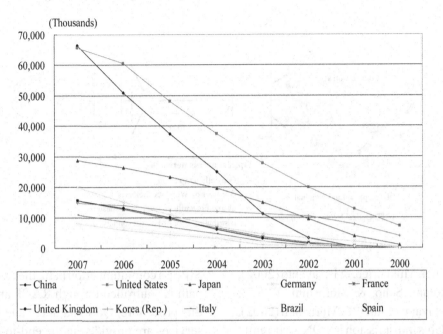

advantages as follows; the reduction of the number of optical fibers, the reduction of a footprint in a central office for each user, and easy providing multicast and broadcast communication. However, it requires an optical burst multiplexing technology for upstream traffic. For this purpose, DBA (Dynamic Bandwidth Assignment) has been

specified and provides similar functions of MAC (Media Access Control) in the token passing LAN. Especially, DBA is the key function to provide various services for each user. Framework of DBA has been standardized in ITU-T (ITU-T G.983.4, 2001). Its detailed specifications and impacts to services have been discussed in many articles,

Figure 4. Evolution of broadband services

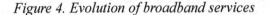

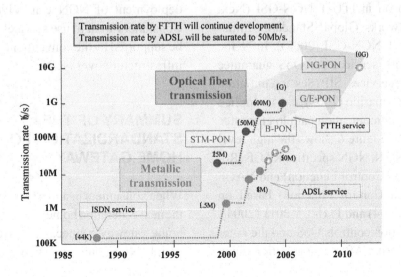

Figure 5. Summary of PON architecture

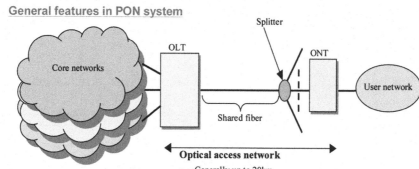

such as: Kramer, Mukherjee, & Pesavento (2002); Mukai, Yokotani, Seno, & Motoshima (2007); Yokotani, Kitayama, Mukai, & Murakami (2002), and so on. 1Gb/s transmission in PON is currently in the mature stage in all aspects such as technology, costs, and services. It is concluded that the next generation PON with 10Gb/s transmission will be deployed within several years.

Enhancement Communication Services by NGN

In addition to high transmission rate, new network services can be provided in NGN. NGN has been summarized including objectives, technologies, services and so on in ITU-T NGN-GSI (Next Generation Networks Global Standards Initiative), (see, ITU-T NGN-GSI (2008)). In NGN, attractive network services are QoS guarantee and high reliable services. SIP (Session Initiation Protocol) (e.g., Camarillo (2002)) can be applied and it controls communication path to guarantee QoS and security. Figure 6 shows the high level architecture in NGN. NGN specifies RACF (Resource Admission Control Function) and NACF (Network Attach Control Function) based on ITU-T Y.2001 (2004) and ITU-T Y.2011 (2004). The former function controls QoS and the latter controls the security. After a negotiation using

SIP on a communication path, the communication path is controlled through RACF and NACF. At this procedure, QoS guarantee and high reliable services are provided in the end-to-end communication path. In this situation, consumers expect the enhancement of existing triple play services and subscription of various new services beyond triple play on this communication path. In short, in the conventional network even in high speed infrastructure, some services cannot be provided, because users and/or service providers worry its quality and security. Thus this case, they have used a dedicated communication line such as phone line, leased line and so on, at the sacrifice of convenience, transmission rate, and costs. The deployment of NGN can solve this situation. Consumers want to use services which could not be supported by the conventional communication infrastructure over NGN.

SUMMARY OF THE STANDARDIZATION FOR HOME GATEWAY

When consumers install new services over NGN, the next generation home gateway is required. This home gateway connects to broadband networks, mainly NGN, and home networks, and provides

Figure 6. Communication control in NGN

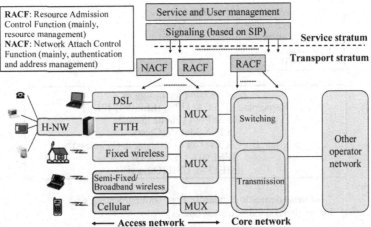

various services such as high quality triple play, and new services beyond triple play services. Therefore, this home gateway is discussed from both aspects of home appliance and telecom infrastructure. This section summarizes a discussion of the next generation home gateway focusing on the standardization.

Overview of the Standardization for Home Gateway

At first, an overview of the standardization for the home gateway is summarized in Figure 7. The detailed information in each organization can be collected using URL described as follows. A key feature of the standardization for the home getaway is transmission functions in a home network. In the home network, it can be concluded that easy and economical connections are important as well as high transmission rate, because a range of users' background, situation and generation is wide and all of users should operate this network. Therefore, consumers are interested in a wireless connection and a connection using conventional cables, such as power line, phone line, and coaxial cable. As the next step, it is important that transmission rate in these methods can be increased as well as a wired communica-

tion, e.g., Ethernet. Moreover, many standards in PLC (Power Line Communication) have been discussed. IEEE1901 manages co-existing among three major candidates, such as HPA (Home Plug Alliance), CEPCA (CE Power line Communication Alliance), and PUA (PLC Utility Alliance). In addition to them, new specifications (ITU-T G.hn, 2008)) are discussed in ITU-T (SG15) recently. By the way, the architecture of the home gateway and the home network has been discussed actively. In particular, ITU-T tries to specify architecture and requirements. Furthermore, it invokes seamless communication between broadband networks and the home network, and provides NGN services for the home network.

URL of each key standardization body

- ITU-T
  ◦ SGs: http://www.itu.int/ITU-T/study-groups/com#/index.asp (#: SG No.)
  ◦ JCA-HN (Joint Coordination Activity on Home Networking)
  ◦ http://www.itu.int/ITU-T/special-projects/jca-hn/
- DSL Forum: http://www.dslforum.org
- HGI (Home Gateway Initiative): http://www.homegateway.org
- OSGi (Open Service Gateway initiative):

Figure 7. Overview of standardization organizations

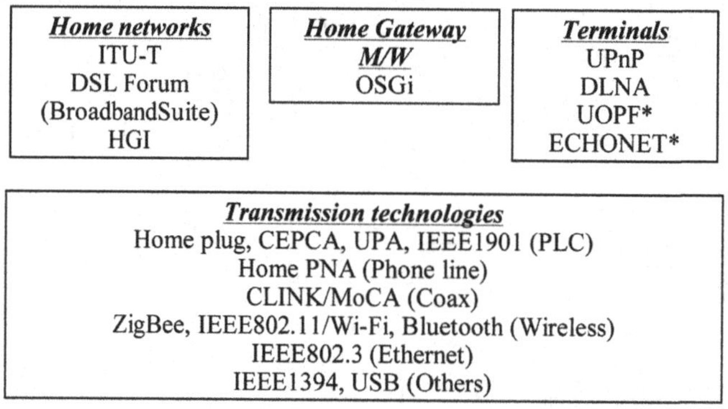

Home networks	*Home Gateway M/W*	*Terminals*
ITU-T	OSGi	UPnP
DSL Forum		DLNA
(BroadbandSuite)		UOPF*
HGI		ECHONET*

Transmission technologies
Home plug, CEPCA, UPA, IEEE1901 (PLC)
Home PNA (Phone line)
CLINK/MoCA (Coax)
ZigBee, IEEE802.11/Wi-Fi, Bluetooth (Wireless)
IEEE802.3 (Ethernet)
IEEE1394, USB (Others)

* Japanese domestic standardization (communicating with International standardization bodies)

http://www.osgi.org

- UPnP (Universal Plug and Play): http://www.upnp.org
- DLNA (Digital Living Network Alliance): http://www.dlna.org
- Home Plug: http://www.homeplug.org
- CEPCA (CE Power line Communication Alliance): http://www.cepca.org
- Home PNA (Home Phone line Network Alliance): http://www.homepna.org
- ZigBee: http://www.zigbee.org
- IEEE802.3, IEEE802.11: http://grouper.ieee.org/groups/802/dots.html

The Standardization in ITU-T for Home Gateway and Home Network

Although ITU-T has been managed by mainly telecom operators, it is interested in the home gateway and the home networks because these are attractive for services over NGN. Therefore, ITU-T is one of the standardization bodies which are most interested in them. Many Study Groups and Questions in ITU-T discuss the home gateway and the home network. Recently, SG15, SG16, SG9 started the next generation home getaway and the home network as the transport aspect associated with optical access networks and NGN

discussed in SGs 12, 13, and 15, service and terminal aspects including IPTV, and CPN (Customer Premise Network) for IP cable network, respectively. Figure 8 summarizes relationship among these study groups.

EVOLUTION OF HOME GATEWAY AND SERVICES

This section describes an evolution of the home gateway and services provided by the home gateway. Functionalities of the home gateway have been expanded according to service evolution.

Figure 9 summarizes an evolution scenario of the home gateway. It can address the roadmap for a future home gateway. This figure can be explained as follows. Originally, the home gateway of STEP 1 works as a router for IP based end devices, mainly PC. It is referred to as a broadband router, and is installed and owned by consumer independent of broadband service operators. In STEP 2, the broadband service operator deploys the home gateway to provide triple play services with "Carrier grade" quality. This home gateway provides terminal adapter functions for IP telephony services, and QoS control, mainly priority control, to provide "Good quality" for voice and

Figure 8. Summary of ITU-T activities for home gateway and home network

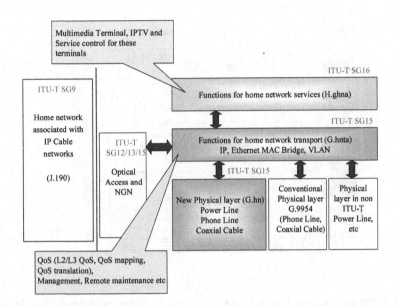

video transmission. In addition to them, it also provides IP-SEC for secure services and a remote maintenance. Moreover, as IPv6 is the maturity in broadband telecom networks, it can handle IPv6 including a translation between IPv4 and IPv6, and an encapsulation by IPv6. In STEP 3, the home network services can be expanded beyond triple play services. For example, the home appliance control, the home security and remote monitoring services can be provided in addition to native IP processing services. In this step, consumers can select services of these home network services on the same hardware. In STEP 4, the home gateway terminates and initiates SIP to connect NGN. In this step, the home gateway controls most of applications by SIP. Controlled communication paths guarantee QoS and security which are features in NGN. However, in this step, the home gateway provides functions for NGN Release 1 and 2 services. Finally, in STEP 5, the home gateway connects the networked tag for ubiquitous services specified in NGN Release 3 services.

KEY TECHNOLOGIES IN HOME GATEWAY

To progress of Home gateway evolution, the following four major technologies should be required; (1) High performance for IP processing, (2) Compliance with WAN interface including NGN, (3) Flexible platform for various services, and (4) Easy management and maintenance. This section describes these technologies.

High Performance for IP Processing

As network transmission rate, the performance of terminals and the bandwidth for applications are increased, the required performance of the home gateway should be enhanced. For example, the interface rate of Ethernet is 10 times every four or five years as shown in Figure 10. Generally, CPU is a key device for the implementation. According to Moore's law (Moore, 1965), the number of packed transistors is twice every 24 months as shown in Figure 11. Moreover, the design process evolves according to Moore's law. This trend is shown in Figure 12. On the other hand, the CPU

Figure 9. Evolution step of home gateway

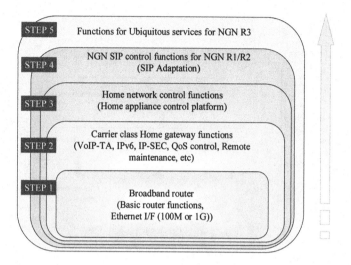

trend is migrated from the increase of a clock rate to a multiple core design, because the clock rate will be saturated in the process of 65nm or the latter by leak current. Therefore, the multiple-core CPU provides the high performance for IP processing in the home gateway. Figure 13 is the typical configuration. Generally, it is feasible that such type of CPU achieves 1Gb/s transmission of IP packets at the best case. To provide the higher performance, the study on offloading architecture

Figure 10. Evolution of interface rate in Ethernet

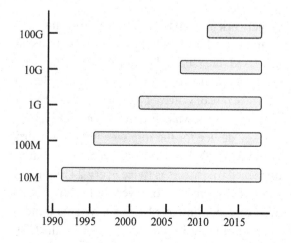

by hardware has been started. However, in the case of offload architecture by hardware, the flexibility such as a modification of functions according to the service situation should be considered carefully. Therefore, for the time being, CPU base architecture will be the mature approach.

Support of NGN Requirements

It is features that QoS guarantee and the secure communication should be provided in the NGN era. In NGN architecture, SIP is a promising candidate of control protocols for these purposes. Communication control in NGN is shown in Figure 14, Narita A (2004) as one of the examples. Communication paths are controlled by SIP as shown in Tsuchida M, Yokotani T, Sato K (2006). QoS and security guarantee can be provided by this method. Although some end devices can support SIP for NGN, most of end devices cannot support SIP. In this case, the home gateway should initiate and terminate SIP in this method instead of these end terminals. However, generally, the home gateway cannot recognize triggers to initiate and terminate SIP sequences. For this purpose, it is proposed that the "SIP adaptation" function is applied as one of NGN requirements. SIP adapta-

Figure 11. Summary of Moore's law

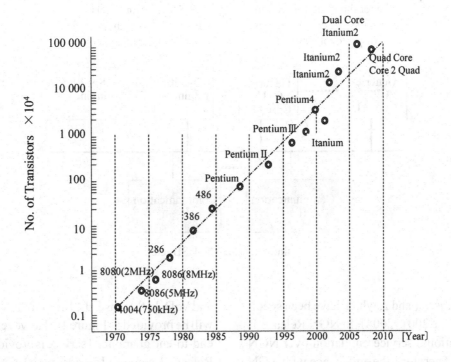

tion has been introduced and prototyped as one of the specific functions for the NGN home gateway. Moreover, reasonability of this function has been verified (reported in Sato, Furuya, Yokotani, Homma, & Sakai (2008) through the NGN field trial refereed to as "the next generation network field trial (e.g., Esaki, Kurokawa, & Matsumoto (2007). Figure 15 shows the relationship between the home gateway and NGN. SIP adaptation can be applied when NGN services are deployed such

Figure 12. Evolution of design process for devices

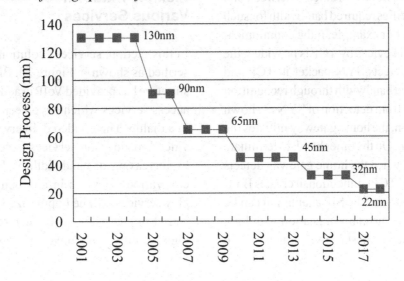

Figure 13 Typical configuration of multiple-core CPU

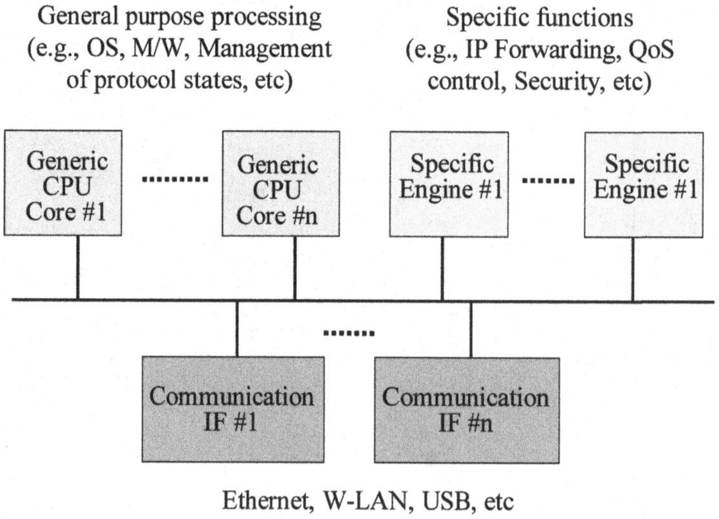

General purpose processing
(e.g., OS, M/W, Management
of protocol states, etc)

Specific functions
(e.g., IP Forwarding, QoS
control, Security, etc)

Generic CPU Core #1 .... Generic CPU Core #n

Specific Engine #1 .... Specific Engine #1

Communication IF #1 .... Communication IF #n

Ethernet, W-LAN, USB, etc

as NGN Releases 1 and 2, which have been speci-fied in ITU-T Y.2001 (2004). In NGN Release 1, only conventional service emulation over NGN is provided. Therefore, the end device with NGN capability does not exist in the home network. In this case, the home gateway initiates and termi-nates SIP to NGN. This operation is illustrated in Figure 16 (See, Sato, Yokotani, Furuya, & Isoda (2005). The home gateway monitors conventional protocols between end devices. It recognizes triggers to initiate and termites SIP through this monitor. Then, the home gateway detects appli-cations over conventional protocols between end devices, and specifies required bandwidth for such communication. For example, if the communica-tion between end devices by TCP is provided, the home gateway detects SYN packet in TCP and estimates required bandwidth through recognition of packet format from detection of SYN as shown in Figure 15. Then, the home gateway initiates the INVITE message. On the other hand, to terminate communication, when the home gateway detects the FIN packet on TCP connection, it creates BYE in SIP for termination. The SIP adaptation can be applied to multimedia communication with long holding time, such as desktop video conference

and VoD. In this case, the SIP adaptation for RTSP will be provided as Figure 16. However, when full end-to-end guaranteed service is provided in NGN Release 2, some end devices with NGN capability appears in the home network. In this stage, SIP adaptation requires the admission control function in addition to functions needed in NGN Release 1. The home gateway controls requirements from end devices to conform the assigned bandwidth in NGN. Figure 18 is illustrated SIP adaptation in this stage.

Flexible Platform for Various Services

In this section, services evolution in NGN is de-scribed as shown in Figure 20. Broadband infra-structure has provided VoIP in addition to Internet access services which have been deployed even in a dialup. Then, video delivery services and IP video broadcasting services have been installed by increasing of bandwidth by FTTH. In NGN era, various services in addition to these triple play services will be popularized over reliability and QoS guarantee infrastructure. Users expect high quality and convenient services beyond

Figure 14. Communication control by SIP in NGN

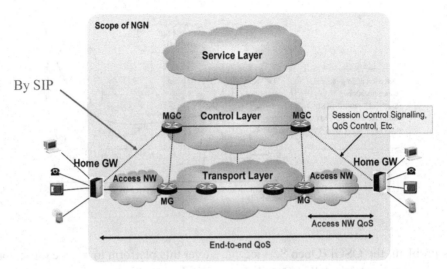

Home Gateway should Control both Access NW QoS and End-to-end NNW QoS

By "ITU-T WS on Home Networking and Home Services, 2004"

triple play services, and customize these services by themselves. However, these home network services are generally deployed by every service provider independently. In this situation, if users try to subscribe these services individually, users have to prepare many home gateways or adopters and dedicated home networks which are provided by every service provider, inconveniently.

Figure 15. Relationship between Architecture of SIP adaptation for NGN Release 1

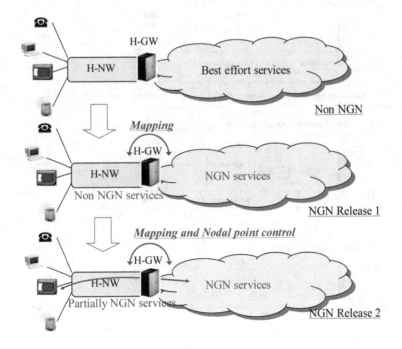

Figure 16 Architecture of SIP adaptation for NGN Release 1

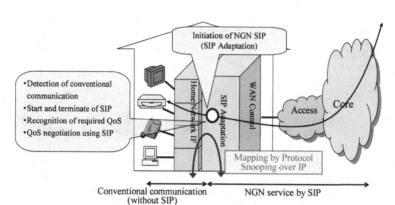

To solve this problem, the OSGi (Open Service Gateway initiative) approach is installed. Detailed description for OSGi has been provided in OSGi Alliance (2003) which can be downloaded from www.osgi.org and OSGi Alliance (2007). OSGi has specified the JAVA based platform in the home gateway as shown in Figure 21. Software blocks, referred to as "Bundles", are provisioned over this platform to realize these home network services. When users try to subscribe some services, bundles for these services are transferred from servers to the home gateway as shown in Figure 22. Therefore, the home gateway can be shared among these services. The home gateway with OSGi approach has been widely introduced. For example, its architecture and implementation

Figure 17. The SIP adaptation function for TCP/UDP communication

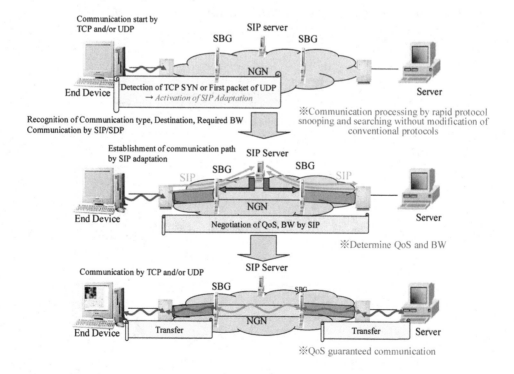

Figure 18. Example of detailed sequences in the SIP adaptation

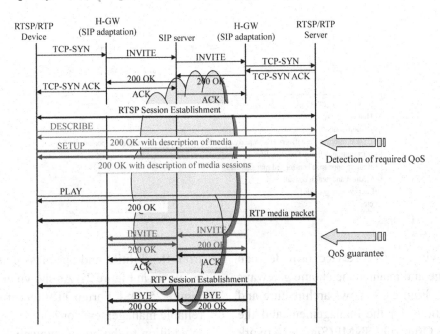

have been reported in the following materials: Moyer (2007), for generic architecture; Kawamura (2007), for dynamic service provisioning using bundle transfer technologies named as OSAP (OSGi Service Aggregation Platform); Yokotani (2007) and Yokotani & Sato (2007), for the implementation of the home gateway using OSGi approach. Users prepare only one home gateway and home network as shown in Figure 23. Service examples provided by the OSGi ap-

proach are shown in Figure 24 and are reported in Soma, Furuya, Sato, Yokotani, & Shimokasa (2008) and so on. The number of these services will be increases according to user preferences.

Easy Management and Maintenance

Finally, one of the key technologies is management and maintenance issues. As the number of subscribers is increasing, it is important that the

Figure 19. Architecture of SIP adaptation for NGN Release 2

Figure 20. Service evolution on broadband infrastructure

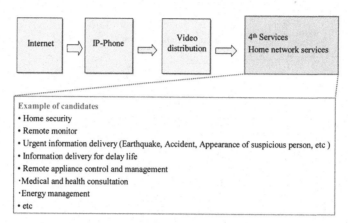

network provider and/or service provider can easily manage and maintenance home gateways and network. Figure 25 shows architecture and typical functions for the management and the maintenance. Originally, SNMP (Simple Network Management Protocol) can be applied to collect MIB data for the management. Moreover, ICMP (Internet Control Message Protocol) is also useful to detect failures. Recently, as the next generation network management and maintenance protocol, DSL Forum TR-069 (2007) as shown in Figure 26 can be highlighted. Since this protocol is implemented over JAVA technologies, it can be provided over OSGi described in the previous section,

from flexibilities and openness points of views, as shown in Figure 27. As shown in Scheme A of Figure 25, DSL Forum TR-069 can provide high reliable management for home network devices in addition to the home gateway. In this case, it is an ideal situation that network operators can distribute software of management functions to these devices independent of their specifications, performance, and implementations. Therefore, Scheme A of Figure 27 is suitable because differences among these devices can be admitted by OSGi framework.

Figure 21. Summary of the OSGi approach

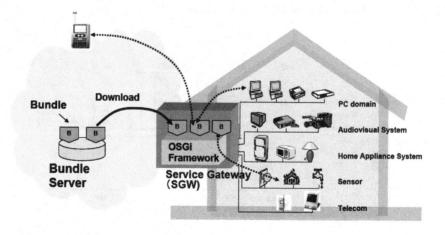

Figure 22. Bundle transfer to provide requested services

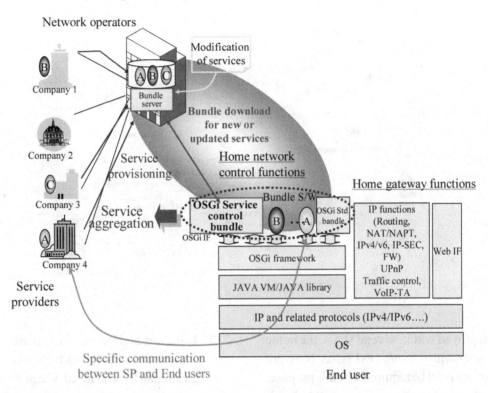

FUTURE ISSUES

The following issues can be nominated as the future evolution for the home gateway and the home network; (1) High performance, (2) Transport architecture, (3) Support of ubiquitous ability, (4) Plug and play, and interoperability, utility, (5)

Security, and (6) Ecology..

1. High performance: As the enhancement of transmission rate in broadband service, enhancement of performance in the home gateway will be continued. Particularly, since 10Gb/s optical access network will

Figure 23. Home gateway and home network configuration

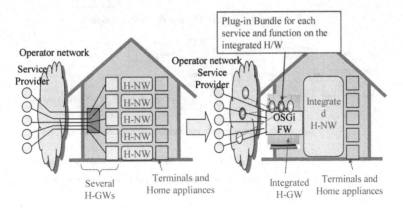

Figure 24. Possible home network services based on the OSGi approach

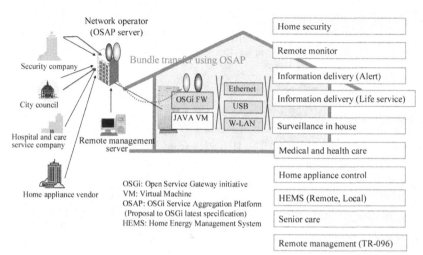

be deployed within several years, the home gateway with 10Gb/s UNI (User Network Interface) will be required. For this purpose, an advanced design process to the development of the key device (e.g., CPU) and multi-core architecture are expected.

2. Transport architecture: To connect various end devices, home work transport technologies should be enhanced. Currently, most of transmission methods expected for Ethernet is less than 100 Mb/s. Performance enhancement of PLC and MoCA will be expected.

Figure 25. Architecture and required functions for management and maintenance

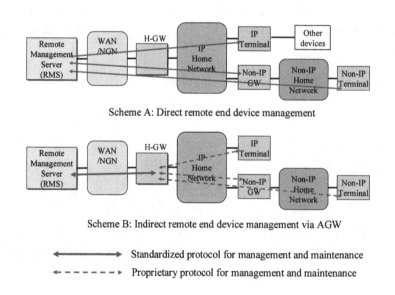

Figure 26. Protocol stack of DSL Forum TR-069

CPE/ACS Management Application
RPC Methods
SOAP
HTTP
SSL/TLS
TCP/IP

3. Support of ubiquitous ability: RF tag, ZigBee, and various sensors should be required for ubiquitous communication. Since ubiquitous services will be specified in NGN Release 3 in ITU-T, the home network also supports ubiquitous services. Therefore, devices with ubiquitous ability should be popularized.

4. Plug and play, and interoperability: In the future, various end devices are connected with the home network. The range of users for these devices is wide, such as from beginner to professional, from child to senior. Therefore, easy connection between these devices and home network should be required. For example, when a small child tries to connect a game machine to the home network, this operation should be simple. For this purpose, plug and play mechanisms and guarantee of interoperability should be considered.

5. Security: Generic security issues, such as DDoS attack, Virus, and Pretender, should be prevented. In addition to them, social guard, such as parental block, should be provided.

6. Ecology: Ecology in the home network consists of two aspects, such as a study on BEMS/HEME (Building Energy Management System / Home Energy Management System), and an energy saving of communication equipment. In the former case, the home gateway controls electric products, especially, lighting and air conditioners which occupy the large percentage of power consumption in a house. In the latter case, the power saving mode, "Sleep mode", should be considered. Foe example, IEEE802.3az (Energy Efficient Ethernet) provides specifications for a reduction of power consumption. In the standardization, ITU-T activates a focus group, ITU-T FG ICT & CC (2008). Moreover, a discussion of device architecture for this purpose should be activated.

CONCLUSION

This chapter has described the popularity of the home network and the home gateway as increasing of the number of broadband services, such as FTTH. Then, new requirements and new required

Figure 27. Implementation of TR-069 over OSGi/JAVA

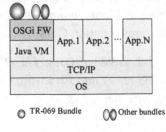

Scheme A: TR-069 over OSGi

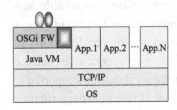

Scheme B: TR-069 over JAVA

technologies in NGN era have been described. NGN provides QoS guarantee and high reliable services. SIP is applied to provide QoS guarantee services. Before user communication across NGN, conformed QoS is negotiated. By the way, in the deployment of NGN, users expect new services beyond triple play services, using high reliable communication capability. When users utilize QoS guarantee and high reliable services provided by NGN, the home gateway should be highlighted. This chapter addresses the home gateway evolution for these NGN services, and proposes required functions of the home gateway according to its evolution scenario. This chapter classifies five steps of the home gateway evolution. Especially, it presents that Steps 3 and 4 are important stages in NGN. In Step3, new home network services beyond triple play services are installed across NGN. Mostly, it is difficult that these new services are provided across the existing Internet by reliability. However, these services can be supported by NGN using high reliable communication capability. Moreover, these services should be programmable according to user preference. This chapter proposes OSGi is suitable approach to provide these services flexibly. By the way, Step 4 provides QoS guarantee services mainly by SIP. In this case, control of SIP should be supported in the home gateway for all of QoS guaranteed communication paths. Although users do not modify end devices in the home network, this control should be provided. For this purpose, this chapter proposes the SIP adaptation function, and presents detailed mechanisms and its suitability.

After this chapter describes remarkable services in NGN and approaches to provide these services, it also proposes required functions in the home gateway. This chapter concludes that the following four technologies should be supported in home gateway for NGN; (1) the high performance home gateway platform using multi-core CPU, (2) SIP adaptation and related technologies, (3) OSGi and its bundle transfer technologies dynamically,

and (4) easy and flexible management.

Finally, this chapter lists future issues for a more flexible and various home network services and the home gateway to provide these services.

ACKNOWLEDGMENT

This work has been in part supported by a project of the Ministry of Internal Affairs and Communications (MIC) of Japan.

REFERENCES

Camarillo, G. (2002). *SIP Demystified*. McGraw-Hill Telecom

Esaki, S., Kurokawa, A., & Matsumoto, K. (2007). Overview of the Next Generation Network. NTT Technical, 5(6).

Forum, D. S. L. TR-069 (2007). *CPE WAN Management Protocol*. DSL Forum Technical Report 069

IEEE802.3 (2005). Carrier Sense Multiple Access with Collision Detection (CSMA/CD) access method and physical layer specifications.

ITU-D. (2008). *ICT statistics*. http://www.itu.int/ITU-D/ict/statistics/

ITU-T FG ICT & CC. (2008). *Focus Group on ICTs and Climate Change (FG ICT&CC)*. http://www.itu.int/ITU-T/focusgroups/climate/index.html

ITU-T G.983.4 (2001). *A broadband optical access system with increased service capability using dynamic bandwidth assignment*. ITU-T Recommendation G.983.4.

ITU-T G.983 (2001). *ITU-T Recommendation G.983 series*. http://www.itu.int/publications/sector.aspx?lang=en&sector=2

ITU-T G.984 (2002). *ITU-T Recommendation G.984 series.* http://www.itu.int/publications/sector.aspx?lang=en&sector=2

ITU-T G.hn (2008). *Generic Home Network Architecture and Transceiver.* ITU-T Draft Recommendation.

ITU-T NGN-GSI. (2008). http://www.itu.int/ITU-T/ngn/

ITU-T Y.2001 (2004). *General overview of NGN.* ITU-T Recommendation Y.2001.

ITU-T Y.2011 (2004). *General principles and general reference model for Next Generation Networks.* ITU-T Recommendation Y.2011.

Kawamura, R. (2007). Network Service Aggregation Platform based on OSGi technologies. *Access '07 Executive Business Forum associated with IEEE Globecom'07, Next Generation Home Network and Home Gateway Session.*

Kramer, G., Mukherjee, B., & Pesavento, G. (2002). IPACT: a dynamic protocol for an Ethernet PON (EPON). *IEEE Communications Magazine, 40,* 74–80. doi:10.1109/35.983911

Maeda, Y., & Okada, F. D. (2001). FSAN OAN-WG and Future Issues for Broadband Optical Access Networks. *IEEE Communications Magazine.*

Moore, G. (1965). Cramming more components onto integrated circuits. *Electronics Magazine, 19.*

Moyer, S. (2007). The OSGi Service Platform: Enabling Web 2.0 for Broadband Services Prepared. *Access '07 Executive Business Forum associated with IEEE Globecom'07, Next Generation Home Network and Home Gateway Session.*

Mukai, H., Yokotani, T., Seno, S., & Motoshima, K. (2007). Enhanced queue status reporting scheme on PON. *Journal of Optical Networking, 6,* 756–766. doi:10.1364/JON.6.000756

Nakagawa, J. (2006). Technical Survey of Next Generation Optical System – 10G-PON. *Fiber of Expo 2006.*

Nakamura, M., Ueda, H., Makino, S., Yokotani, T., & Oshima, K. (2004). Proposal of Networking by PON Technologies for Full and Ethernet Services in FTTx. *IEEE Transaction on Lightwave Technology, 22,* 2631–2640. doi:10.1109/JLT.2004.836765

Narita, A. (2004). Home Gateway for Managing Home Network QoS. *ITU-T Workshop on Home Networking and Home Services,* Session 6-4.

OSGi Alliance (2003). *OSGi Service Platform Release 3.*

OSGi Alliance (2007). *About the OSGi Service Platform*

Sato, K., Furuya, S., Yokotani, T., Homma, H., & Sakai, S. (2008). *Technologies for Next Generation Home Network.* (pp. 19-23).

Sato, K., Yokotani, T., Furuya, S., & Isoda, M. (2005). *Implementation of SIP adaptation function for QoS negotiation in NGN* (pp. 55-60). IEICE Tech. Rep., CS2005-46.

Soma, S., Furuya, S., Sato, K., Yokotani, T., & Shimokasa, K. (2008). *A study on services by home gateway for home appliance control* (pp. 1-6). IEICE Technical Report, CS2008-6.

Tsuchida, M., Yokotani, T., & Sato, K. (2006). *Network Control Technology for Next Generation Network* (pp. 55-59). Mitsubishi Electric Technical Report, 2006.2.

Yokotani, T. (2007). Home Gateway Evolution toward Home Gateway Evolution toward Broadband Service and NGN Broadband Service and NGN. *Access '07 Executive Business Forum associated with IEEE Globecom'07, Next Generation Home Network and Home Gateway Session.*

Yokotani, T., Kitayama, K., Mukai, H., & Murakami, K. (2002). A study on usage of T-CONT in DBA for FTTx. Journal of EXP in Search of Innovation, 2, 64-71.

Yokotani, T., Nakanishi, K., Ogasawara, M., & Maeda, Y. (2006). Next Generation Broadband Access Network Technologies. *Journal of IEICE*, *89*, 1062–1066.

Yokotani, T., & Sato, K. (2007). *Integrated Home Gateway with OSGi Framework*. 3rd Workshop of OSGi User Forum Japan.

Compilation of References

Aagedal, J. O., den Braber, F., Dimitrakos, T., Gran, B. A., Raptis, D., & Stolen, K. (2002). *Model-based risk assessment to improve enterprise security.* Paper presented at the Enterprise Distributed Object Computing Conference EDOC '02 Sixth International

Abrams, M., Becker, P. C., Fujimoto, Y., O'Byrne, V., & Piehler, D. (2005). FTTP deployments in the United States and Japan – Equipment Choices and Service Provider Imperatives. *Journal of Lightwave Technology, 23*(1), 236–246. doi:10.1109/JLT.2004.840340

Adkins, I. (2001). *Strategies for Utilities in the European Telecommunications Market.* London: Reuters & Mason.

Agrawal, P. (2002). *Fiber-Optic Communication Systems.* New York, USA: Wiley-Intersience.

Akanbi, O., Yu, J., & Chang, G. K. (2005, September). *A new bidirectional WDM-PON using DWDM channels generated by optical carrier suppression and separation technique.* Paper presented at European Conference and Exhibition on Optical Communication (ECOC), (We3.3.1), Glasgow, UK.

Akanbi, O., Yu, J., & Chang, G. K. (2006). A new scheme for bidirectional WDM-PON using upstream and downstream channels generated by optical carrier suppression and separation technique. *IEEE Photonics Technology Letters, 18*(2), 340–342. doi:10.1109/LPT.2005.861975

Ali, M., & Deogun, J. S. (2000). Cost-Effective Implementation of Multicasting in Wavelength-Routed Networks. *IEEE Journal of Lightwave Technology, 18*(12), 1628–1638. doi:10.1109/50.908667

Altwegg, L., Azizi, A., Vogel, P., Wang, Y., & Wyler, F. (1994). LOCNET: A fiber in the loop system with no light source at the subscriber end. *IEEE/OSA . Journal of Lightwave Technology, 12*(3), 535–540. doi:10.1109/50.285337

Amendola, G. B., & Pupillo, L. M. (2008). The Economics of Next Generation Access Networks and Regulatory Governance: Towards Geographic Patterns of Regulation. *Communications & Strategies, 69,* 85-105. Retrieved on November 17, 2008, from: http://mpra.ub.uni-muenchen.de/8823/

American National Standards Institute (ANSI) (1991). *ANSI T1.105-1991, Digital Hierarchy - Optical Interface Rates and Formats Specifications.*

An, F., Kim, K. S., Gutierrez, D., Yam, S., Hu, E., Shrikhande, K., & Kazovsky, L. G. (2004). SUCCESS: A Next-Generation Hybrid WDM/TDM Optical Access Network Architecture. *Journal of Lightwave Technology, 22*(11), 2557–2569. doi:10.1109/JLT.2004.836768

Androulidakis, S., Kagklis, D., Doukoglou, T., & Skenter, S. (2004). ADSL2: A sequel better than the original? *IEE Communications Engineering Magazine, June/July Issue* (pp. 22-27).

Arellano, C., Bock, C., Prat, J., & Langer, K. D. (March, 2006). RSOA-based optical network units for WDM-PON. *IEEE/OSA Optical Fiber Communication Conference / National Fiber Optic Engineers Conference (OFC/NFOEC),* Paper OTuC1, Anaheim, California, USA.

Arellano, C., Polo, V., & Prat, J. (2007, July). *Effect of the multiplexer position in Rayleigh-limited WDM-PONs with amplified-reflective ONU.* Paper presented at International Conference on Transparent Optical Networks (ICTON), (Paper Mo.P.22), Rome, Italy.

Arellano, C., Polo, V., Bock, C., & Prat, J. (March, 2005). Bidirectional single fiber transmission based on a RSOA ONU for FTTH using FSK-IM modulation formats. Paper presented at *IEEE/OSA Optical Fiber Communication Conference / National Fiber Optic Engineers Conference (OFC/NFOEC)*, Paper JWA46, Anaheim, California, USA.

Arnaud, B. S. (2001). *Telecom issues and their impact on FTTx architectural designs (FTTH Council).* CANARIE, Inc. Retrieved on August 22, 2004, from: http://www.canarie.ca/canet4/library/customer.html

Assi, C., Ye, Y., Dixit, S., & Ali, M. (2003). Dynamic bandwidth allocation for quality-of-service over Ethernet PONs. *IEEE Journal on Selected Areas in Communications, 21*(9), 1467–1477. doi:10.1109/JSAC.2003.818837

Assi, C., Ye, Y., Dixit, S., & Ali, M. A. (2003). Dynamic Bandwidth Allocation for Quality-of-Service over Ethernet PONs. *IEEE Journal on Selected Areas in Communications, 21*(9), 1467–1477. doi:10.1109/JSAC.2003.818837

Attygalle, M., Anderson, T., Hewitt, D., & Nirmalathas, A. (2006). WDM passive optical network with subcarrier transmission and baseband detection scheme for laser-free optical network units. *IEEE Photonics Technology Letters, 18*(11), 1279–1281. doi:10.1109/LPT.2006.876770

Attygalle, M., Nadarajah, N., & Nirmalathas, A. (2005). Wavelength reused upstream transmission scheme for WDM passive optical networks. *IEE Electronics Letters, 41*(18), 1025–1027. doi:10.1049/el:20052468

Bakaul, M., Nadarjah, N., Nirmalathas, A., & Wong, A. (2008). Internetworking VCSEL-Based Hybrid Base Station towards Simultaneous wireless and wired Transport for Converged Access Network. *IEEE Photonics Technology Letters, 20*(8), 569–571. doi:10.1109/LPT.2008.918875

Banerjee, A., Kramer, G., & Mukherjee, B. (2006). Fair sharing using dual service-level agreements to achieve open access in a passive optical network. *IEEE Journal on Selected Areas in Communications, 24*(8), 32–44.

Banerjee, A., Park, Y., Clarke, F., Song, H., Yang, S., & Kramer, G. (2005). A review of wavelength-division multiplexed passive optical network (WDM-PON) technologies for broadband access. *OSA Journal of Optical Networking, 4*(11), 737–758. doi:10.1364/JON.4.000737

Banerjee, A., Park, Y., Clarke, F., Song, H., Yang, S., & Kramer, G. (2005). Wavelength-division-multiplexed passive optical network (WDM-PON) technologies for broadband access: a review (Invited). *Journal of Optical Networking, 4*(11), 737–758. doi:10.1364/JON.4.000737

Barry, R. A., & Humblet, P. A. (1993). Latin Routers, Design and Implementation. *Journal of Lightwave Technology, 11*(5), 891–899. doi:10.1109/50.233253

Bayvel, P. (2000). Future high-capacity optical telecommunications networks. *Philosophical Transactions of the Royal Society of London. Series A: Mathematical and Physical Sciences, 358,* 303–329. doi:10.1098/rsta.2000.0533

Bbned. (2007). *Subsidiary company of Telecom Italia.* from http://www.bbned.nl/content/english.shtml

Beam Express. (2008). *VCSEL, Long Wavelength VCSEL.* Retrieved in August 2008, http://www.beamexpress.com/

Bell, P., & Trigger, P. (1998). Access network life-cycle costs. *BT Technolology, 16*(4), 165–174. doi:10.1023/A:1009612317200

Bell, P., Walling, R., & Peacok, J. (1996). Costing the access network – an overview. *BT Technology, 14*(2), 128–132.

Bianco, A., Galante, G., Leonardi, E., Neri, F., & Nucci, A. (2003). Scheduling Algorithms for Multicast Traffic in TDM/WDM Networks with Arbitrary Tuning Latencies. *Computer Networks, 41*(6), 727–742. doi:10.1016/S1389-1286(02)00436-X

Bilodeau, F., Johnson, D. C., Theriault, S., Malo, B., Albert, J., & Hill, K. O. (1995). An all-fiber dense wavelength-division multiplexer/demultiplexer using photoimprinted Bragg gratings. *IEEE Photonics Technology Letters, 7*(4), 388–390. doi:10.1109/68.376811

Biraghi, A. M., Cobb, K., Cantone, S. C., Greborio, L., Mercinelli, R., & Walling, R. (2002). *Survey of future broadband services.* Ist for Broadband Europe (pp.1-13).

Blondel, E., & Rubin, R. (2001). Evolution of the telecom powering strategy: moving to a cost effective and reliable solution using remote powering. INTELEC: *Telecommunications Energy Conference, 23*, 532- 539.

Blum, S. (2005). In F. t. t. H. Council (Ed.), *Financial Analysis of FTTH System Proposals: An Operations-Based Approach.* Tellus Venture Associates.

Bock, C., & Prat, J. (2005). Scalable WDMA/TDMA protocol for passive optical networks that avoids upstream synchronization and features dynamic bandwidth allocation. *OSA Journal of Optical Networking, 4*(4), 226–236. doi:10.1364/JON.4.000226

Bock, C., Prat, J., & Walker, S. D. (2005). Hybrid WDM/TDM PON using the AWG FSR and featuring centralized light generation and dynamic bandwidth allocation. *IEEE/OSA . Journal of Lightwave Technology, 23*(12), 3981–3988. doi:10.1109/JLT.2005.853138

Bock, C., Prat, J., Segarra, J., Junyent, G., & Amrani, A. (2004). Scalable Two-stage Multi-FSR WDM-PON Access Network Offering Centralized Dynamic Bandwidth Allocation. *ECOC 2004 Proceedings,* (2), 278-279.

Bogineni, K., Sivalingam, K. M., & Dowd, P. W. (1993). Low-Complexity Multiple Access Protocols for Wavelength-Division Multiplexed Photonic Networks. *IEEE Journal on Selected Areas in Communications, 11*(4), 590–604. doi:10.1109/49.221206

Borella, A., Cancellieri, G., & Chiaraluce, F. (1998). *Wavelength Division Multiple Access Optical Networks.* Norwood, Ohio: Artech House.

Borella, M. S., & Mukherjee, B. (1996). Efficient Scheduling of Nonuniform Packet Traffic in a WDM/TDM Local Lightwave Network with Arbitrary Transceiver Tuning Latencies. *IEEE Journal on Selected Areas in Communications, 14*(5), 923–934. doi:10.1109/49.510916

Borghesani, A., Lealman, I. F., Poustie, A., Smith, D. W., & Wyatt, R. (2007, September). *High temperature, colourless operation of a reflective semiconductor optical amplifier for 2.5Gbit/s upstream transmission in a WDM-PON.* Paper presented at European Conference and Exhibition on Optical Communication (ECOC), (Paper 6.4.1), Berlin, Germany.

Boucart, J., Starck, C., Gaborit, F., Plais, A., Bouche, N., & Derouin, E. (1999). Metamorphic DBR and tunnel-junction injection. A CW RT monolithic long wavelength VCSEL. *IEEE Journal on Selected Topics in Quantum Electronics, 5*(3), 520–529. doi:10.1109/2944.788414

Boucher, Y., Rissons, A., & Mollier, J. C. (2001, April). *Temperature dependence of the near-threshold emission wavelength a linewidth in a Vertical-Cavity Surface-Emitting Laser (VCSEL).* Paper presented at the conference SIOE01, Cardiff, UK.

Briand, J., Payoux, F., Chanclou, P., & Joindot, M. (2007). Forward error correction in WDM PON using spectrum slicing. *Optical Switching and Networking, 4*(2), 131–136. doi:10.1016/j.osn.2006.10.005

Brunnel, H. (1986). Message Delay in TDMA Channels with contiguous output. *IEEE Transactions on Communications, 34*(7), 681–684. doi:10.1109/TCOM.1986.1096608

Byun, H.-J., Nho, J.-M., & Lim, J.-T. (2003). Dynamic bandwidth allocation algorithm in ethernet passive optical networks. *Electronics Letters, 39*(13), 1001–1002. doi:10.1049/el:20030635

Camarillo, G. (2002). *SIP Demystified.* McGraw-Hill Telecom

Cave, M. (2007). *The regulation of access in telecommunications: A European perspective. (Revised, April 2007).* Warwick, UK: Warwick Business School, University of Warwick. Retrieved on October 28, 2008, from: http://www.econ.upf.edu/docs/seminars/cave.pdf.

Cave, M., Stumpf, U., & Valetti, T. (2006, June). *A Review of certain markets included in the Commission's Recommendation on Relevant Markets subject to ex ante Regulation. Report for the European Commission.* Brussels, Belgium: European Commission.

Chae, C. J., & Oh, N. H. (1998). WDM/TDM PON system employing a wavelength-selective filter and a continuous-wave shared light source. *IEEE Photonics Technology Letters, 10*(9), 1325–1327. doi:10.1109/68.705631

Chae, C.-J., Wong, E., & Tucker, R. S. (2002). Optical CSMA/CD media access scheme for Ethernet over passive optical network. *IEEE Photonics Technology Letters, 14*(5), 711–713. doi:10.1109/68.998734

Chan, C. K. (2007). Protection architectures for passive optical networks. In C. Lam (Ed.), *Passive Optical Networks, Principles and Practice*, (pp.243-264). USA: Academic Press, Elsevier Inc.

Chan, L. Y., Chan, C. K., Tong, D. T. K., Tong, F., & Chen, L. K. (2002). Upstream traffic transmitter using injection-locked Fabry–Pérot laser diode as modulator for WDM access networks. *IEE Electronics Letters, 38*(1), 43–45. doi:10.1049/el:20020015

Chang, C.-H., Kourtessis, P., & Senior, J. M. (2006). GPON service level agreement based dynamic bandwidth assignment protocol. *Electronics Letters, 42*(20), 1173–1174. doi:10.1049/el:20062326

Chang, C.-H., Merayo, N., Kourtessis, P., & Senior, J. M. (2007). Dynamic Bandwidth assignment for Multiservice access in long-reach GPON. *Proceedings of the 33rd European Conference and Exhibition on Optical Communication (ECOC 2007)*. Berlin, Germany.

Chang, G. K., Yu, J., Jia, Z., & Yu, J. (2006, March). *Novel optical-wireless access network architecture for simultaneously providing broadband wireless and wired services.* Paper presented at Optical Fiber Communication Conference (OFC), (OFM1), Anaheim, California. USA.

Chang-Hasnain, C. J. (2003). Progress and prospects of long-wavelength VCSEL. *IEEE Optical Communications, 41*(2), 530–534.

Chanlou, P. (2008, April 11). La Fibre: FTTH deployments and plans of France Telecom, *International Workshop on FTTx architectures, technologies, business and economic aspects,* Athens.

Chapuran, T. E., Wagner, S. S., Menendez, R. C., Tohme, H. E., & Wang, L. A. (1991). Broadband multichannel WDM transmission with superluminescent diodes and LEDs. *Proc. of Global Telecommunications Conference (GLOBECOM)* (vol. 1, pp. 612-618).

Chen, L., Shao, Y., Lei, X., Wen, H., & Wen, S. (2007). A novel radio-over-fiber system with wavelength reuse for upstream data connection. *IEEE Photonics Technology Letters, 19*(6), 387–389. doi:10.1109/LPT.2007.891958

Chen, L., Wen, H., & Wen, S. (2006). A radio-over-fiber system with a novel scheme for millimeter-wave generation and wavelength reuse for up-link connection. *IEEE Photonics Technology Letters, 18*(19), 2056–2058. doi:10.1109/LPT.2006.883293

Chia, M. Y. W., Luo, B., Yee, M. L., & Hao, E. J. Z. (2003). Radio over multimode fiber transmission for Wireless LAN using VCSELs. *Electronics Letters, 39*(15), 1143–1144. doi:10.1049/el:20030724

Cho, K. Y., Murakami, A., Lee, Y. J., Agata, A., Takushima, Y., & Chung, Y. C. (February, 2008). Demonstration of RSOA-based WDM PON operating at symmetric rate of 1.25 Gb/s with high reflection tolerance. Paper presented at *IEEE/OSA Optical Fiber Communication Conference / National Fiber Optic Engineers Conference (OFC/NFOEC)*, Paper OTuH4, San Diego, California, USA.

Chochliouros, I. P., & Spiliopoulou, A. S. (2002). Local loop unbundling policy measures as an initiative factor for the competitive development of the European electronic communications markets. [TCN]. *The Journal of the Communications Network, 1*(2), 85–91.

Chochliouros, I. P., & Spiliopoulou, A. S. (2003). Innovative Horizons for Europe: The New European Telecom Framework for the Development of Modern Electronic Networks and Services. [TCN]. *The Journal of The Communications Network, 2*(4), 53–62.

Chochliouros, I. P., Spiliopoulou, A. S., & Lalopoulos, G. K. (2005). Dark Optical Fiber as a Modern Solution for Broadband Networked Cities. In M. Pagani (Ed.), *The Encyclopedia of Multimedia Technology and Networking,* (pp. 158-164). Hershey, PA: IRM Press, Idea Group Inc.

Chochliouros, I.P., & Spiliopoulou, A.S. (2005), Broadband Access in the European Union: An Enabler for Technical Progress, Business Renewal and Social Development. *The International Journal of Infonomics (IJI), 1*, 05-21.

Choi, K. M., & Lee, C. H. (2005, September). *Colorless operation of WDM-PON based on wavelength locked Fabry-Perot laser diode.* Paper presented at European Conference and Exhibition on Optical Communication (ECOC), (We3.3.4), Glasgow, UK.

Choi, S., & Huh, J. (2002). Dynamic bandwidth allocation algorithm for multimedia services over Ethernet PONs. *ETRI Journal, 24*(6), 465–468.

Chow, C. W. (2008). Wavelength remodulation using DPSK down-and-upstream with high extinction ratio for 10-Gb/s DWDM-passive optical networks. *IEEE Photonics Technology Letters, 20*(1), 12–14. doi:10.1109/LPT.2007.911009

Chow, C. W., Talli, G., & Townsend, P. D. (2007). Rayleigh noise reduction in 10-Gb/s DWDM-PONs by wavelength detuning and phase-modulation-induced spectral broadening. *IEEE Photonics Technology Letters, 19*(6), 423–425. doi:10.1109/LPT.2007.892899

Chow, C. W., Talli, G., Ellis, A. D., & Townsend, P. D. (2008). Rayleigh noise mitigation in DWDM LR-PONs using carrier suppressed subcarrier-amplitude modulated phase shift keying. *Optics Express, 16*(3), 1860–1865. doi:10.1364/OE.16.001860

Chow, C. W., Yeh, C. H., Wang, C. H., Shih, F. Y., Pan, C. L., & Chi, S. (2008). WDM extended reach passive optical networks using OFDM-QAM. *Optics Express, 16*(16), 12096–12101. doi:10.1364/OE.16.012096

Chow, W., Choquette, K. D., Crawford, M., Lear, K., & Hadley, G. (1997). Design, Fabrication, and Performance of infrared and Visible Vertical-Cavity Surface-Emitting Lasers. *IEEE Journal of Quantum Electronics, 33*(10), 1810–1824. doi:10.1109/3.631287

Chung, H. S., Kim, B. K., & Kim, K. J. (2008). Effects of upstream bit rate on a wavelength-remodulated WDM-PON based on Manchester or inverse-return-to-zero coding. *ETRI Journal, 30*(2), 255–260.

Chung, H. S., Kim, B. K., Park, H., Chang, S. H., Chu, M. J., & Kim, K. J. (October, 2006). Effects of inverse-RZ and Manchester code a wavelength re-used WDM-PON. Paper presented at *IEEE Lasers and Electro-Optics Annual Meeting (LEOS)*, Paper TuP3, Montréal, Québec, Canada.

Cisco Systems. (2004). *Almere looks to a thriving digital future with real broadband.*

City of Rotterdam. (2007). from http://www.rotterdam.com/

Citynet Amsterdam. (2006). *European municipal Fiber and Fiber backbone projects.* from http://www.citynet.nl/

Clarke, R. N. (2008, April 10). FTTN/VDSL2 Broadband Networks Capabilities and Economics. *Fibre Investment and Policy Challenges OECD Workshop*, Stavanger, Norway.

Cloetens, L. (2001). *Broadband access: the last mile.* Paper presented at the IEEE International Solid-State Circuits Conference.

Coffman, K. G., & Odlyzko, A. M. (2001). *Internet growth: Is there a "Moore's law" for data traffic?* New York, USA: AT&T Labs - Research.

Commission of the European Communities. (2006). *Communication on Market Reviews under the EU Regulatory Framework - Consolidating the internal market for electronic communications [COM(2006) 28 final, 06.02.2006].* Brussels, Belgium: Commission of the European Communities.

Constant, S. B., Le Guennec, Y., Maury, G., Corrao, N., & Cabon, B. (2008). Low-Cost All-Optical Up-Conversion of Digital Radio Signals Using a Directly modulated 1550-nm Emitting VCSEL. *IEEE Photonics Technology Letters, 20*(2), 120–122. doi:10.1109/LPT.2007.912543

Dal Bono, P. (2008, February 28). Telecom Italia domestic NGN2: the first important steps toward FTTH. *FTTx Council Europe Conference*, Paris.

Damien Chew, C. (2006). ING wholesale banking. *European telecoms: Citynet Amsterdam: Fiber-to-the-home is becoming a reality.*

Data Over Cable Service Interface Specifications (DOCSIS) Group. (1999). *DOCSIS 1.1 Interface Specification.* (www.cablemodem.com).

Davey, R. P., & Payne, D. B. (2006). DWDM Reach Extension of a GPON to 135 km. *IEEE/OSA . Journal of Lightwave Technology, 24*(1), 29–31. doi:10.1109/JLT.2005.861140

de la Rosa, P. (2005, October). *The Access Network in the XXI Century.* Broadband World Forum, Madrid, Spain.

Delorme, F. (2000). 1.5-mm tunable DBR lasers for WDM multiplex spare function. *IEEE Photonics Technology Letters, 12*(6), 621–623. doi:10.1109/68.849063

Deng, N., Chan, C. K., & Chen, L. K. (2007). A centralized-light-source WDM access network utilizing inverse-RZ downstream signal with upstream data remodulation. *Optical Fiber Technology, 13*(1), 18–21. doi:10.1016/j.yofte.2006.03.006

Deng, N., Chan, C. K., Chen, L. K., & Tong, F. (2003). Data remodulation on downstream OFSK signal for upstream transmission in WDM passive optical network. *IEE Electronics Letters, 39*(24), 1741–1743. doi:10.1049/el:20031092

Desai, B. N., Frigo, N. J., Smiljanic, A., Reichmann, K. C., Iannone, P. P., & Roman, R. S. (2001). An optical implementation of a packet-based (Ethernet) MAC in a WDM passive optical network overlay. *Proc. of 2001 Optical Fiber Communication Conference (OFC'01), Vol. 3*, WN5-1-WN5-3.

Desurvire, E. (1991). *Erbium-Doped Fiber Amplifiers: Principles and Applications.* New York, NY, USA: Wiley.

Diab, W. W., & Frazier, H. M. (2006). *Ethernet in the First Mile – Access for Everyone.* USA: IEEE Press

Distaso, W., Lupi, P., & Manenti, F. M. (2006). Platform Competition and Broadband Uptake. Theory and empirical evidence from the European Union. *Information Economics and Policy, 18*, 87–106. doi:10.1016/j.infoecopol.2005.07.002

Dixit, S. S. (2003). *IP over WDM, Building the Next Generation Optical Internet.* Hoboken, New Jersey: John Wiley & Sons.

Dixon, R. W., & Joyce, W. B. (1979). Generalized Expressions for the turn-on delay in semiconductor lasers. *Journal of Applied Physics, 50*(7), 4591–4595. doi:10.1063/1.326566

Ecobilan S.A. (2008, February). *Developing a generic approach for FTTH solutions using LCA methodology. Methodological Guide - Final Version.* Neuilly Sur Seine, France: FTTH Council Europe.

Effenberger, F. J., Ichibangase, H., & Yamashita, H. (2001). Advances in broadband passive optical networking technologies. *IEEE Communications Magazine, 39*(12), 118–124. doi:10.1109/35.968822

Effenberger, F., Cleary, D., Haran, O., Kramer, G., Li, R. D., Oron, M., & Pfeiffer, T. (2007). An introduction to PON technologies . *IEEE Communications Magazine, 45*(3), S17–S25. doi:10.1109/MCOM.2007.344582

EICTA. *(2008).* Position on Next Generation Networks (NGN) & Next Generation Access (NGA):"Moving Towards a Very High Speed Europe

Eisenach, R. (2008). *Understanding Mobile Backhaul.* CED Magazine.

Elmirghani, J. M. H., & Mouftah, H. T. (2000). All-Optical Wavelength Conversion: Technologies and Applications in DWDM Networks. *IEEE Communications Magazine, 38*(3), 86–92. doi:10.1109/35.825645

Ericsson AB. (2002). Case Study. *Sollentuna Energi: A Broadband Pioneer.*

Ericsson AB. (2002). Case Study. *Hudiksvallsbostäder: We got fiber all the way.*

Esaki, S., Kurokawa, A., & Matsumoto, K. (2007). Overview of the Next Generation Network. NTT Technical, 5(6).

Eurescom, Project P917, Building and Operating Broadband Access Networks (BOBAN) (2000). *Deliverable 10, DWDM technologies for access networks.* Heidelberg, Germany.

EURESCOM. Project P917, Building and Operating Broadband Access Networks (BOBAN) (2000). *Deliverable 8,* Volume 1 of 2: Main Report, *Broadband cabinet survey, specification and demonstration.* (p. 61)

European Commission. (2000). *Communication on unbundled access to the local loop: Enabling the competitive provision of a full range of electronic communication services, including broadband multimedia and high-speed Internet [COM(2000) 394, 26.07.2000].* Brussels, Belgium: European Commission.

European Commission. *(2003).* Commission Recommendation 2003/311/EC of 11 February 2003 on relevant product and service markets within the electronic communications sector susceptible to ex ante regulation in accordance with Directive 2002/21/EC on a common regulatory framework for electronic communication networks and services. [Official Journal (OJ) C114 (pp. 45-49)]. *Brussels, Belgium: European Commission.*

European Commission. (2005). *Communication on i2010 - A European Information Society for growth and employment [COM(2005) 229 final, 01.06.2005].* Brussels, Belgium: European Commission.

European Commission. (2005). *Commission Staff Working Paper on Communication from the Commission "i2010 - A European Information Society for growth and employment - Extended Impact Assessment" [SEC(2005) 717/2, 01.06.2005].* Brussels, Belgium: European Commission.

European Commission. (2006). *Communication to the Council, the European Parliament, the European Economic and Social Committee and the Committee of the Regions, on Bridging the Broadband Gap [COM(2006) 129 final, 20.03.2006].* Brussels, Belgium: European Commission.

European Parliament & Council of the European Union. (2000). *Regulation (EC) 2887/2000 on unbundled access to the local loop" [Official Journal (OJ) L336, 30.12.2002, pp.04-08].* Brussels, Belgium: European Parliament & Council of the European Union.

European Parliament and Council of the European Union (2002). *Directive 2002/21/EC of March 7 2002 on a common regulatory framework for electronic communications networks and services (Framework Directive), (Official Journal (OJ) L108, 24.04.2002, pp.33-50).* Brussels, Belgium: European Parliament and Council of the European Union.

EXFO Electro-optical Engineering Inc. (2000). *Guide to WDM Technology and Testing.* Quebec City, Canada.

Federal Communications Commission (FCC). (2003). *Triennial Review Order, document FCC-03-36A1: Review of the Section 251 Unbundling Obligations of Incumbent Local Exchange Carriers.* Washington, DC: FCC.

Feldman, R. D. (1997). Crosstalk and loss in wavelength division multiplexed systems employing spectral slicing. *IEEE/OSA . Journal of Lightwave Technology, 15*(10), 1823–1831. doi:10.1109/50.633564

Feldman, R. D., Harstead, E. E., Jiang, S., Wood, T. H., & Zirngibl, M. (1998). An evaluation of architectures incorporating wavelength division multiplexing for broad-band fiber access. *IEEE/OSA .Journal of Lightwave Technology, 16*(9), 1546–1558. doi:10.1109/50.712236

Felten, B. (2008, April 10). Municipal fibre – an overview. *Fibre Investment and Policy Challenges OECD Workshop,* Stavanger, Norway. Yankee Group

Feuer, M. D., Wiesenfeld, J. M., Perino, J. S., Bums, C. A., Raybon, G., Shunk, S. C., & Dutta, N. K. (1996). Single-port laser-amplifier modulators for local access. *IEEE Photonics Technology Letters, 8*(9), 1175–1177. doi:10.1109/68.531827

Feuer, M., Wiesenfeld, J., Perino, J., Burrus, C., Raybon, G., Shunk, S., & Dutta, N. (1996). Single-port laser-amplifier modulators for local access. *IEEE Photonics Technology Letters, 8*(9), 1175–1177. doi:10.1109/68.531827

Foh, C., Andrew, L., Wong, E., & Zukerman, M. (2004). FULL-RCMA: a high utilization EPON. *IEEE Journal on Selected Areas in Communications, 22*(8), 1514–1524. doi:10.1109/JSAC.2004.830459

Forum, D. S. L. (2007, June 5). *More than 200 million customers chose DSL. News Release.* Retrieved on July 23, 2007 from: http://www.dslforum.org/dslnews/pdfs/prbbwfasia060507.pdf

Forum, D. S. L. TR-069 (2007). *CPE WAN Management Protocol.* DSL Forum Technical Report 069

Foxley, D. (2002). *Dark fiber.* TechTarget.com, Inc. Retrieved on August 11, 2004, from: http://searchnetworking.techtarget.com//sDefinition/0,sid7_gci21189,00.html

Frigo, N. J., Iannone, P. P., & Reichmann, K. C. (1998, September 20-24). Spectral slicing in WDM passive optical networks for local access. In *Proceedings of European Conference on Optical Communications (ECOC)* (pp. 119-120). Madrid: Spain.

Frigo, N. J., Iannone, P. P., & Reichmann, K. C. (2004). A view of fiber to the home economics. *IEEE Optical Communications, 42*(8), S16–S23. doi:10.1109/MCOM.2004.1321382

Frigo, N. J., Iannone, P. P., Magill, P. D., Darcie, T. E., Downs, M. M., & Desai, B. N. (1994). A wavelength-division multiplexed passive optical network with cost-shared components. *IEEE Photonics Technology Letters, 6*(11), 1365–1367. doi:10.1109/68.334841

FSAN. (2008). *Full Service Access Network.* Retrieved July 23, 2008, from http://www.fsanweb.org/.

FTTH Council Europe. (2007). *Europe at the Speed of Light - Presentation by J. M. Van Bogaert.* Retrieved on August 15, 2008 from: http://www.ftthcouncil.com.

FTTH Council. (2006). *Definition of Terms.* FTTH Council.

FTTH Council. (2007). *FTTH/FTTP Update April 2007 - Study by RVA LLC.* Retrieved on June 06, 2008 from: http://www.ftthcouncil.org/documents/800832.pdf.

FTTH Council. (2007). *Asia leads the world in Fiber-To-The-Home penetration.* Retrieved on June 06, 2008 from: http://www.ftthcouncil.org/?t=231.

Fujitsu Inc. (2006). *Ethernet in the First Mile Over Point-to-Point Fiber (EFMF).* Fujitsu Inc.

Fujiwara, M., Kani, J., Suzuki, H., & Iwatsuki, K. (2006). Impact of backreflection on upstream transmission in WDM single-fiber loopback access networks. *Journal of Lightwave Technology, 24*(2), 740–746. doi:10.1109/JLT.2005.862429

Fujiwara, M., Suzuki, H., & Yoshimoto, N. (2006). Quantitative loss budget estimation in WDM single-fibre loopback access networks with ASE light sources. *Electronics Letters, 42*(22), 1301–1302. doi:10.1049/el:20062735

Fuller, M. (2006). European municipalities lead FTTH charge. *Lightwave FTTX Direct Newsletter.* Retrieved on July 24, 2007, from: http://www.localret.es/localretnews/bandaampla/num9/docs/5num9.pdf.

Garcés, I., Aguado, J. C., Martínez, J. J., López, A., Villafranca, A., & Losada, M. A. (2007). Analysis of narrow-FSK downstream modulation in colourless WDM PONs. *IEE Electronics Letters, 43*(8), 471–472. doi:10.1049/el:20073908

Gauthey, G. (2007). Regulation must play a role in the breakthrough of fiber. *Lightwave Europe, Q4/2007,* 9-10.

George, J. (2006). Start thinking about 3 to 30 Gbps by 2030! *Broadband Properties, September 2006,* 42-47. Retrieved on July 14, 2007 from: *www.broadbandproperties.com/2006issues/sep06issues/george_sep.pdf*

Gerken, M., & Miller, D. A. B. (2003). Wavelength demultiplexer using the spatial dispersion of multilayer thin-film structures. *IEEE Photonics Technology Letters, 15*(8), 1097–1099. doi:10.1109/LPT.2003.815318

Giles, C. R., Feldman, R. D., Wood, T. H., Zirngibl, M., Raybon, G., & Strasser, T. (1996). Access PON using downstream 1550-nm WDM routing and upstream 1300-nm SCMA combining through a fiber-grating router. *IEEE Photonics Technology Letters, 8*(11), 1549–1552. doi:10.1109/68.541579

Giles, C. R., Zirngibl, M., & Joyner, C. (1997). 1152-subscriber WDM access PON architecture using a sequentially pulsed multifrequency laser. *IEEE Photonics Technology Letters, 9*(9), 1283–1284. doi:10.1109/68.618505

Gillespie, A. (2001). *Broadband Access Technology, Interfaces, and Management.* Norwood, Ohio: Artech House, Inc.

Gillespie, A., Orth, B., Profumo, A., & Webster, S. (1997). Evolving Access Networks: A European Perspective. *IEEE Communications Magazine, 35*(3), 47–54. doi:10.1109/35.581306

Girard, A. (2005). *FTTx PON Technology and Testing.* Tukwila, WA: EXFO Electro-Optical Engineering Inc.

Gislason, H. (2004). Reykjavik: Fiber to Every Home. *BYTE.com.* United Businee Media.

Glance, B., Kamoniw, I. P., & Wilson, R. W. (1994). Applications of the Integrated Waveguide Grating Router. *Journal of Lightwave Technology, 12*(6), 957–962. doi:10.1109/50.296184

Glance, B., Liou, K. Y., Koren, U., Burrows, E. C., Raybon, G., & Burrus, C. A. (1996). A single-fiber WDM local access network based on amplified LED transceivers. *IEEE Photonics Technology Letters, 8*(9), 1241–1242. doi:10.1109/68.531849

Green, P. E. (2004). Fiber to the home: the next big broadband thing. *IEEE Communications Magazine, 42*(9), 100–106. doi:10.1109/MCOM.2004.1336726

Green, P. E. (2005). *Fiber to the Home, The New Empowerment.* Hoboken, New Jersey: John Wiley & Sons.

Gumaste, A., & Anthony, T. (2004). *First Mile Access Networks and Enabling Technologies.* Indianapolis, IN: Cisco Press.

Gumaste, A., & Antony, T. (2002). *DWDM Network Designs and Engineering Solutions.* Indianapolis, USA: Cisco Press.

Gysel, P., & Staubli, R. K. (1990). Statistical properties of Rayleigh backscattering in single-mode fibers.

Journal of Lightwave Technology, 8(4), 561–567. doi:10.1109/50.50762

Hamad, A., Wu, T., Kamal, A. E., & Somani, A. K. (2006). On Multicasting in Wavelength-Routing Mesh Networks. *Computer Networks, 50*(1), 3105–3164. doi:10.1016/j.comnet.2005.12.012

Han, K. H., Kim, H., & Chung, Y. C. (2001). Active Multi-Purpose Fiber-Optic Access Network. In *Asia-Pacific Optical and Wireless Communications Conference* (pp. 31-37). Beijing, China.

Han, K. H., Son, E. S., Lim, K. W., Choi, H. Y., & Chung, Y. C. (2004). Bi-directional WDM PON using light-emitting diodes spectrum-sliced with cyclic arrayed-waveguide grating. *IEEE Photonics Technology Letters, 16*(10), 2380–2382. doi:10.1109/LPT.2004.833865

Han, S., & Yue, W. (2006). *Next-Generation Packet-Based Transport Networks Economic Study.* OFC/NFOEC 2006 Conference Proceedings.

Han, S., Yue, W., & Smith, S. (2006). *FTTx and xDSL: A Business Case Study of GPON versus Copper for Broadband Access Networks.* 2006 FTTH Conference Proceedings.

Hann, S., Kim, T. Y., & Park, C. S. (2005, September). *Direct-modulated upstream signal transmission using a self-injection locked F-P LD for WDM-PON.* Paper presented at European Conference and Exhibition on Optical Communication (ECOC), (We3.3.3), Glasgow, UK.

Hayat, A., Bacou, A., Rissons, A., Mollier, J.-C., Iakovlev, V., & Sirbu, A. (2008, November). *1.3 µm Single-Mode VCSEL-by-VCSEL Optically Injection-Locking for Enhanced Microwave Performance.* Paper Presented at IEEE LEOS Conference, Newport, CA.

Hayat, A., Varon, M., Bacou, A., Rissons, A., & Mollier, J.-C. (2008, September). *2.49 GHz Low Phase-Noise Optoelectronic Oscillator using 1.55 µm VCSEL for Avionics and Aerospace Application.* Presented at the Conference Microwave Photonics 08, Gold Coast, Australia.

He, J.Y., Gary, Chan, S.H., & Danny Tsang, H.K. (2002). Multicasting in WDM Networks. *IEEE Communications Surveys & Tutorials, 4*(1), 2-20.

Healey, P., Townsend, P., Ford, C., Johnston, L., Townley, P., & Lealman, I. (2001). Spectral slicing WDM-PON using wavelength-seeded reflective SOAs. *IEE Electron. Lett.*, *37*(19), 1181–1182. doi:10.1049/el:20010786

Healey, P., Townsend, P., Ford, C., Johnston, L., Townley, P., & Lealman, I. (2001). Spectral slicing WDM-PON using wavelength-seeded reflective SOAs. *IEE Electronics Letters*, *37*(19), 1181–1182. doi:10.1049/el:20010786

Hecht, J. (2005). *Understanding Fiber Optics*. Columbus, Ohio: Prentice Hall.

Heismann, F., Gray, D. A., Lee, B. H., & Smith, R. W. (1994). Electrooptic polarization scramblers for optically amplified long-haul transmission systems. *IEEE Photonics Technology Letters*, *6*(9), 1156–1158. doi:10.1109/68.324697

Hellberg, C., Greene, D., & Boyes, T. (2007). *Broadband Network Architectures: Designing and Deploying Triple Play Services*. Prentice-Hall.

Hinderthur, H., & Friedric, L. (2003). *WDM hybrid transmission based on CWDM plus DWDM*. Lightwave Europe.

Hirosaki, B., Emura, K., Hayano, S., & Tsutsumi, H. (2003). Next-Generation Optical Networks as a Value Creation Platform. *IEEE Communications Magazine*, *41*(9), 65–71. doi:10.1109/MCOM.2003.1232238

Hofmann, W., Wong, E., Böhm, G., Ortsiefer, M., Zhu, N. H., & Amann, M. C. (2008). 1.55-mm VCSEL arrays for high-bandwidth WDM-PONs. *IEEE Photonics Technology Letters*, *20*(4), 291–293. doi:10.1109/LPT.2007.915631

Hogan & Hartson LLP, & Analysys Consulting Ltd. (2006). *Preparing the next steps in regulation of electronic communications. A contribution to the review of the electronic communications regulatory framework. Final Report for the European Commission, July 2006 - Service Contract No.05/48622*. Brussels, Belgium: European Commission.

Holloway, W. T., Keating, A. J., & Sampson, D. D. (1997). Multiwavelength source for spectrum-sliced WDM access networks and LANs. *IEEE Photonics Technology Letters*, *9*(7), 1014–1016. doi:10.1109/68.593384

Hong, J., Kim, H., & Makino, T. (1998). Enhanced wavelength tuning range in two-section complex-coupled DFB lasers by alternating gain and loss coupling. *IEEE/OSA . Journal of Lightwave Technology*, *16*(7), 1323–1328. doi:10.1109/50.701412

Horrocks, J., & Scarr, R. (1994). *The Technology Guide to Telecommunications*. Surrey, UK.

Hossain, A. D., Dorsinville, R., Ali, M. A., Shami, A., & Assi, C. (2006). Ring-based Local Access PON Architecture for Supporting Private Networking Capability. *Journal of Optical Networking*, *5*, 26–39. doi:10.1364/JON.5.000026

Hossain, A. D., Erkan, H., Hadjiantonis, A., Dorsinville, R., Ellinas, G., & Ali, M. A. (2008). Survivable Broadband Local Access PON Architecture: A New Direction for Supporting Simple and Efficient Resilience Capabilities. *Journal of Optical Networking*, *7*, 645–661. doi:10.1364/JON.7.000645

Houbby, R. (2006, September 24-25). FTTH: Where, When, Why and How Much? In *Proceedings of Optical Network Europe (ONE) 2006 - Vision Strategies and Execution*. Cannes, France.

Hsueh, Y. L., Rogge, M. S., Shaw, W. T., Kazovsky, L. G., & Yamamoto, S. (2004). SUCCESS-DWA: Highly scalable and cost effective optical access network architecture. *IEEE Optical Communication Magazine*, *42*(8), S24–S30. doi:10.1109/MCOM.2004.1321383

Hubbard, S. (2008, February). Carrier Ethernet Services: The View From the Enterprise. *Heavy Reading*, *6*(2).

Hudiksvallsbostäder. (2007). from http://www.hudiksvallsbostader.se/

Hung, W., Chan, C. K., Chen, L. K., & Lin, C. L. (September, 2003). System characterization of a robust remodulation scheme with DPSK downstream traffic in a WDM access network. Paper presented at *European Conference on Optical Communications (ECOC)*, Paper We3.4.5, Rimini, Italy.

Hung, W., Chan, C. K., Chen, L. K., & Tong, F. (2003). An optical network unit for WDM access networks with downstream DPSK and upstream remodulated OOK data using injection-locked FP laser. *IEEE Photonics Technology Letters*, *15*(10), 1476–1478. doi:10.1109/LPT.2003.818055

Iannone, P. P., Frigo, N. J., & Darcie, T. E. (March 1995). WDM passive-optical-network architecture with bidirectional optical spectral slicing. Paper presented at *Optical Fiber Communication Conference (OFC)*, Paper TuK2, Anaheim, California, USA.

Iannone, P. P., Reichmann, K. C., Smiljanic, A., Frigo, N. J., Gnauck, A. H., Spiekman, L. H., & Derosier, R. M. (2000). A transparent WDM network featuring shared virtual rings. *IEEE/OSA . Journal of Lightwave Technology*, *18*(12), 1955–1963. doi:10.1109/50.908802

Iannone, P. P., Reicjmann, K. C., & Frigo, N. J. (1998). High-Speed Point-to-Point and Multiple Broadcast Services Delivered over a WDM Passive Optical Network. *IEEE Photonics Technology Letters*, *10*(9), 1328–1330. doi:10.1109/68.705632

Ibaraki, A., Kawashima, K., Furusawa, K., Ishikawa, T., Yamaguchi, T., & Niina, T. (1989). Buried Heterostructure GaAs/GaAlAs Distributed Bragg Reflector Surface Emitting Laser with Very Low Threshold (5.2 mA) under Room Temperature CW Conditions. *Japanese Journal of Applied Physics*, *28*(4), 667–668. doi:10.1143/JJAP.28.L667

IEEE. 802.3ah Ethernet in the First Mile Task Force, (2004). IEEE 802.3ah Standard. Retrieved November 20, 2007, from http://www.ieee802.org/3/efm/public.

IEEE802.3 (2005). Carrier Sense Multiple Access with Collision Detection (CSMA/CD) access method and physical layer specifications.

Iga, K. (2000). Surface-Emitting Laser – Its Birth and Generation of new Optoelectronics Field. *IEEE Journal on Selected Topics in Quantum Electronics*, *6*(6), 1201–1215. doi:10.1109/2944.902168

Iga, K., Koyama, F., & Kinoshita, S. (1988). Surface Emitting Semiconductor Lasers. *IEEE journals of Quantum Electronics, 24*(9), 1845-1855.

Ims, L. A. (1998). *Broadband Access Networks Introduction strategies and techno-economic evaluation, Telecommunications Technology and Applications Series*. Chapman & Hall.

InfoSoc. (2005). *Information Society*. from http://europa.eu/pol/infso/index_el.htm

Inoue, Y., Himeno, A., Moriwaki, K., & Kawachi, M. (1995). Silica-based arrayed-waveguide grating circuit as optical splitter/router. *IEE Electronics Letters, 31*(9), 726–727. doi:10.1049/el:19950497

Inoue, Y., Kaneko, A., Hanawa, F., Takahashi, H., Hattori, K., & Sumida, S. (1997). Athermal silica-based arrayed-waveguide grating multiplexer. *IEE Electronics Letters, 33*(23), 1945–1946. doi:10.1049/el:19971317

Institut de l'Audiovisuel et des Télécommincations en Europe (IDATE) (2006, August). *FTTH Deployment - When and why?* Montpellier, France: IDATE.

Institute of Electrical and Electronic Engineers (IEEE). (2004). *IEEE 802.3ah Standard Ethernet in the First Mile*. New York: IEEE.

Institute of Electrical and Electronic Engineers (IEEE). (2005). *IEEE Standard 802.16e-2005. Mobile 802.16*. New York: IEEE.

Institute of Electrical and Electronic Engineers. (*IEEE*) (2004). *IEEE 802.3ah - Ethernet in the First Mile Task Force archives*. New York: IEEE.

International Telecommunication Union - Telecommunications Standardization Sector (ITU-T). (1999a). *ITU-T Recommendation G.992.1: Asymmetric digital subscriber line (ADSL) transceivers*. Geneva, Switzerland: ITU-T.

International Telecommunication Union - Telecommunications Standardization Sector (ITU-T). (1999b). *ITU-T Recommendation G.992.2: Splitterless asymmetric digital subscriber line (ADSL) transceivers*. Geneva, Switzerland: ITU-T.

International Telecommunication Union - Telecommunications Standardization Sector (ITU-T). (2001). ITU-T Recommendation G.983.3: *A Broadband Optical Access System with Increased Service Capability by Wavelength Allocation.* Geneva, Switzerland: ITU-T.

International Telecommunication Union - Telecommunications Standardization Sector (ITU-T). (2002). *ITU-T Recommendation J.122: Second-generation transmission systems for interactive cable television services - IP cable modems.* Geneva, Switzerland: ITU-T.

International Telecommunication Union - Telecommunications Standardization Sector (ITU-T). (2003). *ITU-T Recommendation G.984.1: Gigabit-capable Passive Optical Networks (GPON): General characteristics.* Geneva, Switzerland: ITU-T.

International Telecommunication Union - Telecommunications Standardization Sector (ITU-T). (2003). *ITU-T Recommendation G.992.5: Asymmetric Digital Subscriber Line (ADSL) transceivers - Extended bandwidth ADSL2 (ADSL2plus).* Geneva, Switzerland: ITU-T.

International Telecommunication Union - Telecommunications Standardization Sector (ITU-T). (2004). *ITU-T Recommendation J.112 Annex B: Data-over-cable service interface specifications: Radio-frequency interface specification.* Geneva, Switzerland: ITU.

International Telecommunication Union - Telecommunications Standardization Sector (ITU-T). (2005). *ITU-T Recommendation G.983.1: Broadband optical access systems based on Passive Optical Networks (PON).* Geneva, Switzerland: ITU-T.

International Telecommunication Union - Telecommunications Standardization Sector (ITU-T). (2006). *ITU-T Recommendation G.993.2: Very high speed digital subscriber line transceivers 2 (VDSL2).* Geneva, Switzerland: ITU.

International Telecommunication Union (ITU-T). (2003). Gigabit-capable Passive Optical Networks (GPON): General characteristics, in Series G: Transmission systems and media, Digital sections and digital line systems-optical line systems for local and access networks, Telecommunication Standardization Sector of ITU. Retrieved February 10, 2008, from http://www.itu.int/rec/T-REC-G.984.1-200303-S /en.

International Telecommunication Union (ITU-T). (2005). Recommendation G.983.1, Broadband optical access systems based on Passive Optical Networks (PON), in Series G: Transmission systems and networks, Digital sections and digital line systems-optical line systems for local and access networks, Telecommunication Standardization Sector of ITU. Retrieved January 15, 2008, from http://www.itu.int/rec/T-REC-G.983.1-200501-I /en.

International Telecommunications Union - Telecommunication Standardization Sector (ITU-T). (2005). *ITU-T Recommendation G.983.2: ONT management and control interface specification for B-PON (with amendments 1 and 2, erratum 1 and an implementer's guide).* Geneva, Switzerland: ITU-T

International Telecommunications Union – Telecommunication Standardization Sector (ITU-T). (2002). *ITU-T Recommendation G.983.5: A broadband optical access system with enhanced survivability.* Geneva, Switzerland: ITU-T.

International Telecommunications Union – Telecommunication Standardization Sector (ITU-T). (2003). *ITU-T Recommendation G.984: Gigabit-capable Passive Optical Networks (GPON).* Geneva, Switzerland: ITU-T.

International Telecommunications Union – Telecommunication Standardization Sector (ITU-T). (2005). *ITU-T Recommendation* G.983.1, Broadband optical access systems based on passive optical networks (PON). Geneva, Switzerland: ITU-T.

Intracom, S. A. (1998). *The Fiber Access System IAS-F, Technical Description.* Athens, Greece.

Ireland's NDP. (2006). Ireland's National Development Plan. *European Union Structural Funds.* from http://www.ndp.ie/viewdoc.asp?fn=/documents/homepage.asp

Ireland's NDP. (2007). *Irelands' Broadband Strategy.* from http://www.ndp.ie/viewdoc.asp?fn=%2Fdocuments%2FNDP2007-2013%2Foverview.htm

ITU-D. (2008). *ICT statistics.* http://www.itu.int/ITU-D/ict/statistics/

ITU-T FG ICT & CC. (2008). *Focus Group on ICTs and Climate Change (FG ICT&CC).* http://www.itu.int/ITU-T/focusgroups/climate/index.html

ITU-T G.694.1 (2002). *Spectral grids for WDM applications: DWDM frequency grid.* Telecommunication Standardization Sector of ITU.

ITU-T G.694.2 (2002). *Spectral grids for WDM applications: CWDM frequency grid.* Telecommunication Standardization Sector of ITU.

ITU-T G.7041 (2003). *Generic Framing Procedure (GFP), in Series G: Transmission Systems and Media, Digital Systems and Networks.* Telecommunication Standardization Sector of ITU.

ITU-T G.983 (2001). *ITU-T Recommendation G.983 series.* http://www.itu.int/publications/sector.aspx?lang=en&sector=2

ITU-T G.983.4 (2001). *A broadband optical access system with increased service capability using dynamic bandwidth assignment.* ITU-T Recommendation G.983.4.

ITU-T G.984 (2002). *ITU-T Recommendation G.984 series.* http://www.itu.int/publications/sector.aspx?lang=en&sector=2

ITU-T G.984.1 (2003). *General characteristics for Gigabit-capable Passive Optical Networks (GPON).* Telecommunication Standardization Sector of ITU.

ITU-T G.984.2. (2003). *Gigabit-capable passive optical networks (GPON): Physical media dependent (PMD) layer specification.* Telecommunication Standardization Sector of ITU.

ITU-T G.984.3. (2003). *Transmission Convergence Layer for Gigabit Passive Optical Networks.* Telecommunication Standardization Sector of ITU.

ITU-T G.hn (2008). *Generic Home Network Architecture and Transceiver.* ITU-T Draft Recommendation.

ITU-T NGN-GSI. (2008). http://www.itu.int/ITU-T/ngn/

ITU-T SG-16. (2002). Full-Service VDSL Focus Group Technical Specification 1 *Part 1: Operator requirements* (p.33)

ITU-T SG-16. (2002). Full-Service VDSL Focus Group Technical Specification 1 *Part 3: Customer Premises Equipment specification.* (p. 49)

ITU-T Y.2001 (2004). *General overview of NGN.* ITU-T Recommendation Y.2001.

ITU-T Y.2011 (2004). *General principles and general reference model for Next Generation Networks.* ITU-T Recommendation Y.2011.

Jang, S., Lee, C. S., Seol, D. M., Jung, E. S., & Kim, B. W. (March, 2007). A bidirectional RSOA based WDM-PON utilizing a SCM signal down-link and a baseband signal for up-link. Paper presented at *IEEE/OSA Optical Fiber Communication Conference / National Fiber Optic Engineers Conference (OFC/NFOEC)*, Paper JThA78, Anaheim, California, USA.

Jang, Y. S., Lee, C. H., & Chung, Y. C. (1999). Effects of crosstalk in WDM system using spectrum-sliced light sources. *IEEE Photonics Technology Letters, 11*(6), 715–717. doi:10.1109/68.766795

Jewell, J. L., Scherer, A., McCall, S. L., Lee, Y. H., Walker, S., Harbison, J. P., & Florez, L. T. (1989). Low threshold electrically pumped vertical-cavity surface-emitting microlaser. *Electronics Letters, 25*(17), 1123–1124. doi:10.1049/el:19890754

Ji, H. C., Yamashita, I., & Kitayama, K. I. (2008). Cost-effective colorless WDM-PON delivering up/down-stream data and broadcast services on a single wavelength using mutually injected Fabry-Perot laser diodes. *Optics Express, 16*(7), 4520–4528. doi:10.1364/OE.16.004520

Jia, Z., Yu, J., Boivin, D., Haris, M., & Chang, G. K. (2007). Bidirectional ROF links using optically up-converted DPSK for downstream and remodulated OOK for upstream. *IEEE Photonics Technology Letters, 19*(9), 653–655. doi:10.1109/LPT.2007.894961

Jia, Z., Yu, J., Yeo, Y. K., Wang, T., & Chang, G. K. (2006, September). *Design and implementation of a low cost, integrated platform for delivering super-broadband dual services simultaneously.* Paper presented at European Conference and Exhibition on Optical Communication (ECOC), (Tu1.6.6), Cannes, France.

Jiang, S., Bristiel, B., Jaouen, Y., Gallion, P., & Pincemin, E. (2007). Bit-error-rate evaluation of the distributed Raman amplified transmission systems in the presence of double Rayleigh backscattering noise . *IEEE Photonics Technology Letters, 19*(7), 468–470. doi:10.1109/LPT.2007.893027

Jiménez, T., Merayo, N., Durán, R. J., Fernández, P., & Lorenzo, R. M. (2008). Adaptive allocation algorithm to support Multi-Service Level Agreements in a Long-Reach EPON. *Proceedings of the 13rd European Conference and Optical Communications* (pp. 101-107). Krems (Austria).

Jones, R. (2007, February). *Opportunities in Fibre to the Home (FTTH) and How to Make a First Assessment.* London, UK: Ventura Team LLP. Retrieved on December 06, 2008 from: http://www.localret.es/localretnews/bandaampla/num18/docs/4num18.pdf.

Jue, J. P., & Mukherjee, B. (1997). The Advantages of Partitioning Multicast Transmissions in a Single-Hop Optical WDM Network. In *ICC'97, 1*, 427-431.

Jung, D. K., Shin, S. K., Lee, C. H., & Chung, Y. C. (1998). Wavelength-division-multiplexed passive optical network based on spectrum-slicing techniques. *IEEE Photonics Technology Letters, 10*(9), 1334–1336. doi:10.1109/68.705634

Kanbach, A., & Körber, A. (1999). *ISDN – Die Technik, Schnittstellen – Protokolle – Dienste – Endsysteme.* Heidelberg, Germany: Hüthig Verlag.

Kaneko, A., Kamei, S., Inoue, Y., Takahashi, H., & Sugita, A. (2000). Athermal silica-based arrayed-waveguide grating (AWG) multi/demultiplexers with new low loss groove design. *IEE Electronics Letters, 36*(4), 318–319. doi:10.1049/el:20000261

Kang, J. M., & Han, S. K. (2006). A novel hybrid WDM/SCM-PON sharing wavelength for up- and downlink using reflective semiconductor optical amplifier. *IEEE Photonics Technology Letters, 18*(3), 502–504. doi:10.1109/LPT.2005.863632

Kang, J. M., Kim, T. Y., Choi, I. H., Lee, S. H., & Han, S. K. (2007). Self-seeded reflective semiconductor optical amplifier based optical transmitter for upstream WDM-PON link. *IET OptoElectronics, 1*(2), 77–81. doi:10.1049/iet-opt:20050116

Kani, J., Teshima, M., Akimoto, K., Takachio, N., Suzuki, H., Iwatsuki, K., & Ishii, M. (2003). A WDM based Optical Access Network for Wide-Area Gigabit Access Services. *IEEE Communications Magazine, 41*(2), 43–48. doi:10.1109/MCOM.2003.1179497

Katsianis, D., Welling, I., Ylonen, M., Varoutas, D., Sphicopoulos, T., & Elnegaard, N. K. (2001). The financial perspective of the mobile networks in Europe. *IEEE Pers. Comm. Mag., 8*(6), 58–64. doi:10.1109/98.972169

Kawamura, R. (2007). Network Service Aggregation Platform based on OSGi technologies. *Access '07 Executive Business Forum associated with IEEE Globecom '07, Next Generation Home Network and Home Gateway Session.*

Keil, N., Yao, H. H., Zawadzki, C., Bauer, J., Bauer, M., Dreyer, C., & Schneider, J. (2001). Athermal all-polymer arrayed-waveguide grating multiplexer. *IEE Electronics Letters, 37*(9), 579–580. doi:10.1049/el:20010406

Keiser, G. (2003). *Optical Communications Essentials.* New York, USA: McGraw-Hill Professional.

Keiser, G. (2006). *FTTX Concepts and Applications.* New Jersey: Wiley Interscience.

Kettler, D., Kafka, H., & Spears, D. (2000). Driving Fiber to the Home. *IEEE Communications Magazine, 38*(11), 106–110. doi:10.1109/35.883497

Killat, U. (1996). *Access to B-ISDN via PONs, ATM Communication in Practice.* Chichester, England: John Wiley and Sons Ltd.

Kim, B. K., Park, H., Park, S. J., Yoon, B. Y., & Kim, B. T. (September, 2006). WDM passive optical networks with symmetric up/down data rates using Manchester coding based re-modulation. Paper presented at *European Conference on Optical Communication (ECOC)*, Paper Tu4.5.5, Cannes, France.

Kim, H. D., Kang, S. G., & Lee, C. H. (2000). A low-cost WDM source with an ASE injected Fabry-Pérot semiconductor laser. *IEEE Photonics Technology Letters*, *12*(8), 1067–1069. doi:10.1109/68.868010

Kim, J., Choi, J. Y., Im, J., Kang, M., & Kevin Rhee, J. K. (2007). Novel Passive Optical Switching Using Shared Electrical Buffer and Wavelength Converter. In *ONDM* (pp. 101-106).

Kim, K. S., Gutierrez, D., An, F. T., & Kazovsky, L. G. (2005). Design and performance analysis of scheduling algorithms for WDM-PON under SUCCESS-HPON architecture. *IEEE/OSA . Journal of Lightwave Technology*, *23*(11), 3716–3731. doi:10.1109/JLT.2005.857729

Kim, S. J., Han, J. H., & Lee, J. S., Park, & C. S. (1999). Intensity noise suppression in spectrum-sliced incoherent light communication systems using a gain-saturated semiconductor optical amplifier. *IEEE Photonics Technology Letters*, *11*(8), 967–1044. doi:10.1109/68.775315

Kim, S. Y., Jun, S. B., Takushima, Y., Son, E. S., & Chung, Y. C. (2007). Enhanced performance of RSOA-based WDM PON by using Manchester coding. *OSA Journal of Optical Networking*, *6*(6), 624–630. doi:10.1364/JON.6.000624

Kim, S. Y., Son, E. S., Jun, S. B., & Chung, Y. C. (March, 2007). Effects of downstream modulation formats on the performance of bidirectional WDM-PON using RSOA. Paper presented at *IEEE/OSA Optical Fiber Communication Conference / National Fiber Optic Engineers Conference (OFC/NFOEC)*, Paper OWD3, Anaheim, California, USA.

Kinoshita, T., Okayasu, M., & Shibata, N. (July 1997). Stable operation condition of optical WDM PDS system employing a rapidly tunable laser diode and wavelength router. Paper presented at *Optoelectronics and Commu-*

nications Conference (OECC), Paper 10A1–5. 368–369. Seoul, Korea.

Kitayama, K., Wang, X., & Wada, N. (2006). OCDMA over WDM PON-solution path to gigabit-symmetric FTTH. *IEEE/OSA . Journal of Lightwave Technology*, *24*(4), 1654–1662. doi:10.1109/JLT.2006.871030

Ko, K.-T., & Davis, B. R. (1984). Delay Analysis for a TDMA Channel with contiguous output and Poisson Message Arrival. *IEEE Transactions on Communications*, *32*(6), 707–709. doi:10.1109/TCOM.1984.1096126

Kopf, R. F., Schubert, E. F., Downey, S. W., & Emerson, A. B. (1992). N- and P- type doping profiles in Distributed Bragg Reflector Structures and their effect on resistance. *Applied Physics Letters*, *61*(15), 1820–1822. doi:10.1063/1.108385

Kouroger, M., Imai, K., Widyatmoko, B., Shimizu, T., & Ohtsu, M. (2000). Continuous tuning of an electrically tunable external cavity semiconductor laser. *OSA Optics Letters*, *25*(16), 1165–1167. doi:10.1364/OL.25.001165

Koyama, F. (2006). Recent advances of VCSEL Photonics. *IEEE Journal Of Lightwave Technology*, *24*(12), 4502–4513. doi:10.1109/JLT.2006.886064

Kramer, G. (2005). *Ethernet Passive Optical Networks*. New York, USA: McGraw-Hill Professional Publishing.

Kramer, G., & Pesavento, G. (2002). Ethernet passive optical network (EPON): building a next-generation optical access network. *IEEE Communications Magazine*, *40*(2), 66–73. doi:10.1109/35.983910

Kramer, G., Mukherjee, B., & Maislos, A. (2003). Ethernet Passive Optical Networks. In Sudhir Dixit (Ed.), *Multiprotocol over DWDM: Building the Next Generation Optical Internet* (pp. 229-275). John Wiley & Sons.

Kramer, G., Mukherjee, B., & Perawnto, G. (2001). Ethernet PON (ePON): design and analysis of an optical access network. *Photonic Network Communications*, *3*(3), 307–319. doi:10.1023/A:1011463617631

Kramer, G., Mukherjee, B., & Pesavento, G. (2002). Interleaved Polling with Adaptive Cycle Time (IPACT): A

Dynamic Bandwidth Distribution Scheme in an Optical Access Network. *Photonic Network Communications, 4*(1), 89–107. doi:10.1023/A:1012959023043

Kramer, G., Mukherjee, B., & Pesavento, G. (2002). IPACT a dynamic protocol for an Ethernet PON (EPON). *IEEE Communications Magazine, 40*(2), 74–80. doi:10.1109/35.983911

Kramer, G., Mukherjee, B., Dixit, S., Ye, Y., & Hirth, R. (2002). Supporting differentiated classes of service in Ethernet passive optical networks. *OSA Journal of Optical Networking, 1*(8/9), 280–298.

Kramer, G., Mukherjee, B., Ye, Y., Dixit, S., & Hirth, R. (2002). Supporting differentiated classes of service in Ethernet passive optical networks. *Journal of Optical Networking, 1*(8), 280–298.

Krishna, M. S., & Subramaniam, S. (2000). *Optical WDM Networks Principles and Practice*. Norwell, Massachusetts: Kluwer Academic Publishers.

Kuznetsov, M., Froberg, N. M., Henion, S. R., Rao, H. G., Korn, J., & Rauschenbach, K. A. (2000). A Next-Generation Optical Regional Access Network. *IEEE Communications Magazine, 38*, 66–72. doi:10.1109/35.815454

Kwon, H. C., Won, Y. Y., & Han, S. K. (2006). A self-seeded reflective SOA-based optical network unit for optical beat interference robust WDM/SCM-PON link. *IEEE Photonics Technology Letters, 18*(17), 1852–1854. doi:10.1109/LPT.2006.881212

Leclerc, O., Brindel, P., Rouvillain, D., Pincemin, E., Dany, B., & Desurvire, E. (1999). 40Gbit/s polarization-insensitive and wavelength-independent InP Mach-Zehnder modulator for all-optical regeneration. *Electronics Letters, 35*(9), 730–732. doi:10.1049/el:19990504

Lee, C.-H., Sorin, V., & Yoon Kim, B. (2006). Fiber to the Home Using a PON Infraestructure. *Journal of Lightwave Technology, 24*(12), 4568–4583. doi:10.1109/JLT.2006.885779

Lee, J. H. Lee, K., & Kim., C. H. (2007, September). *Continuous-wave supercontinuum-based bidirectional long reach WDM-POM incorporating FP-LD-based OLT and RSOA-based ONUs.* Paper presented at European Conference and Exhibition on Optical Communication (ECOC), (Paper 6.4.5), Berlin, Germany.

Lee, J. H., Kim, C. H., Han, Y. G., & Lee, S. B. (2006). Broadband, high power, erbium fibre ASE-based CW supercontinuum source for spectrum-sliced WDM PON applications. *IEE Electronics Letters, 42*(9), 67–68. doi:10.1049/el:20060713

Lee, J. H., Kim, C. H., Han, Y. G., & Lee, S. B. (2006). WDM PON upstream transmission at 1.25 Gb/s using Fabry–Pérot laser diodes injected with spectrum-sliced, depolarized, CW supercontinuum source. *IEEE Photonics Technology Letters, 18*(20), 2108–2110. doi:10.1109/LPT.2006.883288

Lee, J. H., Kim, C. H., Han, Y. G., & Lee, S. B. (2006, September). *1.25 Gbit/s WDM PON upstream transmission using Fabry-Perot laser diodes injected by depolarised CW supercontinuum source.* Paper presented at European Conference and Exhibition on Optical Communication (ECOC), (Tu3.5.5), Cannes, France.

Lee, J. H., Lee, K., Han, Y. G., Lee, S. B., & Kim, C. H. (2007). Single, depolarized, CW supercontinuum-based wavelength-division-multiplexed passive optical network architecture with C-Band OLT, L-Band ONU, and U-Band monitoring. *IEEE/OSA . Journal of Lightwave Technology, 25*(10), 2891–2897. doi:10.1109/JLT.2007.903637

Lee, J. S., Chung, Y. C., & DiGiovanni, D. J. (1993). Spectrum-sliced fiber amplifier light source for multichannel WDM applications. *IEEE Photonics Technology Letters, 5*(12), 1458–1461. doi:10.1109/68.262573

Lee, K. L., & Wong, E. (2006, September). *Directly-modulated self-seeding reflective SOAs in WDM-PONs: performance dependence on seeding power and modulation effects.* Paper presented at European Conference and Exhibition on Optical Communication (ECOC), (Tu4.5.2), Cannes, France.

Lee, S. M., Choi, K. M., Mun, S. G., Moon, J. H., & Lee, C. H. (2005). Dense WDM-PON based on wavelength-locked Fabry–Pérot laser diodes. *IEEE Photonics Technology Letters, 17*(7), 1579–1581. doi:10.1109/LPT.2005.848558

Lee, W., Cho, S. H., Park, M. Y., Lee, J. H., Kim, C., Jeong, G., & Kim, B. W. (2006). Wavelength filter detuning for improved carrier reuse in loop-back WDM-PON. *IEE Electronics Letters*, *42*(10), 596–597. doi:10.1049/el:20060289

Lee, W., Park, M. Y., Cho, S. H., Lee, J., Kim, C., Jeong, G., & Kim, B. W. (2005). Bidirectional WDM-PON based on gain-saturated reflective semiconductor optical amplifiers. *IEEE Photonics Technology Letters*, *17*(11), 2289–2462. doi:10.1109/LPT.2005.858153

Lethien, C., Loyez, C., & Vilcot, J.-P. (2005). Potentials of Radio over Multimode Fiber Systems for the In-Buildings Coverage of Mobile and Wireless LAN Applications. *IEEE Photonics Technology Letters*, *17*(12), 2793–2795. doi:10.1109/LPT.2005.859533

Li, Z. H., Dong, Y., Wang, Y., & Lu, C. (2005). A novel PSK-Manchester modulation format in 10-Gb/s passive optical network system with high tolerance to beat interference noise . *IEEE Photonics Technology Letters*, *17*(5), 1118–1120. doi:10.1109/LPT.2005.845663

Li, Z., Dong, Y., Wang, Y., & Lu, C. (2005). A novel PSK-Manchester modulation format in 10-Gb/s passive optical network system with high tolerance to beat interference noise. *IEEE Photonics Technology Letters*, *17*(5), 1118–1120. doi:10.1109/LPT.2005.845663

Lijbrandt telecom. (2007). from http://www.lijbrandt-telecom.nl/

Lin, C. (2004). *Optical Components for Communications*. Norwell, Massachusetts: Kluwer Academic.

Lin, C. (2006). *Broadband Optical Access Networks and Fiber-to-the-Home. Systems Technologies and Deployment Strategies*. John Wiley & Sons, Ltd.

Lin, C. T., Chen, J., Peng, P. C., Peng, C. F., Peng, W. R., Chiou, B. S., & Chi, S. (2007). Hybrid optical access network integrating fiber-to-the-home and radio-over-fiber systems. *IEEE Photonics Technology Letters*, *19*(8), 610–612. doi:10.1109/LPT.2007.894326

Lin, C. T., Peng, W. R., Peng, P. C., Chen, J., Peng, C. F., Chiou, B. S., & Chi, S. (2006). Simultaneous generation

of baseband and radio signals using only one single electrode Mach-Zehnder modulator with enhanced linearity. *IEEE Photonics Technology Letters*, *18*(23), 2481–2483. doi:10.1109/LPT.2006.887233

Lin, S. C., Lee, S. L., & Liu, C. K. (2008). Simple approach for bidirectional performance enhancement on WDM-PONs with direct modulation lasers and RSOAs. *OSA Optics Express*, *16*(6), 3636–3643. doi:10.1364/OE.16.003636

Liou, K. Y., Koren, U., Burrows, E. C., Zyskind, J. L., & Dreyer, K. (1997). A WDM access system architecture based on spectral slicing of an amplified LED and delay-line multiplexing and encoding of eight wavelength channels for 64 subscribers. *IEEE Photonics Technology Letters*, *9*(4), 517–519. doi:10.1109/68.559407

Liou, K. Y., Koren, U., Dreyer, K., Burrows, E. C., Zyskind, J. L., & Sulhoff, J. W. (1998). A 24-channel WDM transmitter for access networks using a loop-back spectrally-sliced light-emitting diode. *IEEE Photonics Technology Letters*, *10*(2), 270–272. doi:10.1109/68.655381

Lo, Y. L., & Kuo, C. P. (2003). Packaging a fiber Bragg grating with metal coating for an athermal design. *IEEE/OSA .Journal of Lightwave Technology*, *21*(5), 1377–1383. doi:10.1109/JLT.2003.810925

Lowery, A., & Armstrong, J. Orthogonal-frequency-division multiplexing for dispersion compensation of long-haul optical systems. *Optics Express*, *14*(6), 2079–2084.

Lu, G. W., Deng, N., Chan, C. K., & Chen, L. K. (2005, March). *Use of downstream IRZ signal for upstream data re-modulation in a WDM passive network*. Paper presented at Optical Fiber Communication Conference (OFC), (OFI8), Anaheim, California. USA.

Lu, H. H., Ma, H. L., Chuang, Y. W., Chi, Y. C., Liao, C. W., & Peng, H. C. (2007). Employing injection-locked Fabry–Pérot laser diodes to improve bidirectional WDM-PON performances. *Optics Communications*, *270*(2), 211–216. doi:10.1016/j.optcom.2006.09.054

Lu, H.-H., Tzeng, S.-J., Chuang, Y.-W., Chi, Y.-C., & Liao, C.-W. (2007). Bidirectional Radio-over DWDM

Transport Systems Based on Injection-Locked VCSELs and Optoelectronic Feedback Techniques. *IEEE Photonics Technology Letters, 19*(5), 315–317. doi:10.1109/LPT.2007.891627

Luna-Reyes Luis, F., & Andersen, D. F. (2003). Collecting and analyzing qualitative data for systems dynamics: methods and models. *System Dynamics Review, 19*, 271–296. doi:10.1002/sdr.280

Lund, B. (2001). Fiber-to-the-Home Network Architecture: A Comparison of PON and Point-to-Point Optical Access Networks. *Proceedings of the National Fiber Optic Engineers Conference* (pp. 1550-1557).

Lung, B. (1999). PON architecture Futureproofs FTTH. *Lightwave, 16*(10), 104–107.

Luo, Y., & Ansari, N. (2005). Bandwidth allocation for multiservice access on EPONs. *IEEE Communications Magazine, 43*(12), 16–21. doi:10.1109/MCOM.2005.1561907

Luo, Y., & Ansari, N. (2005). Limited sharing with traffic prediction for dynamic bandwidth allocation and QoS provisioning over EPONs. *OSA Journal of Optical Networking, 4*(9), 561–572. doi:10.1364/JON.4.000561

Luo, Y., Yin, S., Ansari, N., & Wang, T. (2007). Resource management for broadband access over time-division multiplexed passive optical networks. *IEEE Network, 21*(5), 20–27. doi:10.1109/MNET.2007.4305168

Ly, K., Rissons, A., Gambardella, E., & Mollier, J.-C. (2008, April). Optimization of An Avionic VCSEL-Based Optical Link Through a Large Signal Characterization. In Proceeding of SPIE, *Photonics Europ Symposium*: Vol.6997.

Ma, J., Yu, J., Yu, C., Xin, X., Zeng, J., & Chen, L. (2007). Fiber Dispersion influence on transmission of the optical millimeter-waves generated using LN-MZM intensity modulation. *Journal of Lightwave Technology, 25*(11), 3244–3256. doi:10.1109/JLT.2007.907794

Ma, M., Zhu, Y., & Cheng, T.-H. (2003). A bandwidth guaranteed polling MAC protocol for ethernet passive optical networks. *Proceedings of the Twenty Second*

Annual Joint Conference of the IEEE Computer and Communications Societies (INFOCOM 2003),1, 22–31, San Francisco.

MacHale, E. K., Talli, G., & Townsend, P. D. (2006, March). *10Gb/s bidirectional transmission in a 116km reach hybrid DWDM-TDM PON.* Paper presented at Optical Fiber Communication Conference (OFC), (OFE1), Anaheim, California, USA.

MacHale, E. K., Talli, G., Chow, C. W., & Townsend, P. D. (2007, September). *Reduction of signal-induced Rayleigh noise in a 10Gb/s WDM-PON using a gain-saturated SOA.* Paper presented at European Conference and Exhibition on Optical Communication (ECOC), (Paper 7.6.3), Berlin, Germany.

MacHale, E. K., Talli, G., Townsend, P. D., Borghesani, A., Lealman, I., Moodie, D. G., & Smith, D. W. (2008, September). *Extended-reach PON employing 10Gb/s integrated reflective EAM-SOA.* Paper presented at European Conference and Exhibition on Optical Communication (ECOC), (Th.2.F.1), Brussels, Belgium.

Maeda, Y. (2004). Overview of Optical Broadband in Japan. *ECOC 2004 Proceedings, 1,* 4-5.

Maeda, Y., & Okada, F. D. (2001). FSAN OAN-WG and Future Issues for Broadband Optical Access Networks. *IEEE Communications Magazine.*

Mahlke, G., & Gössing, P. (2001). *Fiber Optic Cables.* Munich, Germany: Wiley Verlag.

Maier, G., Martinelli, M., Pattavina, A., & Salvadori, E. (2000). Design and Cost Performance of the Multistage WDM-PON Access Networks. *Journal of Lightwave Technology, 18,* 125–143. doi:10.1109/50.822785

Malli, R., Zhang, X. J., & Qiao, C. M. (1998). Benefits of Multicasting in All-optical Networks. In *SPIE Proceedings of All Optical Networking* (pp. 209-220).

Marinkovich, M., & Sybrowsky, J. (2003). *UTOPIA: A Public Network based on FTTP, Layer 2 Ethernet Access and the "OSPN" Model.* Converge Network Digest.

Marki, C. F., Alic, N., Gross, M., Papen, G., Esener, S., & Radic, S. (2007, March) *Performance of NRZ and*

Compilation of References

Duobinary modulation formats in Rayleigh and ASE-dominated dense optical links. Paper presented at Optical Fiber Communication Conference (OFC), (OFD6), Anaheim, California. USA.

Martínez, J. J., Gregorio, J. I. G., Lucia, A. L., Velasco, V. A., Aguado, J. C., & Binué, M. Á. L. (2008). Novel WDM-PON architecture based on a spectrally efficient IM-FSK scheme using DMLs and RSOAs. *IEEE/OSA . Journal of Lightwave Technology, 26*(3), 350–356. doi:10.1109/JLT.2007.909864

Mayer, G., Martinelli, M., Pattavina, A., & Salvadori, E. (2000). Design and cost performance of the multistage WDM PON access networks. *Journal of Lightwave Technology, 18*(2), 125–143. doi:10.1109/50.822785

Mayhew, A. J., Page, S. J., Walker, A. M., & Fisher, S. I. (2002). Fiber to the Home Infrastructure Deployment Issues. *BT Technology Journal, 20*(4), 91–103. doi:10.1023/A:1021374532690

McCabe, J. (2007). *Network Analysis, Architecture and Design.* Amsterdan, The Netherlands, & Boston, US: Elsevier/Morgan Kaufmann Publishers.

McCoy, A. D., Horak, P., Thomsen, B. C., Ibsen, M., Mokhtar, M. R., & Richardson, D. J. (2004, September). *Optimising signal quality in a spectrum-sliced WDM system using SOA-based noise reduction.* Paper presented at European Conference and Exhibition on Optical Communication (ECOC), (Tu4.6.4), Stockholm, Sweden.

McGarry, M. P., Reisslein, M., & Maier, M. (2006). WDM Ethernet passive optical networks. *IEEE Communications Magazine, 44*(2), 15–22. doi:10.1109/MCOM.2006.1593545

Merayo, N., Durán, R. J., Fernández, P., Lorenzo, R. M., de Miguel, I., Aguado, J. C., & Abril, E. J. (2008). EPON algorithm to provide service and client differentiation based on a dynamic weight adaptation. *Proceedings of the 13rd European Conference and Optical Communications* (pp. 129-136), Krems (Austria).

Mestdagh, D. J. G. (1995). *Fundamentals of Multiaccess Optical Fiber Networks.* Norwood, Ohio: Artech House, Inc.

Metro Ethernet Forum (MEF) (2008). *MEF 10. MEF Technical Specifications.*

Michalakelis, C., Varoutas, D., & Sphicopoulos, T. (2008). Diffusion models of mobile telephony in Greece. *Telecommunications Policy, 32*(3-4), 234–245. doi:10.1016/j.telpol.2008.01.004

Mickelsson, H., Sundberg, E., Strömgren, P., & Fujimoto, Y. (2002). *IEEE Contribution on loop lengths.* IEEE EFM. Retrieved July 25, 2008, from http://grouper.ieee.org/groups/802/3/efm/public/jan02/mickelsson_1_0102.pdf.

Mikulla, B., Leng, L., Sears, S., Collings, B. C., Arend, M., & Bergman, K. (1999). Broad-band high-repetition-rate source for spectrally sliced WDM. *IEEE Photonics Technology Letters, 11*(4), 418–420. doi:10.1109/68.752534

Milorad, C. (2004). *Optical Transmission Systems Engineering.* Norwood, Ohio: Artech House.

Mocerino, J. V. (2006). Carrier-Class Ethernet Service Delivery Migrating SONET to IP & Triple Play Offerings. OFC/NFOEC Proceedings 2006 (pp.96-401).

Modiano, E. (1999). Random Algorithms for Scheduling Multicast Traffic in WDM Broadcast-and-Select Networks. *IEEE/ACM Transaction on Networking, 7*(3), 425-434.

Molnár, M. (2008). *Hierarchies for Constrained Partial Spanning Problems in Graphs* (Tech. Rep. No. 1900). Rennes, France: IRISA.

Monroy, I. T., Tangdiongga, E., Jonker, R., & de Waardt, H. (2000). Interferometric crosstalk reduction by phase scrambling . *Journal of Lightwave Technology, 18*(5), 637–646. doi:10.1109/50.842077

Montagne, R. (2006, August). *The market's evolution to very high-speed.* Montpellier, France: Institut de l'Audiovisuel et des Télécommincations en Europe (IDATE). Retrieved on October 06, 2008, from: http://www.idate.org.

Moore, G. (1965). Cramming more components onto integrated circuits. *Electronics Magazine, 19.*

Morgan Securities, J. P. (2006). *The fibre battle – Changing dynamics in European wireline*. European Equity Research.

Moyer, S. (2007). The OSGi Service Platform: Enabling Web 2.0 for Broadband Services Prepared. *Access '07 Executive Business Forum associated with IEEE Globecom'07, Next Generation Home Network and Home Gateway Session*.

Mukai, H., Yokotani, T., Seno, S., & Motoshima, K. (2007). Enhanced queue status reporting scheme on PON. *Journal of Optical Networking, 6*, 756–766. doi:10.1364/JON.6.000756

Mukherjee, B. (1992). WDM-Based Local Lightwave Networks. I. Single Hop Systems. *IEEE Network, 6*(3), 12–27. doi:10.1109/65.139139

Mukherjee, B. (1992). WDM-Based Local Lightwave Networks. II. Multihop Systems. *IEEE Network, 6*(4), 20–32. doi:10.1109/65.145161

Mukherjee, B. (1997). *Optical Communication Networks*. New York, NY, USA: McGraw-Hill.

Mukherjee, B. (2000). WDM Optical Communication Networks: Progress and Challenges. *IEEE Journal on Selected Areas in Communications, 18*(10), 1810–1824. doi:10.1109/49.887904

Multimedia Research Group (MRG). Inc. (2007). *CATV Infrastructure: Assessing Strategies & Forecast - March 2007*. San Jose, CA: MRG, Inc.

Mun, S. G., Moon, J. H., Lee, H. K., Kim, J. Y., & Lee, C. H. (2008). A WDM-PON with a 40 Gb/s (32 × 1.25 Gb/s) capacity based on wavelength-locked Fabry-Perot laser diodes. *Optics Express, 16*(15), 11361–11368. doi:10.1364/OE.16.011361

Murtaza, G., & Senior, J. M. (1996). WDM crosstalk analysis for systems employing spectrally-sliced LED sources. *IEEE Photonics Technology Letters, 8*(3), 440–442. doi:10.1109/68.481143

Nakagawa, J. (2006). Technical Survey of Next Generation Optical System – 10G-PON. *Fiber of Expo 2006*.

Nakamura, H., Suzuki, H., Kani, J., & Iwatsuki, K. (2006). Reliable wide-area wavelength division multiplexing passive optical network accommodating gigabit Ethernet and 10-Gb Ethernet services. *IEEE/OSA . Journal of Lightwave Technology, 24*(5), 2045–2051. doi:10.1109/JLT.2006.871057

Nakamura, M., Ueda, H., Makino, S., Yokotani, T., & Oshima, K. (2004). Proposal of Networking by PON Technologies for Full and Ethernet Services in FTTx. *IEEE Transaction on Light-wave Technology, 22*, 2631–2640. doi:10.1109/JLT.2004.836765

Nakanashi, K., & Maeda, Y. (2002). Standardization Activities of FSAN: International Standardization Trends Concerning the Broadband PON (B-PON). *NTT Review, 14*(2), 108–110.

Narita, A. (2004). Home Gateway for Managing Home Network QoS. *ITU-T Workshop on Home Networking and Home Services*, Session 6-4.

Naser, H., & Mouftah, H. T. (2006). A joint-ONU interval-based dynamic scheduling algorithm for Ethernet passive optical networks. *IEEE/ACM Transactions on Networking, 14*(4), 889-899.

Nesset, D., Davey, R. P., Shea, D., Kirkpatrick, P., Shang, S. Q., Lobel, M., & Christensen, B. (2005, September). *10 Gbit/s bidirectional transmission in 1024-way split, 110 km reach, PON system using commercial transceiver modules, Super FEC and EDC*. Paper presented at European Conference and Exhibition on Optical Communication (ECOC), (Tu1.3.1), Glasgow, UK.

Nowak, D., Perry, P., & Murphy, J. (2004). A Novel Service Level Agreement Based Algorithm for Differentiated Services Enabled Ethernet PONs. In IEICE Press, *3rd International Conference on Optical Internet: Vol. 1* (pp. 598-599), Japan.

NSSG. (2008). General Secretariat of National Statistical Service of Greece (Publication.: www.statistics.gr

Nuss, M. C., Knox, W. H., & Koren, U. (1996). Scalable 32 channel chirped-pulse WDM source. *IEE Electronics Letters, 32*(14), 1311–1312. doi:10.1049/el:19960854

Öberg, M., Rigole, P. J., Nilsson, S., Klinga, T., Back-bom, L., & Streubel, K. (1995). Complete single mode wavelength coverage over 40 nm with a super structure grating DBR Laser. *IEEE/OSA . Journal of Lightwave Technology, 13*(9), 1892–1898. doi:10.1109/50.464740

OECD (Ed.). (2008). *Broadband Growth and Policies in OECD Countries*: An OECD Browse it Edition.

OECD. (2008). *OECD Broadband subscribers per 100 inhabitants, by technology, December 2007*. OECD Broadband Portal. Retrieved July 22, 2008, from http://www.oecd.org/.

OECD. (2008). *Working Party on Communication Infrastructures and Services Policy.* from http://ec.europa.eu/information_society/index_en.htm

Olsen, B. T., Katsianis, D., Varoutas, D., Stordahl, K., Harno, J., & Elnegaard, N. K. (2006). Technoeconomic evaluation of the major telecommunication investment options for European players. *IEEE Network, 20*(4), 6–15. doi:10.1109/MNET.2006.1668398

Ooba, N., Hibino, Y., Inoue, Y., & Sugita, A. (2000). Athermal silica-based arrayed-waveguide grating multiplexer using bimetal plate temperature compensator. *IEE Electronics Letters, 36*(21), 1800–1801. doi:10.1049/el:20001267

Organization for Economic Coordination and Development (OECD) (2007). *OECD Communications Outlook 2007*. Paris, France: OECD. Retrieved on July 30, 2007, from: *http:www//213.253.134.43/oecd/pdfs/browseit/9307021E.PDF*

OSGi Alliance (2003). *OSGi Service Platform Release 3.*

OSGi Alliance (2007). *About the OSGi Service Platform*

Owen, M., & Raj, J. (2003). An Introduction to the New Business Process Modeling Standard. *BPMN and Business Process Management*. Popkin Software.

Pagiatakis, G. C. (2004). *Fiber-Optic Telecommunications*. Salonica, Greece, Tziolas Publications (in Greek). Hellenic Organization for Telecommunications (OTE),

Planning Department (2000). *Description of Digital Technologies used in the Access Network*. Athens, Greece (in Greek).

Papadimitriou, G. I., & Pomportsis, A. S. (1999). Centralized Wavelength Conversion Protocols for WDM Broadcast-and-Select Star Networks. In *Proceedings of the 7th IEEE International Conference on Networks* (pp. 11-18). Washington: IEEE Computer Society.

Park, K. Y., Mun, S. G., Choi, K. M., & Lee, C. H. (2005). A theoretical model of a wavelength-locked Fabry–Perot laser diode to the externally injected narrow-band ASE. *IEEE Photonics Technology Letters, 17*(9), 1797–1799. doi:10.1109/LPT.2005.851886

Park, S. B., Jung, D. K., Shin, D. J., Shin, H. S., Yun, I. K., & Lee, J. S. (2007). Colorless operation of WDM-PON employing uncooled spectrum-sliced reflective semiconductor optical amplifiers. *IEEE Photonics Technology Letters, 19*(4), 248–250. doi:10.1109/LPT.2007.891197

Park, S. J., Kim, G. Y., & Park, T. S. (2006). WDM-PON system based on the laser light injected reflective semiconductor optical amplifier. *Optical Fiber Technology, 12*(2), 162–169. doi:10.1016/j.yofte.2005.07.006

Park, S. J., Lee, C. H., Jeoung, K. T., Park, H. J., Ahn, J. G., & Song, K. H. (2004). Fiber-to-the-home services based on wavelength division multiplexing passive optical network. *IEEE/OSA . Journal of Lightwave Technology, 22*(11), 2582–2591. doi:10.1109/JLT.2004.834504

Parr Rud, O. (2001). *Data Mining Cookbook: Modeling Data for Marketing, Risk, and Customer Relationship Management*. Wiley Computer Publishing.

Payne, D. B., & Davey, R. P. (2002). The future of fibre access systems? *BT Technology Journal, 20*(4), 104–114. doi:10.1023/A:1021323331781

Payoux, F. Chanclou, & P., Genay, N. (2007, March). *WDM-PON with colorless ONUs*. Paper presented at Optical Fiber Communication Conference (OFC), (OTuG5), Anaheim, California, USA.

Payoux, F., Chanclou, P., & Brenot, R. (2006, September). *WDM PON with a single SLED seeding colorless RSOA-*

based OLT and ONUs. Paper presented at European Conference and Exhibition on Optical Communication (ECOC), (Tu4.5.1), Cannes, France.

Payoux, F., Chanclou, P., Moignard, M., & Brenot, R. (2005, September). *Gigabit optical access using WDM PON based on spectrum slicing and reflective SOA.* Paper presented at European Conference and Exhibition on Optical Communication (ECOC), (We3.3.5), Glasgow, UK.

Pepeljugoski, P. K., & Lau, K. Y. (1992). Interferometric noise reduction in fiberoptic links by superposition of high frequency modulation. *Journal of Lightwave Technology, 10*(7), 957–963. doi:10.1109/50.144919

Personick, S. D. (2002). Evolving toward the Next-Generation Internet: Challenges in the Path Forward. *IEEE Communications Magazine, 40*(7), 72–76. doi:10.1109/MCOM.2002.1018010

Pesavento, M., & Kelsey, A. (1999). PONs for the Broadband Local Loop. *Lightwave, 16*(10), 68–74.

Photeon (2008). *Photeon Adds To Extensive Array of AWGs.* Press Release. Retrieved July 24, 2008, from http://www.photeon.com/html/Press/2005 02 28.pdf.

Prat, J. (2002). *Fiber-to-the-Home Technologies.* Dordrecht, Netherlands: Kluwer Academic Publishers.

Prat, J., Arellano, C., Polo, V., & Bock, C. (2005). Optical network unit based on a bidirectional reflective semiconductor optical amplifier for fiber-to-the-home networks. *IEEE Photonics Technology Letters, 17*(1), 250–252. doi:10.1109/LPT.2004.837487

Prat, J., Balaguer, P. E., Gene, J. M., Diaz, O., & Figuerola, S. (2002). *Fiber-to-the-Home Technologies.* The Netherlands: Kluwer Academic Publishers.

Prat, J., Omella, M., & Polo, V. (2007, March). *Wavelength shifting for colorless ONUs in single-fiber WDM-PONs.* Paper presented at Optical Fiber Communication Conference (OFC), (OTuG6), Anaheim, California, USA.

Prat, J., Polo, V., Bock, C., Arellano, C., & Olmos, J. J. (2005). Full-duplex single fiber transmission using FSK downstream and IM remote upstream modulations for fiber-to-the-home. *IEEE Photonics Technology Letters, 17*(3), 702–704. doi:10.1109/LPT.2004.840930

Presi, M., Proietti, R., D'Errico, A., Contestabile, G., & Ciaramella, E. (February, 2008). A full-duplex symmetric WDM-PON featuring OSSB downlink modulation with optical down-conversion. Paper presented at *IEEE/OSA Optical Fiber Communication Conference / National Fiber Optic Engineers Conference (OFC/NFOEC)*, Paper OThT4, San Diego, California, USA.

Pun, S. S., Chan, C. K., & Chen, L. K. (2005). Demonstration of a novel optical transmitter for high-speed differential phase-shift-keying / inverse return-to-zero (DPSK/Inv-RZ) orthogonally modulated signals. *IEEE Photonics Technology Letters, 17*(12), 2763–2765. doi:10.1109/LPT.2005.859412

Qiao, C. M., & Yoo, M. (1999). Optical Burst Switching (OBS) - A New Paradigm for an Optical Internet. *Journal of High Speed Network, 8*(1), 69–84.

Ram Murthy, C. S., & Gurusamy, M. (2002). *WDM Optical Networks: Concepts, Design, and Algorithms.* USA: Prentice-Hall, Inc.

Ramsey, D. (2005). *World's First International Real-time Streaming of 4K Digital Cinema Over Gigabit IP Optical Fiber Networks.* University of California, San Diego. Retrieved July 22, 2008, from http://ucsdnews.ucsd.edu/newsrel/science/iGrid4K.asp.

Rashid, S. (2004). *PON Delivers Optical Access to the Masses.* Alcatel Communications. Retrieved on June 06, 2007, from: www.alcatel.com/bnd/fttu/18282_FTTU_article_final.pdf

Reading, H. (2006). *FTTH in Europe: Forecast & Prognosis, 2006-2011.* New York: Heavy Reading. Retrieved on November 18, 2008, from: http://www.ftthcouncil.eu/documents/studies/Heavy_Reading_FTTH_Europe_2006_final.pdf.

Reed, D. (2003). *Copper Evolution. Federal Communications Commission Technological Advisory Council III.* Retrieved July 23, 2008, from http://www.fcc.gov/oet/tac/TAC_III_04_17_03/Copper_Evolution.ppt.

Reeve, M. H., Hunwicks, A. R., Methley, S. G., Bickers, L., & Hornung, S. (1988). LED spectral slicing for single-mode local loop application. *IEE Electronics Letters*, *24*(7), 389–390. doi:10.1049/el:19880263

Rigole, P. J., Nilsson, S., Backbom, L., Klinga, T., Wallin, J., & Stalnacke, B. (1995). 114-nm wavelength tuning range of a vertical grating assisted co-directional coupler laser with a super structure grating distributed Bragg reflector. *IEEE Photonics Technology Letters*, *7*(7), 697–699. doi:10.1109/68.393177

Rissons, A., Mollier, J.-C., Toffano, Z., Destrez, A., & Pez, M. (2003). Thermal and Optoelectronic Model of VCSEL Arrays for Short Range Communications. In *Proceedings of SPIE, Vertical-Cavity Surface-Emitting Lasers VII, 4994*, 100-111.

Rissons, A., Perchoux, J., Mollier, J. C., & Grabherr, M. (2004). Noise and Signal modeling of various VCSEL structures. In . *Proceedings of SPIE Vertical-Cavity Surface-Emitting Lasers VIII, 5364*, 81–91.

Rosen, E. (2004). Reykjavik leads the way on fiber to the curb. *Network World*.

Rubenstein, R. (2005). GPON versus EPON: the battle lines are drawn. *Lightwave Europe, November 2005 issue*. Retrieved on July 24, 2007, from: http://fibers.org/articles/news/7/12/5/1

Ruderman, K. (2007, May). European trends favour FTTX. *Lightwave Europe*. Retrieved on July 24, 2007, from: http://lw.pennnet.com/display_article/293410/63/ARTCL/none/none/European-trends-favour-FTTX/

Ruffin, A. B., Downie, J. D., & Hurley, J. (2008). *Purely passive long reach 10 GE-PON architecture based on duobinary signals and ultra-low loss optical fiber*. Paper presented at Optical Fiber Communication Conference (OFC), (OThL4), San Diego, California, USA.

Sahasrabuddhe, L. H., & Mukherjee, B. (1999). Light-trees: Optical Multicasting for Improved Performance in Wavelength-Routed Networks. *IEEE Communications Magazine*, *37*(1), 67–73. doi:10.1109/35.747251

Saleh, A. A. M., & Simmons, J. M. (1999). Architectural Principles of Optical Regional and Metropolitan Access Networks. *Journal of Lightwave Technology*, *17*(12), 2431–2438. doi:10.1109/50.809662

Sara. (2005). *Computing & Networking services*. from http://www.sara.nl/news/newsletters/20050105/news_lett_20050105_eng.html

Sato, K., Furuya, S., Yokotani, T., Homma, H., & Sakai, S. (2008). *Technologies for Next Generation Home Network*. (pp. 19-23).

Sato, K., Yokotani, T., Furuya, S., & Isoda, M. (2005). *Implementation of SIP adaptation function for QoS negotiation in NGN* (pp. 55-60). IEICE Tech. Rep., CS2005-46.

Sauer, M., Kobyakov, A., & George, J. (2007). Radio Over Fiber for Picocellular Network Architectures. *Journal of Lightwave Technology*, *25*(11), 3301–3320. doi:10.1109/JLT.2007.906822

Schussmann, J. (2008). *Design and Cost Performance of WDM PONs for Multi-Wavelength Users*. Unpublished doctoral dissertation, Technical University of Ilmenau, Germany.

Scott, J. W., Geels, R. S., Corzine, S. W., & Coldren, L. A. (1993). Modeling Temperature effects And Spatial Hole Burning To Optimize Vertical-Cavity Surface-Emitting Laser Performance. *IEEE Journal of Quantum Electronics*, *29*(5), 1295–1308. doi:10.1109/3.236145

Shami, A., Bai, X., Assi, C., & Ghani, N. (2005). Jitter performance in Ethernet passive optical networks. *IEEE/OSA .Journal of Lightwave Technology*, *23*(4), 1745–1753. doi:10.1109/JLT.2005.844510

Shami, A., Bai, X., Ghani, N., Assi, C., & Mouftah, H. T. (2005). QoS control schemes for two-stage Ethernet passive optical access networks. *IEEE Journal on Selected Areas in Communications*, *23*(8), 1467–1478. doi:10.1109/JSAC.2005.852185

Shannon, C. E. (1948). A Mathematical Theory of Communication. *The Bell System Technical Journal*, *27*, 623–656.

Shau, R., Ortsiefer, M., Rosskopf, J., Boehm, G., Lauer, C., Maute, M., & Amann, M. (2004). Longwavelength InP-Based VCSELs with buried tunnel Junction: properties and applications. In . *Proceedings of SPIE Vertical-Cavity Surface-Emitting Lasers VIII, 5364*, 1–15.

Shea, D. P., & Mitchell, J. E. (2006, March). *Experimental upstream demonstration of a long reach wavelength-converting PON with DWDM backhaul.* Paper presented at Optical Fiber Communication Conference (OFC), (OWL4), Anaheim, California.

Shea, D. P., & Mitchell, J. E. (2007). A 10-Gb/s 1024-way-split 100-km long-reach optical-access network. *Journal of Lightwave Technology, 25*(3), 685–693. doi:10.1109/JLT.2006.889667

Shea, D. P., Ellis, A. D., Payne, D. B., Davey, R. P., & Mitchell, J. E. (2003, September). *10 Gbit/s PON with 100 km reach and x1024 split.* Paper presented at European Conference and Exhibition on Optical Communication (ECOC), (We.P.147), Rimini, Italy.

Sherif, S. R., Hadjiantonis, A., Ellinas, G., Assi, C., & Ali, M. (2004). A novel decentralized Ethernet-Based PON Access Architecture for Provisioning Differentiated QoS . *Journal of Lightwave Technology, 22*(11), 2483–2497. doi:10.1109/JLT.2004.836757

Sheu, S. T., & Huang, C. P. (1997). An Efficient Multicast Protocol for WDM Star-coupler Networks. In *the 2nd Symposium on Computers and Communications* (p. 579). IEEE Computer Society.

Shieh, W., Yi, X., Ma, Y., & Tang, Y. (2007). Theoretical and experimental study on PMD-supported transmission using polarization diversity in coherent optical OFDM systems. *Optics Express, 15*(16), 9936–9947. doi:10.1364/OE.15.009936

Shimada, S., Hashimoto, K., & Okada, K. (1987). Fiber optic subscriber loop systems for integrated services: the strategy for introducing fibers into the subscriber network. *Journal of Lightwave Technology, 5*(12), 1667–1675. doi:10.1109/JLT.1987.1075480

Shin, D. J., Jung, D. K., Lee, J. K., Lee, J. H., Choi, Y. H., & Bang, Y. C. (2003). 155 Mbit/s transmission using ASE-injected Fabry–Pérot laser diode in WDM-PON over 70oC temperature range. *IEE Electronics Letters, 39*(18), 1331–1332. doi:10.1049/el:20030850

Shin, D. J., Jung, D. K., Shin, H. S., Kwon, J. W., Hwang, S., Oh, Y., & Shim, C. (2005). Hybrid WDM/TDM-PON with wavelength-selection-free transmitters. *Journal of Lightwave Technology, 23*(1), 187–195. doi:10.1109/JLT.2004.840031

Shin, D. J., Keh, Y. C., Kwon, J. W., Lee, E. H., Lee, J. K., & Park, M. K. (2006). Low-cost WDM-PON with colorless bidirectional transceivers. *Journal of Lightwave Technology, 24*(1), 158–165. doi:10.1109/JLT.2005.861122

Shin, H. S., Jung, D. K., Kim, H. S., Shin, D. J., Park, S. B., Hwang, S. T., et al. (2005, September). *Spectrally pre-composed ASE injection for a wavelength-seeded reflective SOA in a WDM-PON.* Paper presented at European Conference and Exhibition on Optical Communication (ECOC), (We3.3.7), Glasgow, UK.

Shin, H. S., Jung, D. K., Shin, D. J., Park, S. B., Lee, J. S., Yun, I. K., et al. (2006, March). *16 x 1.25 Gbit/s WDM-PON based on ASE-injected R-SOAs in 60oC temperature range.* Paper presented at Optical Fiber Communication Conference (OFC), (OTuC5), Anaheim, California, USA.

Shinohara, H. (2005). Broadband Access in Japan: Rapidly Growing FTTH Market. *IEEE Communications Magazine, 43*(9), 72–78. doi:10.1109/MCOM.2005.1509970

Shreedhar, M., & Varghese, G. (1996). Efficient fair queuing using deficit round robin. IEEE/ACM Transactions on Networking, 4(3), 375-385.

Siemens ON Training Center. (1998). *Fastlink (document AN4013E6A01).* Munich, Germany.

Sigurdsson, H. M. (2007). *Techno-Economics of Residential Broadband Deployment.* PhD Thesis, Technical University of Denmark, Denmark

Simcoe, M. (2002). A Multiservice Voice, Data, and Video Network Enabled by Optical Ethernet. Proceedings of the National Fiber Optic Engineers Conference, (pp. 119-129).

Simoneaux, C. (2001). Optical Communications: VC-SELs propel 10-Gbit Ethernet. *EETimes*. Retrieved August 2001, from http://www.eetimes.com/story/OEG20010806S0059.

Sivalingam, K. M., & Subramaniam, S. (2005). *Emerging Optical Networks Technologies*. New York, USA: Springer.

Sivalingam, K. M., Bogineni, K., & Dowd, P. W. (1992). Pre-allocation Media Access Control Protocols for Multiple Access WDM Photonic Networks. *ACM Sigcomm Computer Communication Review*, 22(4), 235–246. doi:10.1145/144191.144289

Smit, M. K., & van Dam, C. (1996). PHASAR-Based WDM-Devices: Principles, Design and Applications. *IEEE Journal on Selected Topics in Quantum Electronics*, 2(2), 236–250. doi:10.1109/2944.577370

Söderlund, M., Kapulainen, M., & Koponen, J. (2004). Techno-economic Comparison of Four Alternative FTTH Approaches in a Greenfield Building Site. *ECOC 2004 Proceedings*, 3, 394-395.

Sollentuna Energi. (2002). from www.sollentunaenergi.se

Soma, S., Furuya, S., Sato, K., Yokotani, T., & Shimokasa, K. (2008). *A study on services by home gateway for home appliance control* (pp. 1-6). IEICE Technical Report, CS2008-6.

Song, H., Banerjee, A., Kim, B.-W., & Mukherjee, B. (2007). Multi-Thread Polling: A Dynamic Bandwidth Distribution Scheme in Long-Reach PON. *In Proceedings of the IEEE Globecom '07*, Washington, DC.

Soole, J. B. D., Bhat, R., LeBlanc, H. P., Andreadakis, N. C., Grabbe, P., Caneau, C., Koza, & M. A. (1994). Wavelength precision of monolithic InP grating multiplexer/demultiplexers. *IEE Electronics Letters*, 30(8), 664-666.

Spiekman, L. H., Amersfoort, M. R., De Vreede, A. H., van Ham, F. P. G. M., Kuntze, A., & Pedersen, J. W. (1996). Design and realization of polarization independent phased array wavelength demultiplexers using different array

orders for TE and TM. *IEEE/OSA . Journal of Lightwave Technology*, 14(6), 991–995. doi:10.1109/50.511599

Squire, S. & Dempsey L.L.P. (2002). *Legal study on part II of local loop unbundling sectoral inquiry (Contract No. Comp.IV/37.640)*. Brussels, Belgium: European Commission.

Sreenath, N., Satheesh, K., Mohan, G., & Siva Ram Murthy, C. (2001). Virtual Source Based Multicast Routing in WDM Optical Networks. *Photonic Network Communications*, 3(3), 213–226. doi:10.1023/A:1011443013088

Stark, J. B., Nuss, M. C., Knox, W. H., Cundiff, S. T., Boivin, L., & Dreyer, K. (1997). Cascaded WDM passive optical network with a highly shared source. *IEEE Photonics Technology Letters*, 9(8), 1170–1172. doi:10.1109/68.605539

Starr, T., Sorbara, M., Cioffi, J., & Silverman, P. (1999). *Understanding Digital Subscriber Line Technology*. Upper Saddle River, NJ: Prentice Hall.

Staubli, R. K., & Gysel, P. (1991). Crosstalk penalties due to coherent Rayleigh noise in bidirectional optical communication systems. *Journal of Lightwave Technology*, 9(3), 375–380. doi:10.1109/50.70015

Stern, J. R., Ballance, J. W., Faulkner, D. W., Hornung, S., Payne, D. B., & Oakley, K. (1987). Passive optical local networks for telephony applications and beyond. *Electronics Letters*, 23(24), 1255–1256. doi:10.1049/el:19870872

Stern, J. R., Hoppitt, C. E., Payne, D. B., Reeve, M. H., & Oakley, K. A. (1988). TPON - A passive optical network for telephony. *Fourteenth European Conference on Optical Communication, 1*, 203-206.

Stokab, A. B. (2006b). *Stokab Annual Report*. from http://www.stokab.se/upload/Ladda%20ner/dokument/Stokab%20Annual%20Report_05_ENG_I.pdf

StokabA. B. (2006). from http://www.stokab.se/templates/StandardPage.aspx?id=306

SURFnet. (2007). from http://www.surfnet.nl/info/home.jsp

Syrbu, A. (2005). Wafer fused long-wavelength VCSELs with InP-Based active Cavities. In IEEE, *International Conference on Indium Phosphide and Related Materials* (pp. 670-674).

Szweda, R. (2006). VCSEL applications diversify as technology matures. *III-V Review, 19*(1), 34–38. doi:10.1016/S0961-1290(06)71477-6

Takahashi, H., & Matsuyama, A. (1980). An Approximate Solution for the Steiner Problem in Graphs. *Mathematica Japonica, 24,* 573–577.

Takahashi, H., Hibino, Y., Ohmori, Y., & Kawachi, M. (1993). Polarization-insensitive arrayed-waveguide wavelength multiplexer with birefringence compensating film. *IEEE Photonics Technology Letters, 5*(6), 707–709. doi:10.1109/68.219718

Takahashi, H., Oda, K., Toba, H., & Inoue, Y. (1995). Transmission Characteristics of Arrayed Waveguide NxN Wavelength Multiplexer. *Journal of Lightwave Technology, 13*(3), 447–455. doi:10.1109/50.372441

Takesue, H., & Sugie, T. (2003). Wavelength channel data rewrite using saturated SOA modulator for WDM networks with centralized light sources. *IEEE/OSA . Journal of Lightwave Technology, 21*(11), 2546–2556. doi:10.1109/JLT.2003.819532

Talli, G., & Townsend, P. D. (2006). Hybrid DWDM-TDM long reach PON for next generation optical access. *Journal of Lightwave Technology, 24*(7), 2827–2834. doi:10.1109/JLT.2006.875952

Talli, G., & Townsend, P. D. (2006). Hybrid DWDM-TDM long-reach PON for next-generation optical access. *IEEE/OSA . Journal of Lightwave Technology, 24*(7), 2827–2834. doi:10.1109/JLT.2006.875952

Talli, G., Chow, C. W., & Townsend, P. D. (2008). (accepted for publication). Modeling of modulation formats for interferometric noise mitigation. *Journal of Lightwave Technology.*

Talli, G., Chow, C. W., MacHale, E. K., & Townsend, P. D. (2006, September). *High split ratio 116km reach hybrid DWDM-TDM 10Gb/s PON employing R-ONUs.*

Paper presented at European Conference and Exhibition on Optical Communication (ECOC), (Mo4.5.2), Cannes, France.

Talli, G., Chow, C. W., MacHale, E. K., & Townsend, P. D. (2007). Rayleigh noise mitigation in long-reach hybrid DWDM-TDM PONs. *Journal of Optical Networking, 6*(6), 765–776. doi:10.1364/JON.6.000765

Talli, G., Cotter, D., & Townsend, P. D. (2006). Rayleigh backscattering impairments in access networks with centralised light source. *Electronics Letters, 42*(15), 877–878. doi:10.1049/el:20061546

Tan, A. H. H. (1997). SUPER PON – A Fiber to the Home Cable Network for CATV and POTS/ISDN/VOD as Economical as a Coaxial Cable Network. *Journal of Lightwave Technology, 15*(2), 213–218. doi:10.1109/50.554326

Tanis, D., & Eichenbaum, B. R. (2002). Cost of Coarse WDM Compared with Dense WDM for Wavelength-Addressable Enhanced PON Access. *IEE Seminar on Photonic Access Technologies.*

Tauber, H. (2007). *FTTH: Europe must open its eyes.* Retrieved on July 10, 2007, from: http://optics.org/cws/article/industry/26992.

Telcordia, (2001). *Telcordia GR 253 CORE: SONET Transport Systems: Common Generic Criteria.* Telcordia.

Tera (2008). *Techno Economic Results from ACTS Program from the European Commission, AC364.* Retrieved July 24, 2008, from http://www.telenor.no/fou/prosjekter/tera/index.htm.

Third-Generation Partnership Project. (3GPP) (2007). *Technical Specification 23.402: Architecture Enhancements for Non-3GPP Accesses.* Sophia-Antipolis, France: 3GPP.

Tian, Y., Su, Y., Yi, L., Leng, L., Tian, X., He, H., & Xu, X. (2006, September). *Optical VPN in PON based on DPSK erasing/rewriting and DPSK/IM formatting using a single Mach-Zehnder modulator.* Paper presented at

European Conference and Exhibition on Optical Communication (ECOC), (Tu4.5.6), Cannes, France.

Toffano, Z., Pez, M., Le Brun, C., Desgreys, P., Hervé, Y., & Mollier, J.-C. (2003). Multilevel Behavioral Simulation of VCSEL based Optoelectronic Modules. *IEEE Journal on Selected Topics in Quantum Electronics, 9*(3), 949–960. doi:10.1109/JSTQE.2003.818348

Tonic (2008). *Techno Economics of IP Optimised Networks and Services, Project within the IST Programme (Information Society Technologies) of the European Union.* Retrieved July 24, 2008, from http://www-nrc.nokia.com/tonic/.

Townsend, P. D., Talli, G., Chow, C. W., MacHale, E. M., Antony, C., Davey, R., et al. (2007, October). *Long reach passive optical networks.* Paper presented at IEEE LEOS Annual Meeting, Lake Buena Vista, Florida, USA.

Tridandapani, S., Meditch, J. S., & Somani, A. K. (1994). The Matpi Protocol: Asking Uning Times Through Ipelining in WDM Optical Networks. In *INFOCOM'94* (pp. 1528-1535).

Tse, Y. T., Lu, G. W., Chen, L. K., & Chan, C. K. (November, 2007). Upstream OOK remodulation scheme using injection-locked FP laser with downstream inverse-RZ data in WDM passive optical network. Paper presented at *Asia Pacific Optical Communications Communication Conference (APOC)*, Paper 6784-69, Wuhan, PRC.

Tsuchida, M., Yokotani, T., & Sato, K. (2006). *Network Control Technology for Next Generation Network* (pp. 55-59). Mitsubishi Electric Technical Report, 2006.2.

Tucker, R. (1985). High-Speed Modulation of Semiconductor lasers. *IEEE Transactions on Electron Devices, ED-32*(12), 1180–1192.

Ulm photonics company (2008). VCSEL technology, Retrieved in August 2008, from http://www.ulm-photonics.de/.

Urban, P. J., Koonen, A. M. J., Khoe, G. D., & de Waardt, H. (2007, September). *Coherent crosstalk-suppression in WDM Access networks employing reflective semiconductor optical amplifiers.* Paper presented at European Conference and Exhibition on Optical Communication (ECOC), (Paper 6.4.2), Berlin, Germany.

UTOPIA. (2006). *Connecting Communities.* from http://www.utopianet.org/

Van den Hoven, G. (2007). Why Europe is choosing point-to-point. *Lightwave Europe.* Retrieved on July 24, 2007, from: http://lw.pennnet.com/display_article/293398/63/ARTCL/none/none/Why-Europe-is-choosing-point-to-point/

Van den Hoven, G. (2008). FTTH in Europe, one year later. *Lightwave Europe, Q1/2008*, 9-10.

Van Deventer, M. O., Van Dam, P., Peters, P., & Vermaerke, F. (1997). Evolution Phases to an Ultra-Broadband Access Network: Results from ACTS-PLANET. *IEEE Communications Magazine, 35*(12), 72–77. doi:10.1109/35.642835

Varon, M., Le Kernec, A., & Mollier, J.-C. (2007). Opto-Microwave source using a harmonic frequency generator driven by a VCSEL-Based Ring Oscillator. *Proceeding of the European Microwave Association, 3*(3).

Vassilopoulos, C., & Pagiatakis, G. C. (2002). *Advanced Telecommunications Infrastructures and Services.* Athens, Greece, OTE Publications (in Greek).

Venieris, I. (2007). *Broadband Networks* (Chapter 8). Salonica, Greece, Tziolas Publications (in Greek). Fiber to the premises by country. *Wikipedia, the free encyclopedia.* Retrieved June 18, 2008 from http://en.wikipedia.org/wiki/Fiber_to_the_premises_by_country.

Vertilas (2008), Technology, Retrieved in August 2008, from http://www.vertilas.com/

Very High Speed Digital Subscriber Line 2. *Wikipedia, the free encyclopedia.* Retrieved July 02, 2008 from http://en.wikipedia.org/wiki/*Very_High_Speed_Digital_Subscriber_Line_2*.

VolkerWessels. (2007). from http://www.volkerwessels.com/corporate/bin/en.jsp?enDispWhat=Zone&enPage=HomePage&enDisplay=view&

Wagner, S. S., & Chapuran, T. E. (1990). Broadband high-density WDM transmission using superluminescent diodes. *IEE Electronics Letters, 26*(11), 696–697. doi:10.1049/el:19900454

Wagner, S. S., & Kobrinski, H. (1989). WDM Applications in Broadband Telecommunication Networks. *IEEE Communications Magazine, 27*(3), 22–30. doi:10.1109/35.20264

Wagner, S., & Lemberg, H. (1989). Technology and system issues for a WDM-based fiber loop architecture. *IEEE/OSA . Journal of Lightwave Technology, 7*(11), 1759–1768. doi:10.1109/50.45899

Wagner, S., Kobrinski, H., Robe, T. J., Lemberg, H. L., & Smoot, L. S. (1988). Experimental demonstration of a passive optical subscriber loop. *IEE Electronics Letters, 24*(6), 344–346. doi:10.1049/el:19880234

Wan, P. J. (2000). *Multichannel Optical Networks.* Dordrecht, Netherlands: Kluwer Academic Publishers.

Wang, L. A., Chapuran, T. E., & Menendez, R. C. (1991). Medium-density WDM system with Fabry-Perot laser diodes for subscriber loop applications. *IEEE Photonics Technology Letters, 3*(6), 554–556. doi:10.1109/68.91033

Wang, P., Seah, L. K., Murukeshan, V. M., & Chao, Z. X. (2006). Electronically tunable external-cavity laser diode using a liquid crystal deflector. *IEEE Photonics Technology Letters, 18*(15), 1612–1614. doi:10.1109/LPT.2006.879509

Weldon, M. K., & Zane, F. (2003). The Economics of Fiber to the Home Revisited. *Bell Labs Technical Journal, 8*(1), 181–206. doi:10.1002/bltj.10053

Weller, D. (2008, April 10). A Fiber Future Challenges for markets and policy, *Fibre Investment and Policy Challenges OECD Workshop,* Stavanger, Norway. Verizon.

Wen, Y. J., & Chae, C. J. (2006). WDM-PON upstream transmission using Fabry–Pérot laser diodes externally injected by polarization-insensitive spectrum-sliced supercontinuum pulses. *Optics Communications, 260*(2), 691–695. doi:10.1016/j.optcom.2005.11.029

Wieland, K. (2006). *Voice and Low Prices Drive FTTH in Japan.* Retrieved on July 24, 2007, from: http://www.telecommmagazine.com/newsglobe/article.asp?HH_ID=AR_2570

WIK-Consult GmbH. (2007, June). Possibilities and Prerequisites of a FTTx Strategy in Germany. Bad Honnef, Germany: WIK-Consult GmbH. Retrieved on October 06, 2008, from: http://www.comreg.ie/_fileupload/publications/ComReg0795a.pdf.

WIK-Consult GmbH. (2007, November). *Next Generation Bit Stream Access - Final. Study for the Commission for Communications Regulation (ComReg).* Bad Honnef, Germany: WIK-Consult GmbH. Retrieved on October 06, 2008, from: http://www.comreg.ie/_fileupload/publications/ComReg0795a.pdf.

Wilmsen, C. W., Temkin, H., & Coldren, L. (1999). *Vertical-Cavity Surface-Emitting Lasers, Design, Fabrication, Characterization and applications.* Cambridge University press.

WiMAX Forum (2007). *Deployment of Mobile WiMAX Networks by Operators with Existing 2G & 3G Networks.* WiMAX Forum

Wolf, J., & Zee, N. (2000). *The Last Mile. Broadband and the Next Internet Revolution.* New York, USA: McGraw Hill.

Wong, E., Lee, K. L., & Anderson, T. (2006). Low-cost WDM passive optical network with directly-modulated self-seeding reflective SOA. *IEE Electronics Letters, 42*(5), 299–301. doi:10.1049/el:20060097

Wong, E., Zhao, X., Chang-Hasnain, C. J., Hofmann, W., & Amann, M. C. (2006, March). *Uncooled, optical injection-locked 1.55 mm VCSELs for upstream transmitters in WDM-PONs.* Paper presented at Optical Fiber Communication Conference (OFC), Anaheim, California, USA.

Wong, E., Zhao, X., Chang-Hasnain, C. J., Hofmann, W., & Amann, M. C. (2007). Rayleigh backscattering and extinction ratio study of optically injection-locked 1.55 lm VCSELs . *Electronics Letters, 43*(3), 182–183. doi:10.1049/el:20073446

Wong, E., Zhao, X., Chang-Hasnain, C., Hofmann, W., & Amann, M. C. (2006). Optically injection-locked 1.55-mm VCSELs as upstream transmitters in WDM-PONs. *IEEE Photonics Technology Letters, 18*(22), 2371–2373. doi:10.1109/LPT.2006.885292

Woodward, S. L., Iannone, P. P., Reichmann, K. C., & Frigo, N. J. (1998). A spectrally sliced PON employing Fabry–Perot lasers. *IEEE Photonics Technology Letters, 10*(9), 1337–1339. doi:10.1109/68.705635

Wu, H. T., Ke, K. W., & Huang, S. Y. (2007). A Novel Multicast Mechanism for Optical Local Area Networks. *Computers & Electrical Engineering, 33*(2), 94–108. doi:10.1016/j.compeleceng.2006.08.002

Wu, K. D., Wu, J. C., & Yang, C. S. (2001). Multicast Routing with Power Consideration in Sparse Splitting WDM Networks. In *ICC2001, 2*, 513-517. Helsinki, Finland: IEEE Communication Society.

Xie, J., Jiang, S., & Jiang, Y. (2004). A dynamic bandwidth allocation scheme for differentiated services in EPONs. *IEEE Communications Magazine, 42*(8), S32–S39. doi:10.1109/MCOM.2004.1321385

Xin, Y. F., & Rouskas, G. N. (2004). Multicast Routing Under Optical Layer Constraints. In *INFOCOM2004, 4*, 2731-2742). IEEE Communication Society.

Xu, L., & Tsang, H. K. (2008). Colorless WDM-PON optical network unit (ONU) based on integrated nonreciprocal optical phase modulator and optical loop mirror. *IEEE Photonics Technology Letters, 20*(10), 863–865. doi:10.1109/LPT.2008.921851

Xu, L., & Tsang, H. K. (2008). WDM-PON using differential-phase-shift-keying remodulation of dark return-to-zero downstream channel for upstream. *IEEE Photonics Technology Letters, 20*(10), 833–835. doi:10.1109/LPT.2008.919598

Xu, Z. W., Wen, Y. J., Zhong, W. D., Chae, C. J., Cheng, X. F., & Wang, Y. X. (2007). High-speed WDM-PON using CW injection-locked Fabry-Pérot laser diodes. *OSA Optics Express, 15*(6), 2954–2962.

Xu, Z., Wen, Y. J., Zhong, W., Attygalle, M., Cheng, X., & Wang, Y. (2007). WDM-PON architectures with a single shared interferometric filter for carrier-reuse upstream transmission. *IEEE/OSA .Journal of Lightwave Technology, 25*(12), 3669–3677. doi:10.1109/JLT.2007.909341

Yan, S. G., Deogun, J. S., & Ali, M. (2003). Routing in Sparse Splitting Optical Networks with Multicast Traffic. *Computer Networks, 41*(1), 89–113. doi:10.1016/S1389-1286(02)00345-6

Yeh, C. H., Chien, H. C., & Chi, S. (February 2008). Cost-effective colorless RSOA-based WDM-PON with 2.5-Gbit/s uplink signal. Paper presented at *IEEE/OSA Optical Fiber Communication Conference / National Fiber Optic Engineers Conference (OFC/NFOEC)*, Paper JWA95, San Diego, California.

Yeh, C. H., Lee, C. S., & Chi, S. (2008). Simply Self-Restored Ring-Based Time-Division-Multiplexed Passive Optical Network. *Journal of Optical Networking, 7*, 288–293. doi:10.1364/JON.7.000288

Yokotani, T. (2007). Home Gateway Evolution toward Home Gateway Evolution toward Broadband Service and NGN Broadband Service and NGN. *Access '07 Executive Business Forum associated with IEEE Globecom'07, Next Generation Home Network and Home Gateway Session.*

Yokotani, T., & Sato, K. (2007). *Integrated Home Gateway with OSGi Framework*. 3rd Workshop of OSGi User Forum Japan.

Yokotani, T., Kitayama, K., Mukai, H., & Murakami, K. (2002). A study on usage of T-CONT in DBA for FTTx. Journal of EXP in Search of Innovation, 2, 64-71.

Yokotani, T., Nakanishi, K., Ogasawara, M., & Maeda, Y. (2006). Next Generation Broadband Access Network Technologies. *Journal of IEICE, 89*, 1062–1066.

Yoshida, T., Kimura, S., Kimura, H., Kumozaki, K., & Imai, T. (2006). A new single-fiber 10-Gb/s optical loopback method using phase modulation for WDM optical access networks. *IEEE/OSA .Journal of Lightwave Technology, 24*(2), 786–796. doi:10.1109/JLT.2005.862441

Yoshida, T., Kimura, S., Kimura, H., Kumozaki, K., & Imai, T. (2006). A new single-fiber 10-Gb/s optical loopback method using phase modulation for WDM optical access networks. *Journal of Lightwave Technology, 24*(2), 786–796. doi:10.1109/JLT.2005.862441

Young, G. (2007). Europe at the speed of light: Report of the FTTH Council of Europe Annual Conference. *Broadband Magazine, 291,* 18-19.

Yu, J., Jia, Z., Wang, T., & Chang, G. K. (2007). A novel radio-over-fiber configuration using optical phase modulator to generate an optical mm-wave and centralized lightwave for uplink connection. *IEEE Photonics Technology Letters, 19*(3), 140–142. doi:10.1109/LPT.2006.890087

Yu, J., Kim, N., & Kim, B. W. (2007). Remodulation schemes with reflective SOA for colorless DWDM PON. *OSA Journal of Optical Networking, 6*(8), 1041–1054. doi:10.1364/JON.6.001041

Yue, W. (2006). *How GPON Deployment Drives the Evolution of the Packet-Based Network.* 2006 FTTH Conference Proceedings.

Yue, W. (2006). *The Role of Pseudowires and Emerging Wireless Technologies on the Converged Packet-Based Network.* Proceedings of OFC/NFOEC, Anaheim CA, March 2006 & presented in the QWEST HPN Summit 2006.

Yue, W., & Gutierrez, D. (2003). *Ready for Primetime: MSPPs Can Deliver Digital Video Over Existing Network.* Lightwave North America & Lightwave Europe.

Yue, W., & Gutierrez, D. (2003). *Digital Video Transport over SONET using GFP and Virtual Concatenation.* NFOEC 2003, Conference Proceedings.

Yue, W., & Hunck, B. (2005). *Deploying Multiservice Networks Using RPR over the Existing SONET Infrastructure.* OFC/NFOEC 2005 Conference Proceedings.

Yue, W., & Mocerino, J. V. (2007). *Broadband Access Technologies for FTTx Deployment.* OFC/NFOEC 2007 Conference Proceedings.

Yuen, W., Li, G. S., Nabiev, R. F., Jansen, M., Davis, D., & Chang-Hasnain, C. J. (October 2001). Electrically-pumped directly-modulated tunable VCSEL for metro DWDM applications. Paper presented at *Gallium Arsenide Integrated Circuit (GaAs IC) Symposium,* Paper TuA1.2, Baltimore, Maryland, USA.

Zah, C. E., Favire, F. J., Pathak, B., Bhat, R., Caneau, C., & Lin, P. S. D. (1992). Monolithic integration of a multi-wavelength compressive-strained multi-quantum-well distributed feedback laser array with a star coupler and optical amplifiers. *IEE Electronics Letters, 28*(25), 2361–2362.

Zhang, B., Lin, C., Huo, L., Wang, Z., & Chan, C.-K. (2006, March). *A simple high-speed WDM PON utilizing a centralized supercontinuum broadband light source for colorless ONUs.* Paper presented at OFC'06, Anaheim, CA.

Zhang, C., Kun, Q., & Bo, X. (2007). Passive optical networks based on optical CDMA: Design and system analysis. [English Edition]. *Chinese Science Bulletin, 52*(1), 118–126. doi:10.1007/s11434-007-0020-8

Zhang, X. J., Wei, J., & Qiao, C. M. (2000). Constrained Multicast Routing in WDM Networks with Sparse Light Splitting. *IEEE Journal of Lightwave Technology, 18*(12), 1917–2000. doi:10.1109/50.908787

Zhao, J., Chen, L. K., & Chan, C. K. (2007, March). *A novel re-modulation scheme to achieve colorless high-speed WDM-PON with enhanced tolerance to chromatic dispersion and re-modulation misalignment.* Paper presented at Optical Fiber Communication Conference (OFC), (OWD2), Anaheim, California. USA.

Zheng, J., & Moufah, H. T. (2005). An adaptive MAC polling protocol for Ethernet passive optical networks (EPONs). *Proc. of 2005 IEEE International Conference on Communications (ICC'05), Vol. 3,* 1874-1878.

Zheng, J., & Mouftah, H. T. (2004). *Optical WDM Networks: Concepts and Design Principles.* Hoboken, New Jersey: Wiley-IEEE Press.

Zheng, J., & Mouftah, H. T. (2006). Efficient bandwidth allocation algorithm for Ethernet passive optical networks. *IEE Proceedings. Communications, 153*(3), 464–468. doi:10.1049/ip-com:20050358

Zheng, S. Q., & Gumaste, A. (2006). Smart: An Optical Infrastructure for Future Internet. *The 3rd International Conference on Broadband Communications, Networks and Systems* (pp. 1-12).

Zhou, F., Molnár, M., & Cousin, B. (2008a). Avoidance of Multicast Incapable Branching Nodes for Multicast Routing in WDM Networks. In *the 33rd IEEE Conference on Local Computer Networks* (pp. 336-344). Montreal, Canada: IEEE Computer Society.

Zhou, F., Molnár, M., & Cousin, B. (2008b). Distance Priority Based Multicast Routing in WDM Networks Considering Sparse Light Splitting. In *the 11th IEEE Conference on Communication Systems*. Guangzhou, China: IEEE Communication Society.

Zirngibl, M. (1998). Multifrequency lasers and applications in WDM networks. *IEEE Communications Magazine, 36*(12), 39–41. doi:10.1109/35.735875

Zirngibl, M., Doerr, C. R., & Joyner, C. H. (1998). Demonstration of a splitter/router based on a chirped waveguide grating router. *IEEE Photonics Technology Letters, 10*(1), 87–89. doi:10.1109/68.651116

Zirngibl, M., Doerr, C. R., & Stulz, L. W. (1996). Study of spectral slicing for local access applications. *IEEE Photonics Technology Letters, 8*(5), 721–723. doi:10.1109/68.491607

Zirngibl, M., Joyner, C. H., Stulz, L. W., Dragone, C., Presby, H. M., & Kaminow, I. P. (1995). LARNet, a local access router network. *IEEE Photonics Technology Letters, 7*(2), 215–217. doi:10.1109/68.345927

Zirngibl, M., Joyner, C. H., Stulz, L. W., Koren, U., Chien, M. D., Young, M. G., & Miller, B. I. (1994). Digitally tunable laser based on the integration of a waveguide grating multiplexer and an optical amplifier. *IEEE Photonics Technology Letters, 6*(4), 516–518. doi:10.1109/68.281813

Zukerman, M., Mammadov, M., Tan, L., Ouveysi, I., & Andrew, L. L. H. (2008). To be fair or efficient or a bit of both. *Computers & Operations Research, 35*(12), 3787–3806.

About the Contributors

Ioannis P. Chochliouros is a Telecommunications Electrical Engineer, graduated from the Polytechnic School of the Aristotle University of Thessaloniki, Greece, holding also a M.Sc. (DEA) & a Ph.D. (Doctorat) from the University Pierre et Marie Curie, Paris VI, France. He worked as a Research and Teaching Assistant in the University Paris VI, in cooperation with other European countries. His practical experience as an engineer has been mainly in Telecommunications, as well as in various construction projects in Greece and the wider Balkan area. Since 1997 he has been working at the Competition Department and then as an engineer-consultant of the Chief Technical Officer of OTE (Hellenic Telecommunications Organization S.A.), for regulatory and technical matters. He has been very strongly involved in major OTE's national and international business activities, as a specialist-consultant for technical and regulatory affairs especially for the evaluation and adoption of innovative e-Infrastructures and e-Services. In the same context, he has worked as a consultant in the scope of international projects covering the wider telecommunications sector. He has also served as the Head of Technical Regulations Dept. of the Division for Standardization and Technical Regulations of OTE, involved in an enormous variety of issues regarding European and international standardization, with emphasis on modern technologies. He currently works as the Head of the Research Programs Section of the Labs & New Technologies Division (R&D Dept.) of OTE. Under his supervision, the Section has received several awards by the European Commission, for the successful realization of European research activities. He has been involved in different European and international projects and activities and has participated either as coordinator or scientist-researcher in more than 45 European and national research programs. He is author (or co-author) of more than 125 distinct scientific and business works (i.e. books, book chapters, papers, articles, studies & reports) in the international literature. He has also worked as Lecturer in the Hellenic Academic Sector, in specific areas of modern technologies and/or business and/or regulatory-oriented issues. He has participated in many Conferences, Workshops, Fora and other events, in most of which as an invited speaker. He is also an active participant of various international and national associations, both of scientific and business nature.

George A. Heliotis is a Telecommunications Engineer at the Division of Labs and New Technologies (R&D Department) of the Hellenic Telecommunications Organization (OTE) S.A. and Deputy Head of its Access Networks Lab. He graduated from the University of Crete (Greece) with a B.Sc. in Physics and also holds a M.Sc., and a Ph.D. in Opto-electronics/Photonics from Imperial College London (UK). Prior to joining OTE S.A. he held positions in a variety of academic and industrial institutions, where he was mainly engaged on the design, development and testing of novel opto-electronic devices and systems targeting a wide range of optical communication applications. His current work focuses on the design and evaluation of next-generation access network systems and architectures, mainly involving FTTH and FTTN+VDSL2 implementations. His research interests include photonics devices and systems, all-optical ultrafast networks, FTTx technologies, NGN architectures and services, and xDSL networks. He has participated in several academic and industrial international projects relating to novel optical- and copper-based network systems and infustructures, and has authored or co-authored more than 60 publications in high-profile scientific journals and conferences.

* * *

Evaristo J. Abril received his Telecommunication Engineer and Ph.D. degrees from Universidad Politécnica de Madrid, Spain, in 1985 and 1987, respectively. From 1984 to 1986, he was a Research Assistant at Universidad Politécnica de Madrid, becoming Lecturer in 1987. Since 1995, he has been Full Professor at University of Valladolid, Spain, where he founded the Optical Communications Group. He is currently the Chancellor of the University of Valladolid. His research interests include Integrated Optics, Optical Communication Systems and Optical Networks. Prof. Abril is the author of more than 100 papers in international journals and conferences.

Calvin C. K. Chan received his B.Eng., M.Phil. and Ph.D. Degrees from the Chinese University of Hong Kong, all in Information Engineering. In 1999, he joined Bell Laboratories, Lucent Technologies, Holmdel, NJ, as a Member of Technical Staff where he worked on an optical packet switch fabric with terabit-per-second capacity. In 2001, he joined Department of Information Engineering at the Chinese University of Hong Kong and now serves as an Associate Professor. He has served as members of the Technical Program Committees of many international conferences, including the prestigious Optical Fiber Communication Conference (OFC) (2005-07). Currently, he serves as an Associate Editor for OSA Journal of Optical Networking. Dr. Chan has published more than 180 technical papers in refereed international journals and conferences. He holds one issued US patent. His research interests include optical metro/access networks, optical signal processing and optical performance monitoring.

Aristidis Chipouras was born in Boston, MA, in 1962. He received the B.Sc. degree in physics in 1985, the M.Sc. degree in electronics and radiocommunications in 1987, the M.Sc. degree in information systems in 1994, and the Ph.D. degree in informatics and telecommunications in 2002, all from the University of Athens, Athens, Greece. Since 1993, he has been a Research Associate with the Optical Communications Group, University of Athens, participating in several European R&D projects. His current research interests include the analysis and design of microwave integrated circuits and optical waveguide devices as well as multi-wavelength optical networks. He has several publications in journals and conferences in the area of optical communications.

Chi-Wai Chow received the B.Eng. (First-Class Hons) and Ph.D. Degrees both from the Department of Electronic Engineering, the Chinese University of Hong Kong (CUHK) in 2001 and 2004 respectively. His Ph.D. focused on optical label controlled packet switched networks. After graduation, he was appointed as a Postdoctoral Fellow at the CUHK, working on hybrid integration of photonic components and silicon waveguides. Between 2005-2007, he was a Postdoctoral Research Scientist, working mainly on two European Union Projects: PIEMAN and TRIUMPH in the Tyndall National Institute and Department of Physics, University College Cork (UCC) in Ireland. In August 2007, he joined the Department of Photonics, National Chiao Tung University in Taiwan, as an Assistant Professor. His research interests include WDM-PON, all-optical signal processing, broadband wired and wireless access network, network protection and sensing.

Bernard Cousin is a Professor of Computer Science at the University of Rennes 1, France. Bernard Cousin received in 1987 his Ph.D. Degree in Computer Science from the University of Paris 6. He is, currently, member of IRISA (a CNRS -University-INSA joint research laboratory in computing science located at Rennes). More specifically, he is at the head of a research group on networking. He is the co-author of a network technology book: "IPV6" (Fourth edition, O'Reilly, 2006) and has co-authored a few IETF drafts in the areas of Explicit Multicasting and Secure DNS. His research interests include sensor networks, dependable networking, high speed networks, traffic engineering, multicast routing, network QoS management, network security and multimedia distributed applications.

Ignacio de Miguel received his Telecommunication Engineer degree in 1997, and his Ph.D. degree in 2002, both from the University of Valladolid, Spain. Since 1997 he has worked as a Junior Lecturer at the University of Valladolid. He has also been a Visiting Research Fellow at University College London (UCL), working in the Optical Networks Group. His research interests are the design and performance evaluation of optical networks, especially hybrid optical networks, as well as IP over WDM. Dr. de Miguel is the recipient of the Nortel Networks Prize to the best Ph.D. Thesis on Optical Internet in 2002, awarded by the Spanish Institute and Association of Telecommunication Engineers (COIT/AEIT). He also received the 1997 Innovation and Development Regional Prize for his Graduation Project.

Ramón J. Durán was born in Cáceres, Spain, in 1978. He received his Telecommunication Engineer degree in 2002 and his Ph.D. degree in 2008, both from the University of Valladolid, Spain. Since 2002, he has worked as a Junior Lecturer at the University of Valladolid and currently he is also Secretary of the Faculty of Telecommunication Engineering. He is a member of the Optical Communications Group at the University of Valladolid. His doctoral research focuses on the design and performance evaluation of wavelength-routed optical networks, especially hybrid architectures for optical networks.

Patricia Fernández was born in Madrid, Spain, on April 16, 1973. She received the Telecommunication Engineer degree from the Universidad Politécnica de Cataluña, Barcelona, Spain, in 1997, and the Ph.D. degree from the University of Valladolid, in 2004. Since 1999 she has worked as a Junior Lecturer at the University of Valladolid. Her research interests are passive optical networks and fiberoptic communications components. Dr. Fernández is the author of more than 40 papers in international journals and conferences.

Brian Hunck is Director of Business Access Product Planning at Fujitsu Network Communications. He is responsible for FLASHWAVE® product planning, management and marketing. He has over 20 years of experience in telecommunications and has worked in a variety of management positions as well as in hardware/FPGA design, system engineering and product planning. Before joining Fujitsu, Brian worked with ADC Telecommunications Broadband Access Division on HDSL and optical access products and for Electrospace Systems designing secure communication products for the United States military. He holds an MBA from the University of Texas at Dallas and a B.Sc. in Computer Engineering from Iowa State University.

Dimitris Katsianis received the Informatics Degree, the M.Sc. in Signal Processing and Computational Systems and the Ph.D. Diploma in Network Design with Techno-economics Aspects from the University of Athens, Dept. of Informatics and Telecommunications. He is a research fellow with the Optical Communications Group, participating in several European R&D projects. He has worked as an expert scientific advisor with several firms in the field of techno-economic & network design studies. His research interests include broadband communications and methodology of network design with techno-economic aspects. He has more than 50 publications in journals and conferences in the field of Techno-Economics and telecommunication network design and he serves as a reviewer in journals and conferences.

Vagia Kyriakidou received her degree in Finance and Accounting from the Athens University of Economics and Business (AUEB) in 2003. She holds a M.Sc. degree in Administration and Economics of Telecommunication Networks from the National and Kapodistrian University of Athens (Interfaculty course of the Departments of Informatics and Telecommunications and Economic Sciences). From 2006, she is working with the Laboratory of Optical Communications in the Department of Informatics and Telecommunications (National and Kapodistrian University of Athens). She authored a number of papers that were presented to conferences and project-meetings. Since 2007, she is a Ph.D. candidate in the University of Athens and her research interests include Socio-economic analysis of telecommunication sector, business modeling for optical networks, power laws, game theory, Data Envelopment Analysis, etc.

Rubén M. Lorenzo received his Telecommunication Engineer and Ph.D. degrees from the University of Valladolid, Spain, in 1996 and 1999, respectively. From 1996 to 2000, he was a Junior Lecturer at the University of Valladolid, and joined the Optical Communications Group. Since 2000, he has been a Lecturer. His research interests include Integrated Optics, Optical Communication Systems and Optical Networks. He is currently the Head of the Faculty of Telecommunication Engineering at University of Valladolid and Research Director of CEDETEL (Center for the Development of Telecommunications in Castilla y León).

Noemí Merayo was born in Ponferrada (León), Spain, in 1979. She received her Telecommunication Engineer degree from the University of Valladolid, Spain, in 2004. She is currently working toward her Ph.D. degree in the Optical Communications Group at the University of Valladolid. Her doctoral research focuses on the design and performance evaluation of optical networks, especially passive optical networks.

Jean-Claude Mollier received the Doctorat es Sciences Degree in 1982 from the University of Franche-Comté (Besançon, France). He worked first on stable RF oscillators at the Laboratoire de Physique et Métrologie des Oscillateurs until he joined, in 1984, the Research Institute in Optical and Microwave Communications (X-LIM, France) where he became manager of a research group in the field of broadband and low noise microwave amplification. Since 1991, he has been working about microwave-photonics interactions at SUPAERO-ONERA in Toulouse (France): optical generation of microwave signals, modeling and characterization of semiconductor Lasers (VCSELs and QCLs) for application in optical fiber links and THz active imaging. He is currently Head of the Department of "Electronics-Optronics- Signal" at the Institut Supérieur de l'Aéronautique et de l'Espace (ISAE). Dr

J.C. Mollier is a member of IEEE-LEOS and French Optical Society (SFO) where he created the club "Optics and Microwaves" in 1995.

Miklos Molnar was graduated at the Faculty of Electrical Engineering, Technical University of Budapest in 1976. He received the Ph.D. Degree in Computer Science from the University of Rennes 1. in 1992. Since 1989, he has been with the INSA of Rennes, where he is working as assistant professor, and with the research laboratory IRISA of Rennes. His main research activity is in stochastic control problems, combinatorial optimizing and management algorithms for communication networks. His research results are mainly from the following domains: efficient heuristics for NP-hard optimization problems, heuristics for the Steiner problem, routing algorithms for unicast, incast and multicast communications, dependable communications, routing in optical and in wireless networks, energy aware protocols and optimizations.

Hussein T. Mouftah joined the School of Information Technology and Engineering (SITE) of the University of Ottawa in 2002 as a Tier 1 Canada Research Chair Professor, where he became a *University Distinguished Professor* in 2006. He has been with the ECE Dept. at Queen's University (1979-2002), where he was prior to his departure a Full Professor and the Department Associate Head. He has six years of industrial experience mainly at Bell Northern Research of Ottawa (now Nortel Networks). He served as Editor-in-Chief of the IEEE Communications Magazine (1995-97) and IEEE ComSoc Director of Magazines (1998-99), Chair of the Awards Committee (2002-03), Director of Education (2006-07), and Member of the Board of Governors (1997-99 and 2006-07). He has been a Distinguished Speaker of the IEEE Communications Society (2000-07). He is the author or coauthor of 6 books, 30 book chapters and more than 800 technical papers, 10 patents and 138 industrial reports. He is the joint holder of 8 Best Paper and/or Outstanding Paper Awards. He has received numerous prestigious awards, such as the 2008 ORION Leadership Award of Merit, the 2007 Royal Society of Canada Thomas W. Eadie Medal, the 2007-2008 University of Ottawa Award for Excellence in Research, the 2006 IEEE Canada McNaughton Gold Medal, the 2006 EIC Julian Smith Medal, the 2004 IEEE ComSoc Edwin Howard Armstrong Achievement Award, the 2004 George S. Glinski Award for Excellence in Research of the U of O Faculty of Engineering, the 1989 Engineering Medal for Research and Development of the Association of Professional Engineers of Ontario (PEO), and the Ontario Distinguished Researcher Award of the Ontario Innovation Trust. Dr. Mouftah is a Fellow of the IEEE (1990), the Canadian Academy of Engineering (2003), the Engineering Institute of Canada (2005) and RSC: The Academies of Canada (2008).

Gerasimos C. Pagiatakis was born in Corfu, Greece, in 1961. He received the B.Sc. degree from the National Technical University of Athens, Greece, in 1985 and the Ph.D. degree from the Imperial College of Science, Technology and Medicine, London, UK, in 1990. After his military service, he worked with the Hellenic Organisation for Standardization (ELOT) and the Hellenic Development Company (ELANET). From 1996 to 2005, he was with the Hellenic Organization for Telecommunications (OTE) where he was actively involved in optical networks, including OTE's Olympic network for audio and TV broadcasting (for the Athens 2004 Olympic Games). In 2005, he was elected Associate Professor at the Department of Electronics Instructors, School for Pedagogical and Technological Education (ASPETE). Dr. Pagiatakis is the author or co-author of 11 papers and 2 books. His research interests are on optical communications and the application of numerical techniques in electromagnetics and optoelectronics. Dr. Pagiatakis is a member of the Technical Chamber of Greece (TEE), the IET, the IEEE and the FITCE.

Angelique Rissons was born in Reims, France, in 1975. She obtained the M.S. Degree in Microwave and Optoelectronics in 2000 and the Ph.D. Degree in 2003 from the "Grande Ecole" Supaero (National School of Aeronautics and Space), Toulouse, France. Her thesis work was on the optoelectronic modelling and characterisation of 850nm VCSELs. Since 2004, she has been with the MOSE (microwave and optoelectronics for embedded systems) research group. She is in charge of the VCSELs activities of DEOS (Department of "Electronics-Optronics-Signals") at the "Institut Supérieur de l'Aéronautique et de l'Espace" (ISAE). She is currently working on modelling and characterisation of long wavelength VCSELs in optical subassembly. She is also interested by the various applications of the VCSEL in avionic and optical access network.

Thomas Schirl developed digital X-ray flat-panel imagers for more than 8 years for Siemens Health Care. He received his Ph.D. in Applied Physics from the Friedrich-Alexander-University of Erlangen in 1999. Since 2001 he is Professor in Klagenfurt, Austria. In 2007 he passed from the Telematics (RF) Department to the Department of Medical Technology of the Carinthian University of Applied Sciences. His interests comprise medical imaging, thermography and neurosciences.

Jürgen Schussmann received the Ph.D. Degree in electrical engineering from the Technical University of Ilmenau, Germany, in 2008 and the Dipl.-Ing. Degree in Telematics / Network Engineering from the Carinthia University of Applied Sciences, Klagenfurt, Austria in 2002. He previously worked for Siemens over 10 years. His current research interests include photonic networking and the design of passive optical access networks based on WDM.

Thomas Sphicopoulos received the Physics Degree from Athens University in 1976, the D.E.A. Degree and Doctorate in Electronics both from the University of Paris VI in 1977 and 1980 respectively, the Doctorat Es Science from the Ecole Polytechnique Federale de Lausanne in 1986. From 1976 to 1977 he worked in Thomson CSF Central Research Laboratories on Microwave Oscillators. From 1977 to 1980 he was an Associate Researcher in Thomson CSF Aeronautics Infrastructure Division. In 1980 he joined the Electromagnetism Laboratory of the Ecole Polytechnique Federal de Lausanne where he carried out research on Applied Electromagnetism. Since 1987 he is with the Athens University, engaged in research on Broadband Communications Systems. In 1990 he was elected as an Assistant Professor of Communications in the Department of Informatics & Telecommunications, in 1993 as Associate Professor and since 1998 he is a Professor in the same Department. His main scientific interests are Optical Communication Systems and Networks and Techno-economics. He has lead about 50 National and European R&D projects. Professor Sphicopoulos has more than 150 publications in scientific journals and conference proceedings and he is an advisor in several organizations.

Anastasia S. Spiliopoulou is a Lawyer, possessing a Post-Graduate Diploma in Internet-related studies from the Athens University Law School, where she currently exercises research activities as a Ph.D. candidate. As a lawyer, she is Member of the Athens Bar Association in Athens, Greece. During the latest years, she had a major participation in matters related to telecommunications & broadcasting policy, in Greece and abroad, within the wider framework of the Information Society. She has been involved in current legal, research and business activities, as a specialist for e-Commerce and e-Businesses, Electronic Signatures, e-Contracts, e-Procurement, e-Security and other modern Information Society applications. She has published more than 90 scientific and business papers in the international literature (books, book chapters, journals, conferences and workshops), with specific emphasis given on regulatory, business, commercial and social aspects. Many of the above publications have been internationally recognized for their validity. She also works as Lawyer-Partner of the Hellenic Telecommunications Organization S.A. (OTE) in the General Directorate for Regulatory Affairs, Dept. for Regulatory Framework Issues, where she has been efficiently involved in a great variety of regulatory issues, with impact on business and market related matters.

Christos Vassilopoulos was born in Athens Greece (1961) and studied Electronic Engineering in U.K. obtaining the Degrees of B.Sc. (1982) from the University of Sussex, M.Sc. in Telecommunications (1983) and Ph.D. in Optical Communications (1987) from Imperial College of Science and Technology (University of London). In the period 1989-1992 he followed Post-Doctoral Research at the Institute of Telecommunications and Informatics of the NCPR "Demokritos", Athens, Greece, participating in numerous international EC funded research projects on optical communication and sensing. For the period 1992-1994 he worked as a specialist at the Computer Centre of the Greek Ministry of Health (KHYKY) on networks developed for public hospitals. In the period 1994-1996 he worked as a consultant technical evaluator for ELANET S.A., a subcontractor of the Greek Ministry of Finance, responsible for the technical evaluation of the Telematics program for the Greek Private Sector, funded by the EC. He joined OTE (Hellenic Telecommunications Organization) S.A. in 1996 as a telecommunication specialist on optical communications and was promoted section-head for wireline access in 2002 and division-head for access network planning in 2005. He is currently responsible for OTE access network planning, in the areas of xDSL, optical access (FTTx technologies), wireless access (WiFi, WiMax) and network infrastructure. Dr. Vassilopoulos is co-author in two books, has participated in numerous conferences and has published extensively in scientific magazines and conference proceedings. Over the last 15 years he also holds a part time teaching job at the Technical Educational Institution of Piraeus, where he lectures on Tele-communications, Digital Networks and Optical Communications.

Tetsuya Yokotani obtained his B.S., M.Sc., and Ph.D. Degrees in information science from the Tokyo University of Science in 1985, 1987 and 1997, respectively. He joined Mitsubishi Electric Corporation in 1987. Since then, he has studied high-speed data communication, and system performance evaluation. Currently, his interests include optical access and home networks, and standardization of these protocols. He is also a section manager in Information Technology R&D Center of Mitsubishi Electric Corporation. He has joined standardization activity in ITU-T SG15 for FTTH and Home networks from 2000. He obtained ITU-T Japanese associates award for this activity on 2003. He has also been worked for the Tokyo University of Science as a part-time lecturer from 2008.

William Yue is senior product planner of access product planning at Fujitsu Network Communications. He is responsible for planning the company's broadband wireless access products that focus on WiMAX, IP and Ethernet technologies. William has over 15 years of experience in telecommunications and prior to joining Fujitsu, worked in software and product design at Nortel Networks. William holds a bachelor's degree in computer engineering and a master's degree in electrical engineering from McMaster University, Canada.

Jun Zheng is a research scientist with the School of Information Technology and Engineering, University of Ottawa, Canada. He serves as an associate technical editor for IEEE Communications Magazine, an editor for IEEE Communications Surveys & Tutorials, and an associate editor for IEEE/OSA Journal of Optical Communications and Networking. He has served as Lead Guest Editor of eight special issues for different refereed journals and magazines, including IEEE Network and IEEE Journal on Selected Areas in Communications (J-SAC). He has served as General Chair, TPC Chair, and Symposium Chair for several international conferences and symposia, and has served on the technical program committees of a number of international conferences and symposia, including IEEE GLOBECOM and ICC. Dr. Zheng has conducted extensive research in the field of communications and computer networks. His research interests include optical networks, wireless networks, and IP networks, focused on network architectures and protocols. He has coauthored (first author) a book on optical WDM networks published by Wiley-IEEE, and has published about 80 technical papers in refereed journals, magazines, and conference proceedings.

Fen Zhou was born in HeFei, P. R. China in 1982. He received the B.S. and M.S. Degree in Communication and Information System respectively in 2004 and 2007, from Northwestern Polytechnical University in P. R. China. Currently, he is preparing for his Ph.D. Degree in the Lab. IRISA/INSA Rennes (Institut de Recherche en Informatique et Systèmes Aléatoires), France. His research interest includes multicast routing in optical networks.

Index